IoT and Spacecraft Informatics

Aerospace Engineering

IoT and Spacecraft Informatics

Edited by

K.L. YUNG

Department of Industrial and Systems Engineering, The Hong Kong Polytechnic University, Hong Kong, P.R. China

ANDREW W.H. IP

Department of Mechanical Engineering, University of Saskatchewan, Saskatoon, SK, Canada; Department of Industrial and Systems Engineering, The Hong Kong Polytechnic University, Hong Kong SAR, P.R. China

FATOS XHAFA

Universitat Politècnica de Catalunya, Barcelona, Spain

K.K. TSENG

Harbin Institute of Technology, Shenzhen, P.R. China

ELSEVIER

Elsevier
Radarweg 29, PO Box 211, 1000 AE Amsterdam, Netherlands
The Boulevard, Langford Lane, Kidlington, Oxford OX5 1GB, United Kingdom
50 Hampshire Street, 5th Floor, Cambridge, MA 02139, United States

ISBN: 978-0-12-821051-2

Publisher: Matthew Deans
Acquisitions Editor: Brian Guerin
Editorial Project Manager: Fernanda A. Oliveira
Production Project Manager: Nirmala Arumugam
Cover Designer: Greg Harris

Typeset by MPS Limited, Chennai, India

Prof. Xhafa dedicates this book to the memory of his late mother.

Contents

List of contributors

Yim Shan Au
Department of Industrial and Systems Engineering, The Hong Kong Polytechnic University, Hong Kong, P.R. China; Laboratory for Artificial Intelligence in Design, Hong Kong Science Park, New Territories, Hong Kong SAR, P.R. China

Shing Chow Chan
Department of Electrical and Eletronic Engineering, The University of Hong Kong, Hong Kong, P.R. China

Yachin Chang
Harbin Institute of Technology (Shenzhen), Shenzhen, P.R. China

Song Chen
Tianjin University of Science and Technology, Tianjin, P.R. China

Jingyi Dong
School of Management Science and Engineering, Key Laboratory of Big Data Management Optimization and Decision of Liaoning Province, Dongbei University of Finance and Economics, Dalian, P.R. China

Na Dong
School of Electrical and Information Engineering, Tianjin University, Tianjin, P.R. China

Ming Gao
School of Management Science and Engineering, Key Laboratory of Big Data Management Optimization and Decision of Liaoning Province, Dongbei University of Finance and Economics, Dalian, P.R. China

Tan Guo
Faculty of Information Technology, Macau University of Science and Technology, Taipa, Macau, P.R. China; State Key Laboratory of Lunar and Planetary Sciences, Macau University of Science and Technology, Taipa, Macau, P.R. China; School of Communication and Information Engineering, Chongqing University of Posts and Telecommunications, Chongqing, P.R. China

Bin Hu
Changsha Normal University, Changsha, P.R. China

Andrew W.H. Ip
Department of Mechanical Engineering, University of Saskatchewan, Saskatoon, SK, Canada; Department of Industrial and Systems Engineering, The Hong Kong Polytechnic University, Hong Kong SAR, P.R. China; College of Engineering, University of Saskatchewan, Saskatoon, SK, Canada

Muhammad Irshad
Department of Electronic and Information Engineering, The Hong Kong Polytechnic University, Hong Kong, China

Ghadeer Khader
Diligent Trust Inc., IT Solutions, Toronto, Canada

Ang Li
School of Management Science and Engineering, Key Laboratory of Big Data Management Optimization and Decision of Liaoning Province, Dongbei University of Finance and Economics, Dalian, P.R. China

Bo Li
School of Management Science and Engineering, Key Laboratory of Big Data Management Optimization and Decision of Liaoning Province, Dongbei University of Finance and Economics, Dalian, P.R. China

Donghui Li
School of Electrical and Information Engineering, Tianjin University, Tianjin, P.R. China

Jun Li
Harbin Institute of Technology (Shenzhen), Shenzhen, P.R. China

Wenqiang Li
Department of Industrial and Systems Engineering, The Hong Kong Polytechnic University, Hong Kong SAR, P.R. China

Bin Liao
College of Elecronics and Information Engineering, Shenzhen University, Shenzhen, P.R. China

Xinyu Liu
School of Electrical and Information Engineering, Tianjin University, Tianjin, P.R. China

Xiao-Ping Lu
Faculty of Information Technology, Macau University of Science and Technology, Taipa, Macau, P.R. China; State Key Laboratory of Lunar and Planetary Sciences, Macau University of Science and Technology, Taipa, Macau, P.R. China

Fulin Luo
State Key Laboratory of Information Engineering in Surveying, Mapping and Remote Sensing, Wuhan University, Wuhan, P.R. China

Aparna Murthy
EIT, PEO, Toronto, ON, Canada

Sohail M. Noman
Shantou University Medical College, Shantou, Guangdong, P.R. China

Xilang Tang
Air Force Engineering University, Xi'an, P.R. China

Yuk Ming Tang
Laboratory for Artificial Intelligence in Design, Hong Kong Science Park, New Territories, Hong Kong SAR, P.R. China; Department of Industrial and Systems Engineering, The Hong Kong Polytechnic University, Hong Kong SAR, P.R. China

K.K. Tseng
Harbin Institute of Technology (Shenzhen), Shenzhen, P.R. China

Fatos Xhafa
Universitat Politècnica de Catalunya, Barcelona, Spain

Keping Yu
Global Information and Telecommunication Institute, Waseda University, Tokyo, Japan

K.L. Yung
Department of Industrial and Systems Engineering, The Hong Kong Polytechnic University, Hong Kong, P.R. China

Yong-Xiong Zhang
Faculty of Information Technology, Macau University of Science and Technology, Taipa, Macau, P.R. China; State Key Laboratory of Lunar and Planetary Sciences, Macau University of Science and Technology, Taipa, Macau, P.R. China

Zhiguo Zhang
School of Biomedical Engineering, Health Science Center, Shenzhen University, Shenzhen, P.R. China

About the editors

Prof. K.L. Yung, the Hong Kong Polytechnic University, China

Professor K.L. Yung is an Associate Head and Chair Professor of the Department of Industrial and Systems Engineering of The Hong Kong Polytechnic University. He received his BSc in Electronic Engineering at Brighton University in 1975, MSc, DIC in Automatic Control Systems at Imperial College of Science, Technology & Medicine, University of London in 1976, and PhD in Microprocessor Applications in Process Control at Plymouth University in the United Kingdom in 1985. He became a Chartered Engineer (C.Eng., MIEE) in 1982. After graduation, he worked in the United Kingdom for companies such as BOC Advanced Welding Co. Ltd., the British Ever Ready Group, and the Cranfield Unit for Precision Engineering (CUPE). In 1986, he returned to Hong Kong to join the Hong Kong Productivity Council as a Consultant and subsequently switched to academia to join the Department of Industrial and Systems Engineering of The Hong Kong Polytechnic University. He has a wealth of experience in making sophisticated space tools for deep space exploration missions. These include Space Holinser Forceps for the MIR Space Station, the "Mars Rock Corer" for the European Space Agency's Mars Express Mission (2003), the "Soil Preparation System" for the Sino-Russian Phobos-Grunt Mission (2011), and advanced precision robotic systems for the China Lunar Exploration Missions (Chang'e 3, 4, 5, and 6) such as the Camera Pointing Systems (CPS), and the Mars Exploration Mission (Tianwen-1), such as the Mars Surveillance Camera.

Prof. Andrew W.H. Ip, University of Saskatchewan, Department of Mechanical Engineering, Canada

Prof. Andrew W.H. Ip received his PhD from Loughborough University (UK), MBA from Brunel University (UK), MSc in Industrial Engineering from Cranfield University (UK), and LLB (Hons) from the University of Wolverhampton (UK). In 2015, Dr. Ip was awarded "Gold Medal with the Congratulations of the Jury" and "Thailand Award for Best International Invention" in the 43rd International Exhibition of Geneva, and in 2013, he was awarded the "Natural Science Award" of the Ministry of Education Higher Education Outstanding Scientific Research Output Awards by the Ministry of Education of Mainland China. Dr. Ip has been appointed as a distinguished professor and visiting professor of various universities in China including Shenzhen University, Civil Aviation University of China, South China Normal University, City University of Macau, and the Hunan University of Finance and Economics. He is a Professor Emeritus and an adjunct professor of Mechanical Engineering, University of Saskatchewan, Canada; Honorary Fellow of the University of Warwick (Warwick Manufacturing Group), and a senior research fellow of the Hong Kong Polytechnic University. He is a chartered engineer, senior member of IEEE, member of Hong Kong Institution of Engineers, and member of various professional bodies in mechanical and electrical engineering.

He is the Editor-in-Chief of the *International Journal of Enterprise Information System* (SCI index), Taylor & Francis; the Founder & Editor-in-Chief of the *International Journal of Engineering Business Management* (ESCI & SCOPUS index), SAGE; and Editor-in-Chief of the *International Journal of Software Science and Computational Intelligence* of IGI Global. He also serves as associate editor and guest editor of various SCI international journals on engineering and technology.

Prof. Fatos Xhafa, PhD,
UPC-BarcelonaTech,
Barcelona, Spain

Fatos Xhafa, PhD in Computer Science, is a full professor at the Technical University of Catalonia (UPC), Barcelona, Spain. He has held various tenured and visiting professorship positions. He was a visiting professor at the University of Surrey, UK (2019/2020), visiting professor at the Birkbeck College, University of London, UK (2009/2010), and a research associate at Drexel University, Philadelphia, USA (2004/2005). He was a distinguished guest professor at Hubei University of Technology, China, for 3 years (2016–19). Prof. Xhafa has widely published in peer-reviewed international journals, conferences/workshops, book chapters, edited books, and proceedings in the field (H-index 55). He is awarded teaching and research merits by the Spanish Ministry of Science and Education, in IEEE conferences, and best paper awards. Prof. Xhafa has extensive editorial experience. He is the founder and Editor-in-Chief of *Internet of Things*—Journal—Elsevier (Scopus and Clarivate WoS Science Citation Index) and of *International Journal of Grid and Utility Computing*, (Emerging Sources Citation Index) and AE/EB Member of several indexed international journals. He is the founder and Editor-in-Chief of two books series: the Springer Book Series Lecture Notes in Data Engineering and Communication Technologies (SCOPUS, EI Compendex, ISI WoS) and the Elsevier Book Series Intelligent Data-Centric Systems (SCOPUS, EI Compendex). Prof. Xhafa is a member of IEEE Communications Society, IEEE Systems, Man & Cybernetics Society, and Founder Member of the Emerging Technical Subcommittee of Internet of Things.

His research interests include IoT and Cloud-to-thing continuum computing, massive data processing and collective intelligence, optimization, security, and trustworthy computing and machine learning. He can be reached at fatos@cs.upc.edu. Please visit also http://www.cs.upc.edu/~fatos/ and at http://dblp.uni-trier.de/pers/hd/x/Xhafa:Fatos

Prof. K.K. Tseng,
Harbin Institute of
Technology, China

K.K. Tseng is a tenured associate professor and Shenzhen Peacock B-level talent. He was born in 1974 and received his doctoral degree in computer information and engineering from the National Chiao Tung University of Taiwan in 2006. He has many years of research and development experience and has long engaged in deep learning architecture and algorithms research. His recent research focus is on brain-like processor design, unmanned driving, and biological signal research.

He has published more than 80 articles, of which about 40 have a high SCI impact factor including the famous ACM/IEEE series of journals. In addition, he has registered more than 40 patents for inventions.

Foreword

Space exploration is one of the most exciting and contemporary topics both in academia and industries in recent years. Many countries such as Russia, United States, Japan, and India are deploying new and innovative technologies involving AI, drones, robotics, and machine learning, and so on into their deep space exploration missions. Moreover, considering the high complexity, high cost, and high risk involved in spacecraft, advanced technologies in information processing, simulation, optimization, and decision-making are required to improve the effectiveness, efficiency, reliability, and safety of space exploration. The emerging Internet of Things (IoT) or Internet of Planets and informatics offer the possibility of integration in the areas of spacecraft design, development, and implementation regarding in-orbit spacecraft, satellites, space stations, stations on Moon, Mars, and other planets, which is imperative. For example, in 2020 China completed two historical deep space missions. Chang'e-5 successfully returned the first Chinese acquired lunar regolith sample, and Tiawen-1-Zhurong completed the first Chinese soft landing on Mars. Tiawen-1-Zhurong undertook an ambitious mission that sent an orbiter, lander, and rover in one go which no other mission had attempted before. The Hong Kong Polytechnic University team lead by Prof. K.L. Yung and a group of scientists was honored to play a key role in these missions through the development of the landing site selection methodology, the Lunar Surface Sample Acquisition and Packaging System for returning the lunar regolith, and the Mars Landing intelligent surveillance camera on board the Mars landing platform. Over the years, the university has also participated in the Chang'e-3 and Chang'e-4 missions, where intelligent instruments, sensors, and IoT are located on the near and far sides of the Moon. In the near future, the team will be designing a device for Chang'e-6 to return a sample from the far side of the moon, Chang'e-7 searching for water-ice at the lunar south pole, an asteroid sample return from Chang'e-8 experimental lunar base, and a sample return from Mars. This book, written by Prof. K.L. Yung, Prof. Andrew W.H. Ip, Prof. Fatos Xhafa, and Prof. K.K. Tseng provide straightforward concepts as the starting point without overlooking their limitations and address many of the implications of successful cases and examples. The book follows a very practical approach, dedicated to all readers who would like to immediately understand all the insights of this interdisciplinary research area which integrates and innovates in key deep space exploration technologies. This book would also be a good choice for academicians and industrialists who want to bring specific theory and case studies into their classrooms and working environment.

Prof. Wenjun (Chris) Zhang
University of Saskatchewan, Saskatoon, SK, Canada

Preface

Internet of Things (IoT) and Spacecraft Informatics are the applications of IoT systems and theory in the design, development, and operation of spacecraft. A spacecraft is a complex system that involves the integration of hardware and software, requiring different IoT architectures with sensors, networks, applications, and so on for information modeling, simulation, optimization, and decision support methods and techniques. Such a spacecraft system represented by theory and techniques can be used to describe in-orbit spacecraft, satellites, space stations of any types in deep-space exploration missions from ground control, user payload, space weather and conditions, remote sensing and telemetry, and many more spaceflight missions and activities in designing, forecasting, planning, and control. The motivation of this book is to provide the fundamentals and theory in the area of IoT and spacecraft informatics with the goal of directly contributing to the present and future space exploration and spacecraft development. It aims to bring the latest advances, findings, and state of the art in research, case studies and examples, development, and implementation of IoT key technologies for spacecraft systems. The book consists of 10 chapters and is the first in a series of books with Elsevier Aerospace Engineering.

In the first chapter of our book, "Artificial intelligence approach for aerospace defect detection using a single shot detector", we describe an inspection system that has been designed and implemented with the support of advanced artificial intelligence (AI) technologies. In aerospace engineering, ultrasonic testing is a reliable method to examine the integrity of composite components in an aircraft. The development of a practical and operational system using the latest AI technology for defect detection in an aircraft with the convolutional neural network is illustrated and demonstrated, which can be used to detect defects in the composite laminates to increase the accuracy and efficiency of ultrasonic inspection. This chapter provides a simple and easy-to-understand introduction to AI and IoT sensors. In Chapter 2, we investigate the composition and mineralogical characteristics of asteroids which is very significant for understanding the physical and chemical evolution of the solar system. To overcome the unexpected noise of observation systems and the ever-changing external observation environment, the observed asteroid spectral data always contain noise and outliers exhibiting inseparable patterns, which will bring great challenges to the precise classification of asteroids. To alleviate the problem and improve the separability and classification accuracy for different kinds of asteroids, this chapter introduces a Neighboring Discriminant Component Analysis (NDCA) model for asteroid spectrum feature learning in a data-driven supervised machine learning fashion. In Chapter 3, we discuss the problem of the capturing, removal, and maintenance of old satellites. Many satellites launched each year become inoperative ahead of time because of failing to enter orbit or are considered obsolete because of the expiry date. The vision-based navigation system has become the popular detection method for space missions in short distances due to the advantages of higher precision, lower power consumption, and cost. Various computer vision methods and techniques are discussed and the approaches to satellite maintenance are proposed. In Chapter 4, we provide more examples of IoT, from wearable medical devices to home appliances. Various components play a role in the communication infrastructure, and issues such as security and privacy are considered particularly important. The specific usage of IoT is in terms of digital twin (DT) which is a representation of the physical system in terms of variables. Using DT substantially reduces the cost of prototyping and testing time and wastage. The readers can relate everyday lives with those of an astronaut living in a space station who needs medical and support devices through various IoT and sensors. In Chapter 5, a more advanced treatment of AI and a computational algorithm are provided. Subspace estimation and tracking are of great benefits for high-resolution sensor array signal processing in aerospace and defense applications. Several subspace tracking algorithms with different arithmetic complexities and tracking abilities are introduced. The application of these algorithms to time-varying direction-of-arrival (DOA) estimation is presented with two modified methods, namely modified PAST (MPAST) and modified orthonormal PAST (MOPAST). Numerical examples are given to demonstrate the flexibility, effectiveness, and robustness of these algorithms for subspace and DOA tracking.

In Chapter 6, "Optimization problems and resolution methods in satellite scheduling and spacecraft operation: Description, modeling, and applications," we discuss the state of the art in satellite scheduling regarding spacecraft design, operation, and satellite deployment system. With heuristics methods, the constraint features in satellite mission planning, including window accessibility and visibility requirements, can be addressed for producing small and low-cost satellites; some proposed algorithms that improve the accuracy and efficiency are illustrated. In Chapter 7, a simulation approach with colored Petri net (CPN) modeling is presented to describe the resource type and execution logic in the manufacturing workflow of a spacecraft instrument called SOil Preparation SYStem (SOPSYS). In this study, we applied the CPN model in simulating the manufacturing process, planning, and controlling the various resources. In Chapter 8, a more managerial approach to the innovation aspects of IoT and spacecraft information is given. Many successful products are developed through innovative ideas, and the design and development of space equipment is one such example. This chapter explains and evaluates the impact of five major factors that affect the success: the team of individuals, team technology, funding and resources, human resources system, and government support. Through a case study of the SOPSYS space instruments, we can understand the five factors on the design and development of space equipment products and their respective subfactors. In Chapter 9, "Monocular SLAM for a space rover application," a novel simultaneous localization and mapping system are presented to track the unconstraint motion of the mobile robot on the Moon or Mars. Through the integration with the ellipse search algorithm and MVEKF filter, the proposed system enhances the grid-based feature point extraction with satisfactory performance and a low error rate for the Lunar rover's locating tasks. In the final chapter, we describe an important topic on the reliability and health management of spacecraft which is the health condition estimation of spacecraft key components using the belief rule with a semi-quantitative decision science method for examining the health status of a spacecraft. It is compared with the traditional optimization method in training an expert's knowledge, and the Markov Chain Monte Carlo technique is embedded in the proposed method to overcome the problem of overfitting in a backpropagation network. It assists human users to deal with the uncertainties in the spacecraft during deep-space exploration to minimize the risks and unexpected events.

Prof. K.L. Yung
Prof. Andrew W.H. Ip
Prof. Fatos Xhafa
Prof. K.K. Tseng

Acknowledgment

We would like to express our gratitude and appreciation for the hard work and support of many people who have been involved in the development and writing of this book. The contributions and participations of them have helped in completing the book satisfactorily. Among many of them, we wish to thank all the authors who despite their busy schedules devoted so much of their time in preparing and writing the chapters, they have also shown great enthusiasm to respond to many comments made by the reviewers and editors. Special thanks goes to Prof. Chris Zhang from the Faculty of Engineering, the University of Saskatchewan, Canada, for writing the forward to this book, and also to the reviewers for their constructive feedbacks and suggestions for improvements. We are also thankful for the support of the publisher who has an experienced and dedicated team to guide us throughout the development of this book.

Last but not least, we wish to thank the Research Center for Deep Space Explorations, the Hong Kong Polytechnic University, China, to provide the resources and support during the development of this book. We look forward to the further development of next book on Spacecraft Informatics to be carried out shortly.

Finally, we are deeply indebted to our families for their love, patience, and support throughout this rewarding experience.

Prof. K.L. Yung
Prof. Andrew W.H. Ip
Prof. Fatos Xhafa
Prof. K.K. Tseng

CHAPTER 1

Artificial intelligence approach for aerospace defect detection using single-shot multibox detector network in phased array ultrasonic

Yuk Ming Tang[1,3], Andrew W.H. Ip[1,2] and Wenqiang Li[1]
[1]Department of Industrial and Systems Engineering, The Hong Kong Polytechnic University, Hong Kong SAR, P.R. China
[2]Department of Mechanical Engineering, University of Saskatchewan, Saskatoon, SK, Canada
[3]Laboratory for Artificial Intelligence in Design, Hong Kong Science Park, New Territories, Hong Kong SAR, P.R. China

1.1 Introduction

1.1.1 Ultrasonic inspection in aircraft

Ultrasonic testing (UT) is one of the most reliable methods to examine the serviceability of structural components (Khaira, Srivastava, & Suhane, 2015), such as welds, composite lightweight material, and structures. UT has been applied in many different applications in manufacturing, aviation, building structures, etc. (Katnam, Da Silva, & Young, 2013).

In aircraft inspection, the Federal Aviation Administration (FAA) requires a series of aircraft parts traceability and trackability for inventory management. Recently, many technologies are adopted in aviation industries to enhance working accuracy and efficiency. Ho, Tang, Tsang, Tang, and Chau (2021) adopted the blockchain-based system to enhance aircraft parts' traceability and trackability. AI-empowered approaches are used for aircraft inspections to provide reasonable assurance that the aircraft is functioning properly. Despite AI can be used in many different forecasting, prediction applications, etc. (Tang, Chau, Li, & Wan, 2020), in this chapter, we focus on the AI approach for aerospace defect detection and the basic inspection requirements for aircraft inspection. Such requirements usually differ with the usage of the aircraft. For example, aircraft being used for compensation or hire must have a thorough

IoT and Spacecraft Informatics
DOI: https://doi.org/10.1016/B978-0-12-821051-2.00008-8

inspection every 100 h, while other aircraft are required to have a complete inspection every year (Chen, Ren, & Bil, 2014).

Traditionally, UT is a manual task that is usually labor-intensive and time-consuming. It is because the ultrasonic inspection usually needs a high degree of operator skill and is subject to specimen geometric and equipment limitations. Although some commercial software has been developed for defect sizing, the software requires the use of expensive specific ultrasonic flaw detectors and probes. The discontinuities and defects usually have no specific shapes, positions, and orientations. As such, aircraft defects are usually inspected by skilled operators and qualified inspectors who are certified to perform inspection and evaluation. However, due to the testing equipment limitations, it is difficult for scanning components with complex geometry, such as the large curved surfaces of aircraft.

1.1.2 Autonomous inspection

Nowadays, due to the revolution of advanced Industry 4.0 and Internet of Things (IoT) technologies, the maintenance, repair, and operation industry (MRO) is being transformed due to the use of the intelligent predictive maintenance approach. Predictive maintenance is an important task to ensure safety and reliability. Taking aircraft inspection as an example, inspectors need to locate, search, make decisions, record the defect and prepare the repair plan. The predictive maintenance approaches adopt the latest sensors and computer vision technologies to analyze image data effectively and efficiently in the whole maintenance process. The system can shorten the measurement time for evaluating large areas of delamination and debonding and is suitable for a great variety of composite materials in the aviation industry. To meet the strict standards of the aircraft industry and improve the current inspection procedure, a new ultrasonic inspection system with image analysis function using single-shot multibox detector (SSD) network and computer vision approach has been designed and developed to assist the inspector to improve the efficiency and accuracy for identifying the defect from interpreting the ultrasonic scanning image, thereby reduce inspector's workload (Du, Shen, Fu, Zhang, & He, 2019). An effective inspection procedure not only shortens the maintenance time and reduces worker fatigue, but compared to conventional manual inspection, the novel approach of an automated inspection system can also enhance the speed and accuracy of the whole inspection process.

It can also improve the efficiency for inspection of delamination with higher precision. Therefore, this chapter focuses on illustrating the AI algorithm for defects inspection based on the UT.

1.2 Literature review

In this section, the inspection technologies applied in the industries and the case studies of inspection on aircraft composite materials and defects detection by using AI approaches are reviewed.

1.2.1 Composite material for the aerospace industry

In the aerospace industry, weight is a key criterion in materials selection. Many types of research are devoted to striving for improving the thrust-to-weight ratio of materials. Composite materials are widely used and frequently used in many aircraft and spacecraft structural applications and components. Composite materials have played an important role in reducing weight. Not only and carbon fiber reinforced polymer (CFRP) and glass-fiber reinforced plastic (GFRP) commonly used, but many fiber and metal laminates are also used.

Glass laminate aluminum reinforced epoxy (GLARE) is a fiber metal laminate (FML) that is commonly used in the aviation industry. Kakati and Chakraborty (2020) applied finite element (FE) analysis for delamination in GLARE laminates under impact. GLARE laminates have better impact properties compared to those of both solely aluminum and glass/epoxy composites.

In general, the material properties of composite materials are more complicated than monolithic materials, such as metal and plastic, and the composite material can be modified depending on the component usage.

1.2.2 Defects on composite materials

Delamination is the major failure mechanism of composite laminates and ultimately leads to material failure. Since delamination is a common failure mode in CFRP, attention has been paid to the mechanism of delamination crack propagation due to the inherent weakness between layers. Through nondestructive assessment, this phenomenon can be discovered early, with a better understanding of the appropriate in-situ repair method, to better assess its impact on the residual strength of lightweight structural components.

During repairs, composite parts may suffer fatigue damage, which may cause delamination and may probably decrease the rigidity of the laminate. The crack density and delamination rate are the indicators of structural damage. Transverse cracks, delamination, and fiber failure are the ultimate failure welds of laminate structures (Carraro, Maragoni, & Quaresimin, 2019). To discover these failure modes is the major function of the inspection system.

To effectively measure the functionalities of the inspection systems, Wong (2013) prepared layering defects. Before processing, up to 16 carbon epoxy composite prepreg laminates were stacked, and a 15-mm thick aluminum foil was inserted between the intermediate layers to provide initial delamination cracks and cured by the autoclave "mini-bonder." Compared with the source of a defective workpiece in industry, and the artificially produced standard defect can more easily control and assess the inspection equipment.

1.2.3 Defect inspection of composite materials

There are many nondestructive testing (NDT) methods for evaluating composite materials, for example, the visual test (VT), ultrasonic test (UT), thermal imaging, radiographic test (RT), electromagnetic test (ET), acoustic emission (AE) and shear imaging test, etc.

The UT configuration includes transmitter and receiver and display equipment. According to the information carried by the signal, features such as crack location, defect size, and direction can be inspected (Lu, Shi, Li, & Yu, 2010). Fast scanning speed, good resolution, and portability are the advantages, tolerates the disadvantages are the difficult setup, the skills required to accurately scan parts, and the need to test samples for calibration.

Pulse echo and pass-through transmission methods are two commonly used methods, in which both methods use high-frequency sound waves of the order of 1−50 MHz to detect internal defects in materials (Gholizadeh, 2016). UT is performed in three modes, transmission, reflection, and backscatter (Stonawski, 2008).

Pulse-echo ultrasound methods can easily locate defects in homogeneous materials. In this method, the operator focuses on the wave propagation time and energy loss due to defect attenuation and wave scattering. Whether the material is homogeneous or heterogeneous, helps to identify inconsistencies in the material (Warnemuende, 2006). For large defect inspection, positioning and imaging, and quality control, ultrasonic pulse inspection is a suitable solution (Oguma, Goto, & Sugiura, 2012).

Apart from the aviation industry, infrared thermography was developed for detecting delamination in CFRP-jacketed concrete members to analyze the accuracy of the NDT method. Artificial delamination was made for experiments with the test parameters includes size and depth, presence of surface cover mortar, and water content in the delamination void. Infrared thermography was applied to the specimens with active heating and could successfully identify delamination regions of 50 mm × 50 mm in size. However, it failed to identify the depth. A recent paper (Gholizadeh, 2016), states that UT is common for the different industries for inspection. For construction and building applications, maintenance technicians use UT to locate welding defects in structural steel members.

Zhao, Wang, Wang, Hao, and Luo (2019), Zhang et al. (2020), and Colombo, Bhujangrao, Libonati, and Vergani (2019) applied the ultrasonic method to detect the characteristics and effects of delamination on different composites components. B-scan images of ultrasonic and IR-thermography are both able to identify damage.

To conclude, in the aviation industry, nowadays, UT is one of the most reliable nondestructive examination methods for finding the delamination and debonding of composite materials in the aviation industry. The UT detector, on the other hand, must be used in combination with scanners or an automated system to allow for the recording of results in "C-scan" form, which shows a cross-section of the test sample parallel to the scanning surface. The processing time is relatively long, and the scanner systems are less practical for onsite and in-service inspections.

1.3 Defect detection algorithm

Haar-like features, local binary pattern, histogram of gradient boosting functions, and Cascade classifier for detection are commonly used in conventional target detection. The defect inspection methods based on deep learning can be divided into two categories including two-stage detector and one-stage detector. In the two-stage detection, the inspection task firstly generates the bounding boxes and predicts the boxes. R-CNN is a popular example of two-stage detectors (Wang, Li, Ji, & Wang, 2017). In the one-stage detection, there is no intermediate box generation. The prediction results are obtained directly from the pictures. You only look once (YOLO) and single-shot mulibox detector (SSD) are the common methods of a one-stage detector (Hong, Han, Kim, Lee, & Kim, 2019).

For the inspection of the two-stage detector, the first stage is to generate a set of frames, then the second stage applies the deep convolutional neural network (DCNN) to encode the additional frames with feature vectors and then predict the image class. In the case of the one-stage detector, it does not have an additional frame generator and an intermediate region. The prediction results are obtained directly from the pictures, which is a region-free method. Further details of the inspection algorithm are illustrated as follows.

1.3.1 R-convolutional neural network

The two-level detector divides the detection task into two stages which are generating a prototype box and predicting the title box. In the stage of generating the position frame, the detector may mark the area that may be an object. Second, the model classifies these regions where maybe backgrounds, or categories are defined in advance correctly. In addition, the model also redefines the localized location of the predetermined area.

The steps of R-CNN are to first, input the test image and use a selective search algorithm to extract from the bottom of the image about 2000 prospective areas that may contain objects, with different withdrawn areas, it is necessary to expand each area proposal to a uniform size of 227×227 grid and input it to CNN, using the output of CNN's fc7 layer as a feature. After that, input the CNN features extracted from each regional proposal are inputted to the SVM for classification.

The calculation information cannot be shared due of the characteristics of each box are separated by DCNN. The double calculations make R-CNN time-consuming in training and inference. The three steps of R-CNN, which include proposal generation, feature extraction, and region classification, are independent and cannot be optimized end-to-end for object detection. Selective searches are not accurate and effective as applied in complex scenes. In addition, Selective search cannot be accelerated by GPU processing.

1.3.2 You only look once

YOLO is the first example of using a one-stage approach and makes the detection task a unified, end-to-end regression problem. The object is named after processing the image only once to get the position and classification simultaneously.

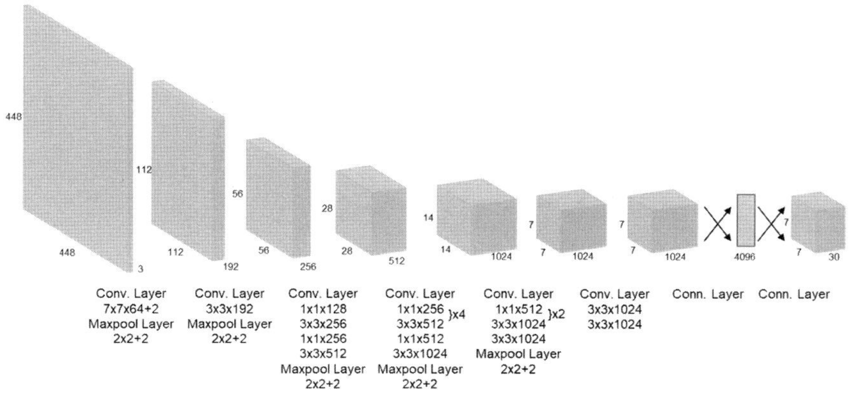

Figure 1.1 The architecture of you only look once.

Redmon et al. (2016) proposed the real-time detector YOLO. YOLO regards object detection as a regression problem, dividing the image into a fixed number of grid cells (e.g., 7×7 grid). Fig. 1.1 shows the network structure of YOLO. Each cell predicts whether one or more objects exist and is considered to contain a central location of up to two objects. Prediction is made that includes the location for an object, the bounding box's coordinates, size (width/height), and category of the object.

The entire framework is a single network, minus the proposal generation process, so the end-to-end manner can be optimized. YOLO is designed to give lightweight architecture that can achieve an inference speed of 45 frames per second (FPS), and even can reach 155 FPS in lighter structures.

It is relatively fast compared to the general two-stage detector. Global processing makes background errors relatively small, compared to local (regional) based methods such as Fast R-CNN. The generalization performance of YOLO performs better than fast R-CNN.

However, YOLO also has some defects. First, only two objects can be detected in a given place at most. It is difficult to detect small objects or crowded objects because only the feature map of the last layer is used for prediction, and it is not suitable for the detection of multisize objects.

1.3.3 Single-shot mulibox detector

Liu et al. (2016) proposed a one-stage detector called SSD to solve the shortcomings of YOLO. Fig. 1.2 shows the architecture of SSD. Similar to YOLO, SSD also divides the image into grid cells but generates a series

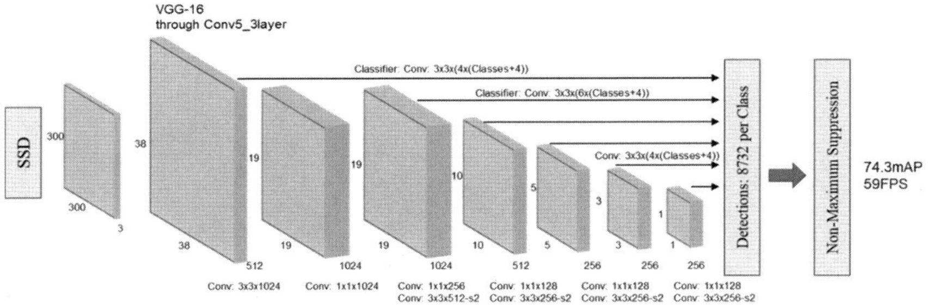

Figure 1.2 The architecture of single-shot mulibox detector.

of anchors of different sizes and ratios in each grid cell. YOLO makes predictions from fixed grid cells. At the same time, SSD predicts objects on multiple feature maps. Each layer of feature maps must detect objects of a certain size based on the receptive fields.

To detect more object categories, additional convolutional feature maps are added. In the training process, the localization loss and classification loss of all predicted maps are added together as a total loss function to optimize the entire network. Combining all detection results of different feature maps as the final prediction, SSD achieves similar detection accuracy as Faster R-CNN in the inference stage, but the speed is faster.

The SSD output discretizes bounding boxes, which are generated on the feature maps of different layers with different aspect ratios. In the prediction stage, to deal with different sizes of the same object, SSD combines the prediction of different feature maps. The detection model of the proposed object, such as the faster R-CNN. SSD method has eliminated the generation of pixel resampling and feature resampling. It is easier to use SSD to optimize training and integrate the detection model into the system.

1.3.4 Single-shot mulibox detector versus you only look once

Fig. 1.3 shows the performance comparison of SSD and YOLO. Compared to YOLO, SSD has a multiscale function graph based on different convolutional segments of VGG and is capable of outputting the function graph to the regressor. It improves the detection accuracy of small objects for SSD. Second, SSD has more anchor boxes. Each grid point generates boxes of different sizes and length-to-width ratios, divided into category prediction probabilities based on box prediction. However, YOLO is on the grid. To summarize, SSD is an effective single-stage

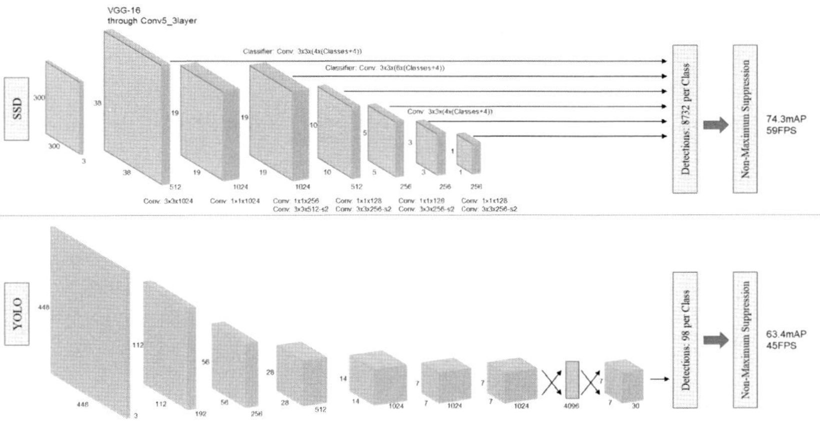

Figure 1.3 The comparison between single-shot mulibox detector and you only look once.

model which can achieve accuracy close to that of two-stage models and faster than two-stage models.

1.3.5 Convolutional neural network-based object detection in nondestructive testing

Ultrasonic flaw classification in welding is an active area of research. Lots of artificial intelligence (AI) approaches have been applied to reality. During the ultrasonic inspection, the signals are required to reduce any environmental noise. Automatic intelligent defect classification algorithms show relatively low classification performance.

Munir, Kim, Park, Song, and Kang (2019) applied the CNN to separate the noisy ultrasonic signatures to improve the classification performance of welding defects and applicability. The result shows that CNN is robust, does not require specific feature extraction methods, and gives considerate high defect classification accuracies, even for noisy signals.

Moreover, CNN image understanding technology is widely used in autonomous driving, industry, and medicine. In the diagnosis of hydronephrosis, some pathological symptoms and causes are difficult to judge. To improve the ultrasonic diagnosis of hydronephrosis, based on CNN neural network technology, the traditional image resolution processing algorithm was analyzed through a comparative analysis method, and a fast image superresolution reconstruction algorithm was proposed by Zhang et al. (2020).

This method improves the efficiency of image processing and combines CNN neural networks to effectively treat ultrasound images of hydronephrosis. In addition, Zhang et al. (2020) analyzed the clinical effects of the algorithm by establishing controlled trials. The analysis showed that this study can obtain effective diagnostic results, that is, the diagnostic results can provide an effective reference for the diagnosis of hydrocephalus ultrasound images and have certain clinical significance.

Nighttime vehicle recognition based on driver assistance systems (DASs), is targeted at enhancing the driver perception of the environment. Kang, Zhao, Yang, Ahmed, and Ma (2020) applied thermal infrared cameras in the field of DAS and studied the performance of nighttime vehicle recognition. A dataset of thermal infrared images comprising four vehicle classes was generated: bus, truck, van, and car. To achieve robust vehicle recognition, a CNN was developed to classify the four vehicle types. The recognition time of a single picture was only 0.52 Ms on a desktop with a common CPU. Thermal cameras were useful to detect pedestrians during the nighttime with low external light. However, thermal cameras are currently expensive and thus difficult to install in many places. Kim, Batchuluun, and Park (2018) proposed a method of pedestrian detection at nighttime using a visible-light camera and a faster R-CNN.

For ancient buildings' protection and maintenance, it is crucial to identify the damaged components. However, the current identification and statistics work is mainly carried out by the human eye, which is time-consuming and labor-intensive. Zou, Zhao, Zhao, Qi, and Wang (2019) proposed a methodology to identify and count components based on CNN for the Forbidden City. In Beijing, the algorithm was structured by faster R-CNN, which is effective in object detection. In addition, the positions of the missing components can be labeled in the images.

Nasiri, Taheri-Garavand, Omid, and Carlomagno (2019) proposed an accurate radiator condition monitoring system using a deep CNN. The CNN model was constructed with VGG-16 and batch normalization layer, dropout layer, and dense layer. This model is directly applied the infrared thermal images to classify the conditions of the radiator. The results proved that the CNN model is better than traditional methods with higher performance and accuracy.

Liu et al. (2019) developed a novel noncooperative Target Detection of Spacecraft Objects based on the Artificial Bee Colony (ABC) Algorithm. ABC method and proposed a multipeak optimization algorithm named Species-based Artificial Bee Colony (SABC) which was applied to detect an NCT. Experiments were conducted using real cases of "Shenzhou 8" and "Apollo 9" space missions as well as the "Change" Camera Point System (CPS) developed by the Hong Kong Polytechnic University.

In summary, objective detection using CNN-related technologies is commonly and maturely applied to autonomous-related applications. It is still a new technology and there is still large room to develop the technology especially in a labor-intensive task such as inspection. Applying the SDD network to detect defects on composite materials is a brand-new technology, that can contribute to the industries.

1.4 Deployment of defect detection

1.4.1 Setting up of the deep learning environment

Before installing the packages, a library version is needed to run the platform to be clearly understood. The package version is critical to the whole system environment, and an incompatible library version leads the system damaged. There are no official references and instructions. However, the latest version is likely to have compatibility issues. For example, TensorFlow 1.3 supports up to Python 3.6 (as of August 1, 2017) but Theano only supports up to python 3.5. Anaconda's latest version runs on Python 3.6 but a Python Conda environment can be created with Python 3.5.

1.4.1.1 NVidia Tensorflow Object Detection API

The NVidia TensorFlow Object Detection application programming interface (API) is an open-source framework built on top of the TensorFlow. Since building a new and valid convolutional neuron network is time-consuming, the TensorFlow Object Detection API provides a fast and convenient way for detection. However, building a defect detection model from scratch can be difficult and can take a very long time to train the model. The TensorFlow Object Detection API offers a very useful framework to construct, train and deploy object detection models. It also allows the developer to use pretrained object detection models or create and train a new model by making use of transfer learning. This enables us to create accurate machine learning models that are capable of localizing and identifying multiple objects in a single image to deal with a core challenge in computer vision.

TensorFlow Object Detection API requires the following libraries:

- Protobuf 3.0.0
- Python-tk
- Pillow 1.0
- lxml
- tf Slim (which is included in the "tensorflow/models/research/ "checkout)
- Jupyter notebook
- Matplotlib
- Tensorflow (> = 1.12.0)
- Cython
- contextlib2
- cocoapi

TensorFlow Object Detection API and the libraries are installed in the Nvidia to train the object detection algorithm.

1.4.1.2 TensorRT

We propose to set up defect detection in the NVidia environment. The TensorRT can be used to optimize the neural network model trained in all major frameworks, calibrate with lower accuracy to achieve high accuracy, and finally deploy on different platforms. Reduced precision inference can significantly reduce application latency, which is a real-time service for many automated and embedded applications. During the reasoning process, TensorRT-based applications run 40 times faster than CPU-based platforms.

The trained models in each deep learning framework can be imported into TensorRT. After application optimization, TensorRT selects platform-specific cores to maximize the performance on the GPU of the Jetson TX2 platform.

TensorRT uses a trained network consisting of network definitions and a set of trained parameters and generates a highly optimized runtime engine to facilitate high-performance inference of NVIDIA graphics. TensorRT applies graph optimization, layer fusion, and other optimization approaches, while also using various highly optimized kernels to find the fastest implementation of the model.

1.4.1.3 OpenCV

OpenCV is a cross-platform computer vision and machine learning software library to perform real-time image processing, computer vision, pattern recognition programs and supports certain models in the deep learning framework based on the defined list of supported layers, such as TensorFlow, and Caffe. The library has more than 2500 optimization algorithms, including a complete set of classic and latest computer vision and machine learning algorithms (Culjak, Abram, Pribanic, Dzapo, & Cifrek, 2012). OpenCV is mainly intended for real-time vision applications and utilizes Multi-Media eXtension (MMX) and Streaming SIMD Extensions (SSE) instructions when available.

1.4.2 Model training

Before the model training, the detection method is first identified. In the defect inspection, the inspection speed is one of the core concerns. The speed of defect detection is calculated based on the FPS. Faster R-CNN can only run at 7 fps in higher-level hardware (Liu et al., 2016). Even if there is network fine-tuning to speed up the detection, the detection accuracy can be lowered. Although the faster R-CNN method has higher accuracy, it also has the problem of excessive calculation. To balance between the detection performance and deployment configuration, SSD is a better choice to apply in this project rather than R-CNN.

Secondly, the SSD Common Objects in Context (COCO) (i.e., ssd_mobilenet_v1_coco) is selected for object detection (Liu et al., 2016). The object detection model aims to localize and identify multiple objects in a single image at the same time. This model is a TensorFlow.js port of the COCO-SSD model and is defined in the COCO dataset, which is large-scale object detection, segmentation, and captioning dataset. The model can detect 90 classes of objects. It can also take inputs of image elements (e.g., <img>, <video>, <canvas> elements) and returns an array of bounding boxes with class name and confidence level.

Thirdly, the TFRecords for training are required to be generated. The TFRecords are created with the labeled images which can serve as input data for training. To create the TFRecords we use two scripts which are xml_to_csv.py and generate_tfrecord.py files to transform the created xml files to csv correctly.

Fourthly, the object detection training pipeline is configured. The TensorFlow Object Detection API uses Protocol Buffers (Protobuf) to configure the training and evaluation process. The schema for the training pipeline can be split into five parts, model configuration, training configuration, evaluation configuration, train input configuration and evaluation output configuration. Once the pipeline configuration is done, the training can start anytime by running the train.py file.

Eventually, after the previous steps, training a model may take over a day to complete. After completion, the trained model is exported and an inference graph is generated. To obtain a complete training result, the highest training step number is recorded to navigate the training directory and find the model.ckpt file with the largest index, and save finally for validation and deployment.

1.4.3 Deployment in NVidia jetson TX2

1.4.3.1 Program structure

Training interference is the output of the SSD network. Utilizing these outputs for the defect classification and the coordinate of the boundary box is the core of defect detection. Since the inspection process is a labor-intensive task that relies on a qualified inspector, building an automatic inspection algorithm and generating a report from the inspection system not only reduces the inspectors' workload but also boosts the inspection speed and even the whole maintenance time.

A program is written to put the image and information in the blank report. The user only needs to input the name of the ultrasonic C-scan image file, and the program then automatically generates the full report. Fig. 1.4 shows the function involved in the deployment process.

To test the program, the original image in 173 × 173 pixels is resized to 500 × 500 pixels for better arrangement and calculation. Then, the

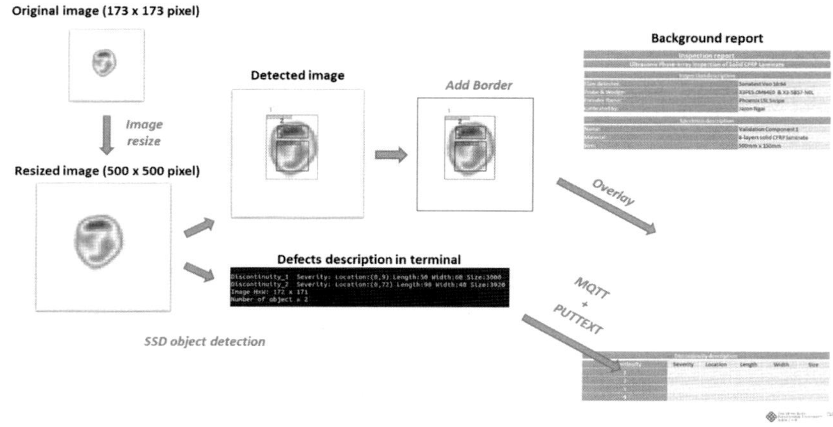

Figure 1.4 The functions involved in the deployment process.

resized image is used to perform the SSD object detection. The output detected images have several boundary boxes with the color depending on the cataloges and number of the defects found. Another output is the description of the defects in the terminal. The image output will have a border to clearly show the boundary of the images and then overlay to the blank report. The text output is sent and displayed on the report by using the self-design protocol, MQTT, which can transmit the message to any device connected internet.

1.4.3.2 OpenCV

In the program structure, there are at least four OpenCV functions that are applied to modify the report. Initially, the image resized function is applied to change the size of the image. Then, the output image is added to a border. Since the background of C-scan images is sometimes in white which is easily confused with the inspectors about the image sizes are the border-image can make it clearer. After that, two images will be combined by using the overlay function. Finally, once the terminal message of defect descriptions is received, the text related to the inspection result is added by using the "puttext" function.

1.4.3.3 MQTT

The MQTT protocol is a machine-to-machine (M2M)/IoT connectivity protocol for messaging. The nature of MQTT is an extremely lightweight publish/subscribe messaging transport approach which is useful for connections with remote locations where a small code footprint is required and/or network bandwidth is at a premium.

The source code for the Eclipse Paho MQTT Python client library applies versions 3.1 and 3.1.1 of the MQTT protocol. It offers a client class that allows applications to connect to an MQTT broker to publish messages and to subscribe to topics and collect published messages. It also makes publishing one-off messages to an MQTT server very straightforward.

1.4.4 Validation

Finally, validation is important for neuron network training. Validation is the process of making sure that the model generalizes well. The generalization of the model means that it is built using one set of data and it performs well on a completely different set of data. The validation set is then improved to minimize the generalization error.

After the program is done, the validation test sets, maybe several pieces of C-scan images which are out of the original training and testing sets, are used to validate the trained model and program.

1.5 Implementation

To implement the defect detection of composite materials in aircraft or spacecraft, some procedures and processes are required, including dataset preparation, scanning setup and configuration, model training, and final deployment.

1.5.1 Dataset preparation

Since the project focuses on composite delamination defects, the samples can be classified into two types. The first type is a standard defect. The delamination defects inside the sample are artificially made of the Teflon strips. The second type is deflected due to improper repair work. The random delamination defects in repair work are due to improper techniques and procedures. The preparation of the diverse samples and the dataset is important to improve the accuracy of predictions.

1.5.2 Defect scanning

To test the algorithm, two repaired work samples and one standard defect was prepared for the phased array ultrasonic (PAUT). PAUT is an advanced nondestructive examination technique that utilizes a set of UT probes made up of numerous small elements, each of which is pulsed individually with computer-calculated timing ("phasing"). When these elements are excited using different time delays, the beams can be steered at different angles, focused on different depths, or multiplexed over the length of a long array, creating an electronic movement of the beam. To operate the scanning procedures, the wedge delay, probe sensitivity, and encoder require calibration. The sample size is scanned based on the index distance, which is 20 mm in our settings.

After setup, shown in Fig. 1.5, the artificial defects are scanned by using the recording function. The results and the C-scan images are output into the jpeg files. The intensity of the image signal reflection was adjusted and saved into different files for increasing the training data size. Then, the C-scan image is cropped into lots of small square images for better organization. Fig. 1.6 shows the dataset ultrasonic C-scan images postprocessing.

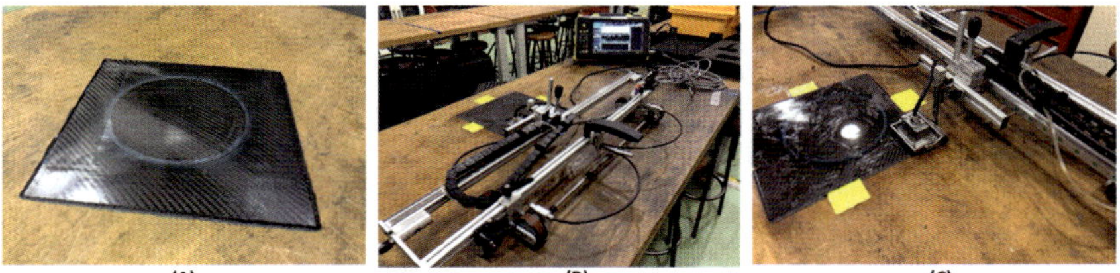

(A) (B) (C)

Figure 1.5 (A) Artificial defect–repair; (B) ultrasonic inspection equipment; (C) ultrasonic scanning.

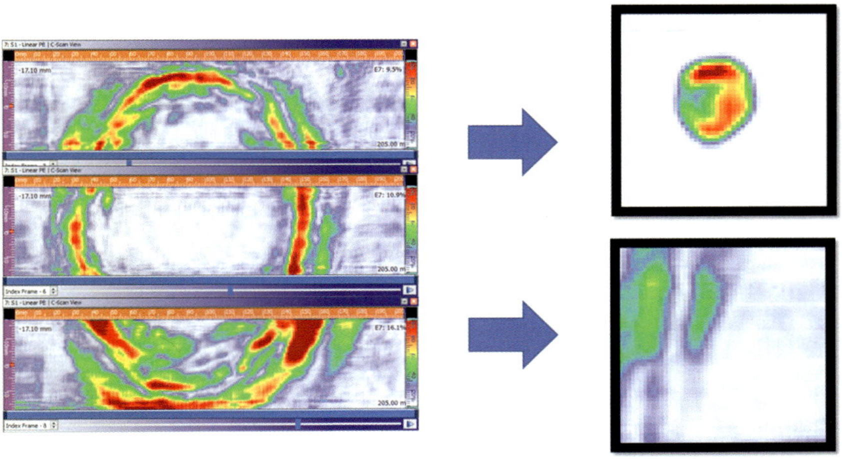

Figure 1.6 Dataset ultrasonic C-scan images postprocessing. The intensity of the image signal reflection was adjusted and saved into different files for increasing the training data size.

1.5.3 Image augmentation

After the image scanning, the original C-scan images are cropped into small jpeg format images in a square shape with 173×173 pixels as shown in Fig. 1.7. Batch-Image-Cropper (BIC) a free image cropping program for Windows, is used to boost the process for resizing and cropping the images by cropping the c-scan image in a batch. Image augmentation techniques, such as resizing, cropping and rotating, are applied in the C-scan images dataset. In our case study, 58 images are captured from the composite standard defects and 431 images are captured from the composite repair works. There are 489 images in total for the dataset.

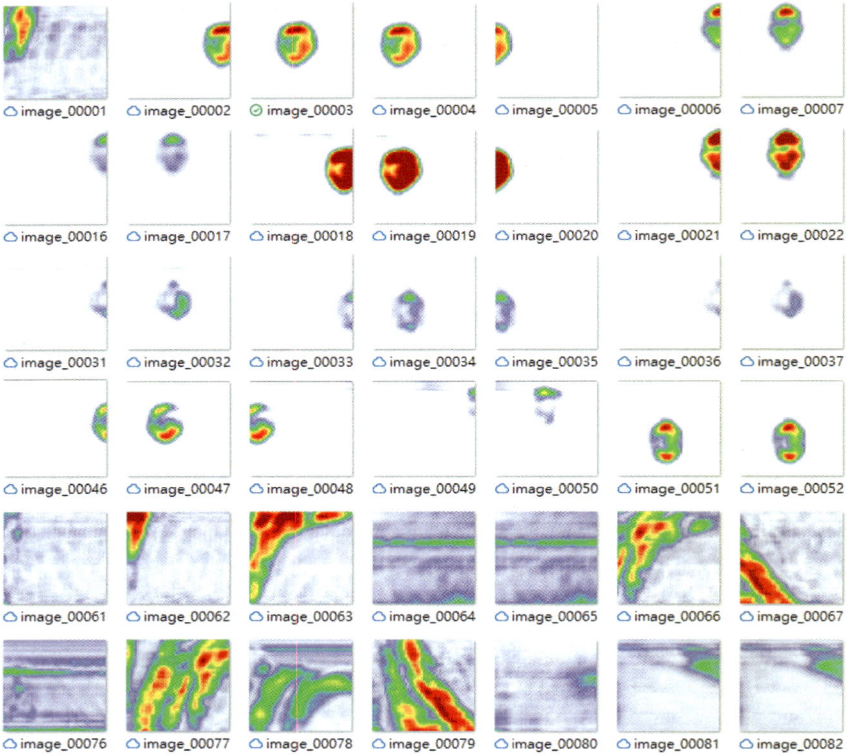

Figure 1.7 Processed training images.

1.5.4 Image annotation

After preparing the dataset of delamination, download labelImg. Image annotation is the next step to handle the dataset. First, download labelImg (Windows_v1.5.0). The latest version of labelImg (Windows_v1.6.0) is not recommended due to software bugs. LabelImg is a graphical image annotation tool and labels object bounding boxes in images. It is written in Python and uses Qt for its graphical interface. Annotations are saved as XML files in PASCAL VOC format, the format used by ImageNet. Besides, it also supports the YOLO format.

After downloading the labelImg, unzip the downloaded file, open the data/predefined_classes.txt in the same folder as labelImg.exe (use Notepad + + to open) and then edit the name of the class to be marked. In this project, there are three categories of defects to be classified by the SSD network, which are discontinuity, high severity discontinuity, and low severity discontinuity, depending on the intensity of the ultrasonic sound. After opening labelImg.exe. Insert the image file to be marked and place it in the same folder. Click "Open Dir" in the left column menu and select the folder where the image file is placed. Click "Change Dir" in the left column menu to set the storage path of the annotation mark (XML file). Click "Create RectBox" and select the object to be marked by dragging the mouse on the image, and then select the category of this object. After marking, click "Save" to save the annotation. Click "Next Image" to mark the next image. After completing the tagging of all images, an XML file will be created in the folder where the tags are stored.

1.6 Results

1.6.1 Loss

Loss is the penalty for a bad prediction. That is, the loss is a number indicating how bad the prediction of the model is on a single example. If the prediction of the model is perfect, the loss is zero; otherwise, the loss is greater. The goal of training a model is to find a set of weights and biases that have low loss, on average, across all examples.

1.6.1.1 Classification loss and localization loss

The classification task aims to recognize the object in the given box, while the localization task aims to predict the precise bounding box of the object (Chen et al., 2019). Loss used for training is a sum of classification and localization losses, shown in Figs. 1.8 and 1.9, with the latter calculated only for boxes with the same class label as the ground truth.

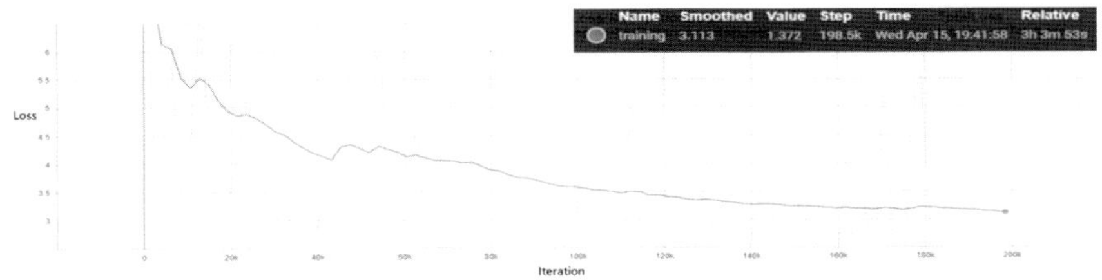

Figure 1.8 Classification loss.

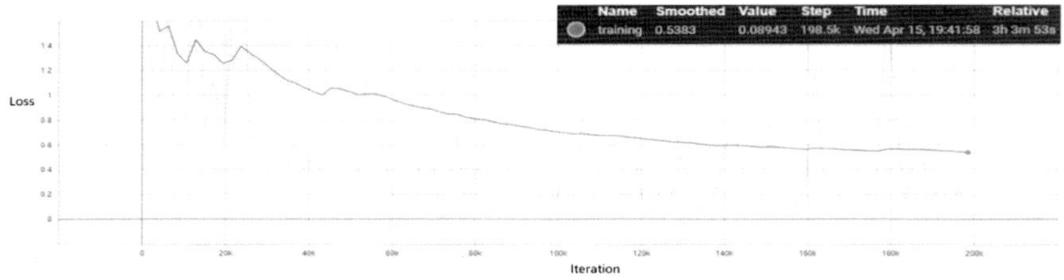

Figure 1.9 Localization loss.

1.6.1.2 Network configuration comparison and improvement

The results in Table 1.1 show that the batch size, losses, and duration of the network interference.

It has been observed in practice that when using larger batches of products, the size quality of the model will be reduced, measured by the generalization ability of the model. Large batch methods tend to converge to the minimum of training and testing functions, and it is well known that minimum values lead to poor generalization.

Therefore, the choice of small batch size affects the training time until convergence. Fig. 1.10 shows the relation between batch size and losses, and Fig. 1.11 shows the relation between batch size and training duration. The smaller value allows the learning process to quickly converge at the expense of noise during the training process. If the batch size is large, then it will also exceed the minimum value. The larger value makes the learning process converge slowly and accurately estimates the error gradient.

Batch size = 32 is a good default value, recommended by Bengio (2012) and Masters and Luschi (2018). However, in this case, batch size = 24 is the best value to achieve the minimum both in classification loss and localization loss. The interference of batch size = 32 outputted for deployment.

1.6.2 Validation of the defect detection system

To implement the deflect detection, hyperparameter optimization is performed to offer minimum loss to the interference for the defect detection system. Then, the validation test sets are prepared to validate the functionality of the system.

1.6.2.1 Validation test sets

The validation test sets were generated by using the market-available graphics editing tools. To ensure the accuracy of the validation test, the 489

Table 1.1 Batch size, losses, and duration of training.

Batch size	Step	Classification loss	Localization loss	Duration
2	200k	1.866	0.1815	03:03:53
8	200k	1.415	0.1074	04:09:52
16	200k	1.236	0.09979	06:05:45
24	200k	0.9381	0.04811	07:31:41
32	200k	1.127	0.0784	09:37:37
64	200k	1.117	0.06038	16:25:47

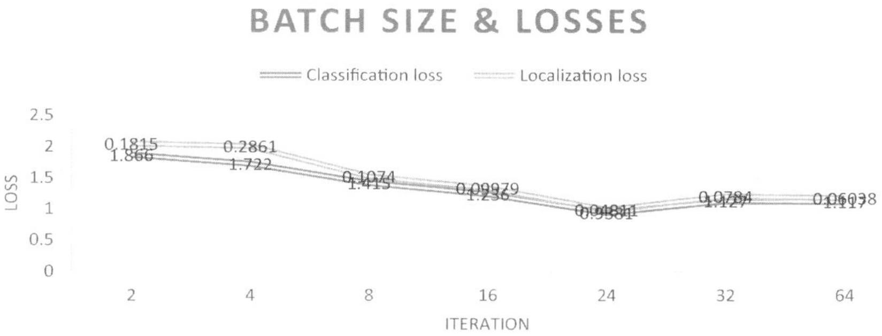

Figure 1.10 Relation between Batch size and Losses.

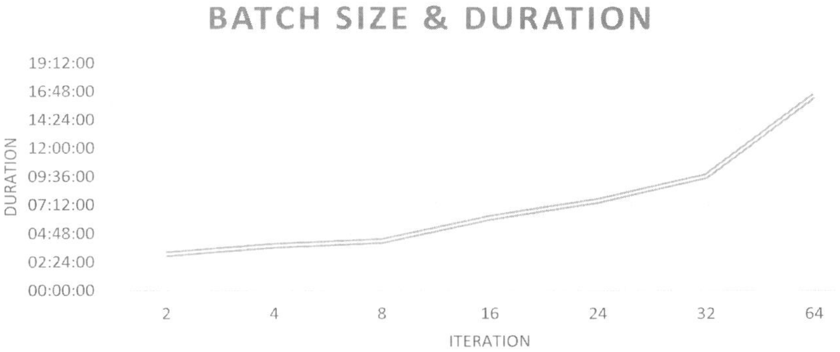

Figure 1.11 Relation between Batch size and Training duration.

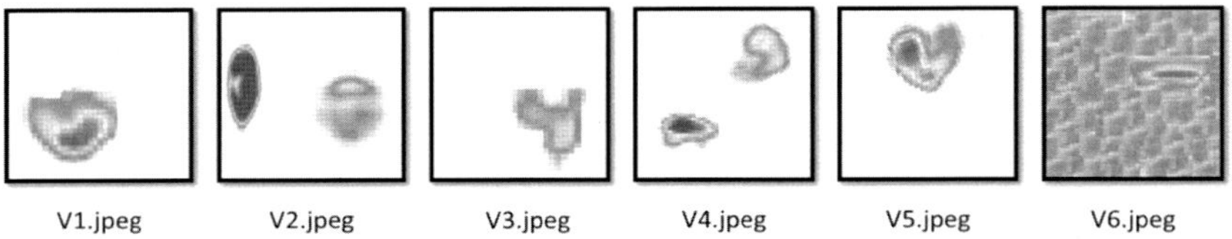

Figure 1.12 Validation image from V1 to V6.

images already used for training and testing are not used again for validation. There is a total of six validation test sets, which are shown in Fig. 1.12. The defect severity and location are random. The details of the validation test sets are shown below.

1.6.2.2 Manual labeling

Then, the validation test sets are classified and measured manually. A bounding box represented by the rectangle to surround the discontinuities of the severity is first identified. Then, the coordinates of the rectangle and the size of the bounding box are determined in terms of cartesian coordinates and millimeters (mm). All the jpeg files are labels with a suffix "m" which represents manually marked, as shown in Fig. 1.13. Table 1.2 summarizes the manual labeling results.

1.6.2.3 Preliminary result of system and improvement

The testing images were inputted into the defect detection system, however, some errors were found in V1.jpg shown as shown in Fig. 1.14. An unexpected boundary box, with a low confidence level of 0.48 exists in the image. To eliminate the unexpected boundary box for the robust result, boundary boxes a lower than 80% confidence levels were eliminated. The visualization.py was edited and an "if" function to visualize the boundary boxes is added. "***if cf > 0.8***" prevents a boundary box lower the 80% confidence level from appearing in the result.

1.6.2.4 Automatic inspection

After preliminary adjustment, the six validation images were inputted into the defect detection system again. Six reports were generated and are shown in Fig. 1.15. The images and defect information are all printed on the report. Table 1.3 is generated to record the automatic inspection results. All the report images are listed below.

The results show that the automatic system cannot classify the low severity discontinuity in the V6.jpg. In the V6.jpg, noises are added in the

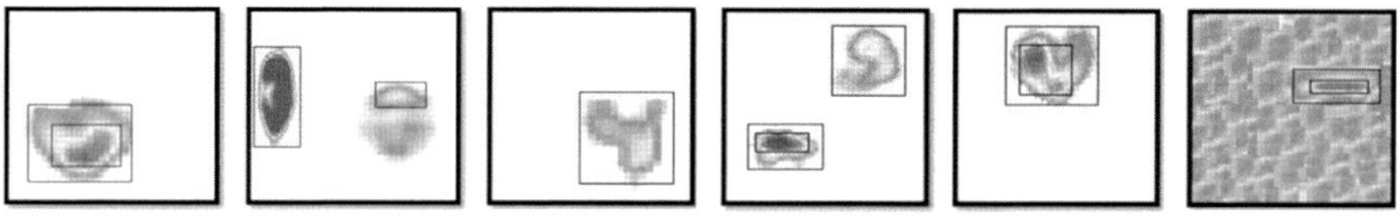

Figure 1.13 Manual inspection on validation images.

Table 1.2 Manual inspection results.

	Discontinuity	Severity	Location *x*-axis	Location *y*-axis	Length (mm)	Width (mm)	Size (mm^2)
V1.jpg	1	LO	8.6	19.5	17.5	21.9	383.25
	2	HI	11.5	20	12.5	14.5	181.25
V2.jpg	1	LO	21	10	17	17.5	297.5
	2	HI	0.7	12	22.4	9.9	221.76
V3.jpg	1	LO	18.5	5.2	18	19.4	349.2
V4.jpg	1	LO	3	6.5	9.5	20	190
	2	LO	22.5	23.4	15.2	16	243.2
	3	HI	6.5	11.1	4	12.2	48.8
V5.jpg	1	LO	9	18	18	19.5	351
	2	HI	13.2	20	13	12	156
V6.jpg	1	LO	21.2	21.2	7.5	17.6	132
	2	HI	24.3	23.5	3.2	11.9	38.08

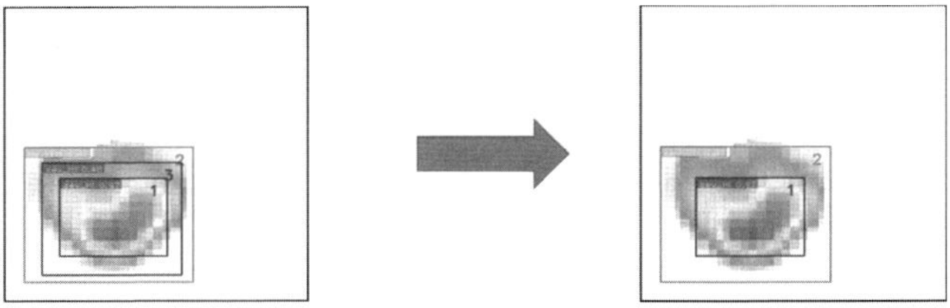

Figure 1.14 The problem caused by the low confidence level.

background, and maybe a reason that affects the performance of the system. The reason is not just simply due to the threshold for the confidence level interval, but also due to the sample size of the dataset.

1.6.2.5 Comparison between automatic and manual inspection

Table 1.4 shows the results between the automatic and manual inspection. In the manual inspection, 11 defects were classified from 6 images manually, while the automatic inspection has classified 10 defects. In terms of

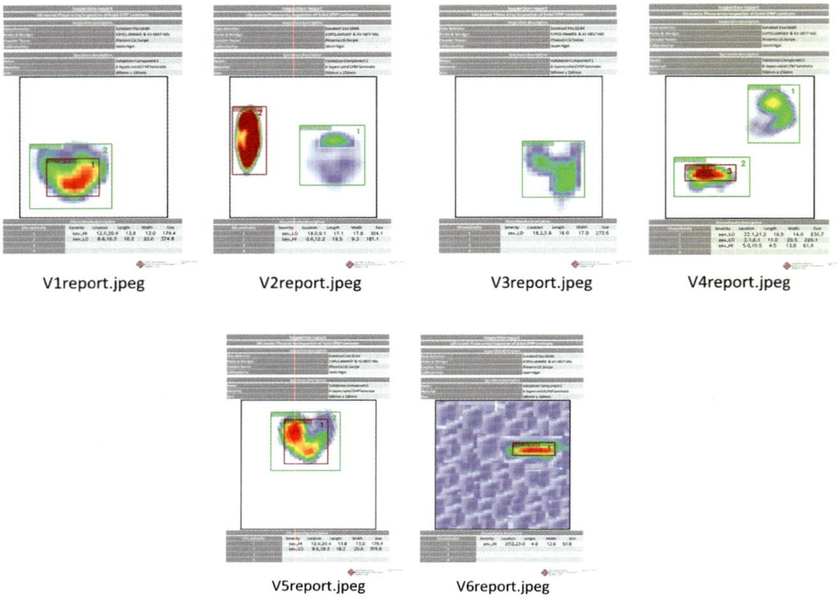

Figure 1.15 Auto-generated reports from V1 to V6.

Table 1.3 Auto-inspection results.

	Discontinuity	Severity	Location *x*-axis	Location *y*-axis	Length (mm)	Width (mm)	Size (mm^2)
V1.jpg	1	LO	8.6	18.3	18.2	20.6	374.92
	2	HI	12.6	20.4	13.8	13	179.4
V2.jpg	1	LO	19	9	17.1	17.8	304.38
	2	HI	0.6	12.2	22.4	9.3	208.32
V3.jpg	1	LO	18.2	5.6	16	17	272
V4.jpg	1	LO	2.1	6.1	11	20.5	225.5
	2	LO	22.1	21.2	16.5	14	231
	3	HI	5	10.5	4.5	13.8	62.1
V5.jpg	1	LO	8.6	18.3	18.2	20.6	374.92
	2	HI	12.6	20.4	13.8	13	179.4
V6.jpg	1	—	—	—	—	—	—
	2	HI	23	23	4	12.6	50.4

the total 50 defects, there was only one unrecognized defect which counts for a 5% difference compared with the manual inspection results. The system is reliable for assisting the inspector to classify the discontinuities in the ultrasonic phased-array inspection.

Table 1.4 Auto-inspection results.

	Discontinuity	Severity	Location x-axis	y-axis	Length	Width	Size
V1.jpg	1	LO	0.00%	−3.00%	1.75%	−3.25%	−0.52%
	2	HI	2.75%	1.00%	3.25%	−3.75%	−0.12%
V2.jpg	1	LO	−5.00%	−2.50%	0.25%	0.75%	0.43%
	2	HI	−0.25%	0.50%	0.00%	−1.50%	−0.84%
V3.jpg	1	LO	−0.75%	1.00%	−5.00%	−6.00%	−4.83%
V4.jpg	1	LO	−2.25%	−1.00%	3.75%	1.25%	2.22%
	2	LO	−1.00%	−5.50%	3.25%	−5.00%	−0.76%
	3	HI	−3.75%	−1.50%	1.25%	4.00%	0.83%
V5.jpg	1	LO	−1.00%	0.75%	0.50%	2.75%	1.50%
	2	HI	−1.50%	1.00%	2.00%	2.50%	1.46%
V6.jpg	1	—	—	—	—	—	—
	2	HI	−3.25%	−1.25%	2.00%	1.75%	0.77%

1.7 Conclusions

In this chapter, the framework of automatic defect detection for the ultrasonic inspection for composite materials was introduced. The approach can be used to detect discontinuities in the C-scan images and generate an inspection report. The time for scanning and creating the inspection result is less than 30 s. The SSD network is applied to automatically detect the defects in the pictures. Defects with different severities and uneven shapes in the noisy background can be detected and be correctly sized. A bounding box is created and integrated into the images to identify the locations of the defects and provide the inspection information in the report. The defect detection report can improve the effectiveness and efficiency of the inspection.

However, there are limitations in the current system, which in general is not able to handle complex defect geometries. In addition, the stability of the inspection performance can be improved to avoid any environmental interruption. In the validation, it is shown that if the background of the image is noisy, the system may not detect the discontinuities precisely. This is related to the training dataset and the generalization of the network. In general, and well-prepared and labeled images can reduce the losses of the network. Avoiding overtraining and overfitting can optimize network performance. It is expected that a larger sample size of the ultrasonic scanning datasets can be used to improve the accuracy of the networks further.

Acknowledgment

We acknowledge the support of the Laboratory for Artificial Intelligence in Design (Project Code: RP2-1), Innovation and Technology Fund, Hong Kong Special Administrative Region, and the Research Committee of The Hong Kong Polytechnic University (project account code: RK3N), in preparing this chapter.

References

Bengio, Y. (2012). *Practical recommendations for gradient-based training of deep architectures. Neural networks: Tricks of the trade* (pp. 437–478). Berlin, Heidelberg: Springer.

Carraro, P. A., Maragoni, L., & Quaresimin, M. (2019). Characterisation and analysis of transverse crack-induced delamination in cross-ply composite laminates under fatigue loadings. *International Journal of Fatigue*, *129*, 105217.

Chen, K., Li, J., Lin, W., See, J., Wang, J., Duan, L., ... Zou, J. (2019). Towards accurate one-stage object detection with ap-loss. In *Proceedings of the IEEE/CVF conference on computer vision and pattern recognition* (pp. 5119–5127).

Chen, X., Ren, H., & Bil, C. (2014). Inspection intervals optimization for aircraft composite structures considering dent damage. *Journal of Aircraft*, *51*, 303–309. Available from https://doi.org/10.2514/1.C032377.

Colombo, C., Bhujangrao, T., Libonati, F., & Vergani, L. (2019). Effect of delamination on the fatigue life of GFRP: A thermographic and numerical study. *Composite Structures*, *218*, 152–161.

Culjak, I., Abram, D., Pribanic, T., Dzapo, H., & Cifrek, M. (2012). A brief introduction to OpenCV. In *2012 proceedings of the 35th international convention MIPRO* (pp. 1725–1730). IEEE.

Du, W., Shen, H., Fu, J., Zhang, G., & He, Q. (2019). Approaches for improvement of the X-ray image defect detection of automobile casting aluminum parts based on deep learning. *NDT & E International*, *107*, 102144.

Gholizadeh, S. (2016). A review of nondestructive testing methods of composite materials. *Procedia Structural Integrity*, *1*, 50–57.

Ho, G. T. S., Tang, Y. M., Tsang, K. Y., Tang, V., & Chau, K. Y. (2021). A blockchain-based system to enhance aircraft parts traceability and trackability for inventory management. *Expert Systems with Applications*, *179*, 115101. Available from https://doi.org/10.1016/j.eswa.2021.115101.

Hong, S. J., Han, Y., Kim, S. Y., Lee, A. Y., & Kim, G. (2019). Application of deep-learning methods to bird detection using unmanned aerial vehicle imagery. *Sensors*, *19* (7), 1651.

Kakati, S., & Chakraborty, D. (2020). Delamination in GLARE laminates under low velocity impact. *Composite Structures*, *240*, 112083.

Kang, Q., Zhao, H., Yang, D., Ahmed, H. S., & Ma, J. (2020). Lightweight convolutional neural network for vehicle recognition in thermal infrared images. *Infrared Physics & Technology*, *104*, 103120.

Katnam, K. B., Da Silva, L. F. M., & Young, T. M. (2013). Bonded repair of composite aircraft structures: A review of scientific challenges and opportunities. *Progress in Aerospace Sciences*, *61*, 26–42.

Khaira, A., Srivastava, S., & Suhane, A. (2015). Analysis of relation between ultrasonic testing and microstructure: A step towards highly reliable fault detection. *Engineering Review: Međunarodni časopis namijenjen publiciranju originalnih istraživanja s aspekta analize konstrukcija, materijala i novih tehnologija u području strojarstva, brodogradnje, temeljnih tehničkih znanosti, elektrotehnike, računarstva i građevinarstva*, *35*(2), 87–96.

Kim, J. H., Batchuluun, G., & Park, K. R. (2018). Pedestrian detection based on faster R-CNN in nighttime by fusing deep convolutional features of successive images. *Expert Systems with Applications*, *114*, 15–33.
Liu, W., Anguelov, D., Erhan, D., Szegedy, C., Reed, S., Fu, C. Y., & Berg, A. C. (2016). Ssd: Single shot multibox detector. In *European conference on computer vision* (pp. 21–37). Cham: Springer.
Liu, X., Li, D., Dong, N., Ip, W. H., & Yung, K. L. (2019). Noncooperative target detection of spacecraft objects based on artificial bee colony algorithm. *IEEE Intelligent Systems*, *34*(4), 3–15.
Lu, R. S., Shi, Y. Q., Li, Q., & Yu, Q. P. (2010). *AOI techniques for surface defect inspection*, . *Applied mechanics and materials* (Vol. 36, pp. 297–302). Trans Tech Publications Ltd.
Masters, D. & Luschi, C. (2018). *Revisiting small batch training for deep neural networks*. arXiv preprint arXiv:1804.07612.
Munir, N., Kim, H. J., Park, J., Song, S. J., & Kang, S. S. (2019). Convolutional neural network for ultrasonic weldment flaw classification in noisy conditions. *Ultrasonics*, *94*, 74–81.
Nasiri, A., Taheri-Garavand, A., Omid, M., & Carlomagno, G. M. (2019). Intelligent fault diagnosis of cooling radiator based on deep learning analysis of infrared thermal images. *Applied Thermal Engineering*, *163*, 114410.
Oguma, I., Goto, R., & Sugiura, T. (2012). Ultrasonic inspection of an internal flaw in a ferromagnetic specimen using angle beam EMATs. *Przeglad Elektrotechniczny*, *88*(7B), 78–81.
Redmon, J., Divvala, S., Girshick, R., & Farhadi, A. (2016). You only look once: Unified, real-time object detection. In *Proceedings of the IEEE conference on computer vision and pattern recognition* (pp. 779–788).
Stonawski, O. (2008). *Non-destructive evaluation of carbon/carbon brakes using air-coupled ultrasonic inspection systems*. Southern Illinois University at Carbondale.
Tang, Y. M., Chau, K. Y., Li, W., & Wan, T. W. (2020). Forecasting economic recession through share price in the logistics industry with artificial intelligence (AI). *Computation*, *8*(3), 70. Available from https://doi.org/10.3390/computation8030070.
Wang, H., Li, Z., Ji, X., & Wang, Y. (2017). Face r-cnn. arXiv preprint arXiv:1706.01061.
Warnemuende, K. (2006). *Amplitude modulated acousto-ultrasonic nondestructive testing: Damage evaluation in concrete*. Wayne State University.
Wong, K. J. (2013). *Moisture absorption characteristics and effects on mechanical behaviour of carbon/epoxy composite: Application to bonded patch repairs of composite structures* (Doctoral dissertation). Dijon.
Zhang, Z., Guo, S., Li, Q., Cui, F., Malcolm, A. A., Su, Z., & Liu, M. (2020). Ultrasonic detection and characterization of delamination and rich resin in thick composites with waviness. *Composites Science and Technology*, *189*, 108016.
Zhao, G., Wang, B., Wang, T., Hao, W., & Luo, Y. (2019). Detection and monitoring of delamination in composite laminates using ultrasonic guided wave. *Composite Structures*, *225*, 111161.
Zou, Z., Zhao, X., Zhao, P., Qi, F., & Wang, N. (2019). CNN-based statistics and location estimation of missing components in routine inspection of historic buildings. *Journal of Cultural Heritage*, *38*, 221–230.

CHAPTER 2

Classifying asteroid spectra by data-driven machine learning model

Tan Guo[1,2,3], Xiao-Ping Lu[1,2], Keping Yu[4], Yong-Xiong Zhang[1,2] and Fulin Luo[5]

[1]Faculty of Information Technology, Macau University of Science and Technology, Taipa, Macau, P.R. China
[2]State Key Laboratory of Lunar and Planetary Sciences, Macau University of Science and Technology, Taipa, Macau, P.R. China
[3]School of Communication and Information Engineering, Chongqing University of Posts and Telecommunications, Chongqing, P.R. China
[4]Global Information and Telecommunication Institute, Waseda University, Tokyo, Japan
[5]State Key Laboratory of Information Engineering in Surveying, Mapping and Remote Sensing, Wuhan University, Wuhan, P.R. China

2.1 Introduction

Deep space exploration is the focus of space activities around the world, whose aims are to explore the mysteries of the universe, search for extra-terrestrial life and acquire new knowledge (Dorsky, 2001; Guo et al., 2020a, 2020b; Wu et al., 2012; Zhang et al., 2021; Guo et al., 2021). Planetary science plays an increasingly important role in the high-quality and sustainable development of deep space exploration (Kerr, 2004; Liou, 2006). As a kind of special celestial bodies rotating around the sun, asteroids are of great scientific significance for human beings to explore the origin and evolution of the universe, seek new living space, and protect the safety of the earth due to their large number, different individual characteristics and special orbit (Carry, 2012; Keil, 2000; Lu & Jewitt, 2019). Previous studies have shown that the thermal radiation from asteroids mainly depends on its size, shape, albedo, thermal inertia, and roughness of the surface (Bus & Binzel, 2002; Xu et al., 1993). The asteroids with different types (such as S-type, V-type, etc.) and different regions (such as the Jupiter trojans, Hungarian group, etc.) show different spectral characteristics, which establishes the foundation for identifying different kinds of asteroids via remote spectra observation (Binzel et al., 2001; Howell et al.,1994; Zhang et al., 2014). For example, the near-infrared data reveals diagnostic compositional information because of the presence of

IoT and Spacecraft Informatics
DOI: https://doi.org/10.1016/B978-0-12-821051-2.00003-9

the features at 1 and 2 μm bands primarily owing to the existence of olivine and pyroxene. The astronomers have developed many remote observation methods for asteroids, such as spectral and polychromatic photometry, infrared and radio radiation methods (Burbine & Binzel, 2002; Vilas & Mcfadden, 1992; Xu et al., 1995; Zellner et al., 1985). Therefore, a large volume of asteroid visible and infrared data has been gathered with the development of space and ground-based telescope observation technology, which have made great progress in the field of asteroid taxonomy through their spectral characteristics (Bus, 1999; Bus & Binzel, 2002; Demeo et al., 2009; Tholen, 1984).

2.1.1 Asteroid spectroscopic survey

The Eight-Color Asteroid Survey (ECAS), begun in 1979, is the most remarkable ground-based asteroid observation survey, which gathered the spectrophotometric observations of about 600 large asteroids (Zellner et al., 1985). The survey is aimed at providing high-quality reflection spectra for asteroids and natural satellites in a photometric system using eight filter passbands ranging from 0.34 to 1.04 μm wavelength, as shown in Table 2.1. All observations were made with the eight-color photometer at the 1.54-m Catalina reflector and the 2.29-m Steward reflector of the University of Arizona on 106 nights. Most of the work was done with the full set of eight filters, which requires the use of the Varian VPM-159A (InGaAsP) detector for the p and z filters (Bus & Binzel, 2002). Some of the faintest objects were observed in only five colors using the quieter, more sensitive RCA C31034−04 (GaAs) detector in the red channel. Except for a few of the faintest high-priority objects, an attempt was made to keep the noise level less than a few percent for all data (Bus & Binzel, 2002). However, very few small main-belt asteroids have been observed due to their faintness, and the observations of small main-belt asteroids may help to solve some fundamental problems in asteroid science.

With the appearance of charge-coupled device (CCD) detectors, it has been possible to study the large-scale spectra of small main-belt asteroids with a diameter less than 1 km (Xu, 1994). The first phase of the Small Main-belt Asteroid Spectroscopic Survey (SMASSI) was implemented from 1991 to 1993 at Michigan-Dartmouth-MIT Observatory (Xu et al., 1995). The main goal of SMASSI was to measure the spectral properties of the asteroids with small and medium size, and primarily focuses on the

Table 2.1 Effective wavelengths of the filter passbands.

Filter	Effective wavelength (μm)	Full width half maximum (μm)
s	0.337	0.047
u	0.359	0.060
b	0.437	0.090
v	0.550	0.057
w	0.701	0.058
x	0.853	0.081
p	0.948	0.080
z	1.041	0.067

objects in the inner main belt to study the correlations between meteorites and asteroids. One of the most significant achievements of SMASSI was the discovery of more than 20 small asteroids resembling basaltic achondrite (HED) meteorites (Bus & Binzel, 2002). Based on the survey, abundant spectral measurements for 316 different asteroids have been observed.

In view of the successes of SMASSI, the second phase of the SMASSII mainly concerns collecting an even larger and internally consistent dataset with the spectral observations and reductions which were carried out in as consistent away as possible. All SMASSII observations were made between August 1993 and March 1999, using either the 2.4-m (f/7.5) Hiltner or 1.3-m (f/7.6) McGraw-Hill reflecting telescopes (Bus & Binzel, 2002). These telescopes are operated by the MDM Observatory located on the southwest ridge of Kitt Peak in Arizona (Bus & Binzel, 2002). The data were obtained using the Mark III long-slit CCD spectrograph, along with one of two CCD cameras, all facility instruments of the MDM Observatory. For each asteroid, a set of 2–5 spectral images were usually taken, with exposure times ranging from a few seconds up to 900 s for the faintest objects. Longer exposures were avoided to limit the number of cosmic rays strikes that accumulate on the detector. The procedures used during SMASSII were roughly parallel to those used for SMASSI (Xu et al., 1995), though several minor changes were made in instrumentation and portions of the data reduction process. The primary observational goals of SMASSII were the sampling of planet-crossing asteroids and the continued study of dynamical families, with particular emphasis on the Vesta family, and an in-depth study of several families located in the middle of the main belt between 2.7 and 2.8 AU (Bus, 1999, Bus &

Binzel, 2002). Thus, the SMASSII survey provides the largest internally consistent sample of asteroid spectra ever obtained revealing a greater range of spectral diversity among asteroids than has been previously shown, and a new basis for studying the composition and structure of the asteroid belt (Bus & Binzel, 2002).

2.1.2 Asteroid taxonomy

The spectral reflectivity of asteroid surfaces can be used to classify these objects into several broad groups with similar spectral characteristics. The minimal tree method has been applied through combination with the principal component analysis (PCA) model to classify nearly 600 asteroid spectra from the ECAS (Zellner et al., 1985). Since principal component scores provide a fundamental representation of the data that can be used for a variety of analyses, including the assignment of taxonomic classifications. The near-infrared data range reveals diagnostic compositional information. Some classification systems created using near-IR data include Howell et al. (1994) by created a neural network-based taxonomy method. An S-complex taxonomy of olivine- and pyroxene-rich asteroids was created based on near-infrared data (Gaffey et al., 1993). For a more comprehensive and accurate classification of asteroids, DeMeo et al. developed an extended taxonomy to characterize visible and near-infrared wavelength spectra (DeMeo et al., 2009). The asteroid spectral data used for the taxonomy are based on reflectance spectrum characteristics measured over the wavelength range 0.45–2.45 μm containing 379–688 bands. In summary, the dataset was comprised of 371 objects with both visible and near-infrared data. SMASSII dataset provided the most visible wavelength spectra, and the near-infrared spectral measurements from 0.8 to 2.5 μm were obtained using SpeX, the low- to medium-resolution near-infrared spectrograph, and the imager on the 3-m NASA IRTF located on Mauna Kea, Hawaii (DeMeo et al., 2009). A description for the dataset is illustrated in Table 2.2. Based on this dataset, DeMeo et al., have presented the taxonomy, as well as the method and rationale for the class definitions of different asteroids. Specifically, three main complexes, that is, S-complex, C-complex, and X-complex, were defined according to some empirical spectral characteristics/features, including spectral curve slope, absorption bands, and so on.

Nevertheless, how to automatically discover the key category-related spectral characteristics/features is still an open problem (Bus & Binzel,

Table 2.2 Description of the asteroid spectral datasets for 371 asteroids with 24 classes.

Class	"A"	"B"	"C"	"Cb"	"Cg"	"Cgh"	"Ch"	"D"
#Samples	6	4	13	3	1	10	18	16
Class	"K"	"L"	"O"	"Q"	"R"	"S"	"Sa"	"Sq"
#Samples	16	22	1	8	1	144	2	29
Class	"Sr"	"Sv"	"T"	"V"	"X"	"Xc"	"Xe"	"Xk"
#Samples	22	2	4	17	4	3	7	18

2002; Imani & Ghassemian, 2014; Taskin et al., 2017). Meanwhile, owing to the noise of the system and ever-changing external observation conditions, the observed spectral data usually contain noise and distortions, which will cause spectrum mixture due to the random perturbation of electronic devices. As a result, the observed asteroid spectra data often show inseparable pattern characteristics (Gaffey et al., 2012, Wood & Kuiper, 1963). Furthermore, most of the spectral absorption features have very broadbands, such as the ultraviolet and near-infrared absorption bands. Thus, the reflectance at one wavelength is usually correlated with the reflectance of adjacent wavelengths (Sun & Du, 2019). Accordingly, the adjacent spectral bands are usually redundant, and some bands may not carry discriminant information for certain applications. Besides, the abundant spectral information will result in high data dimensionality with undesired information and brings about the "curse of dimensionality" problem, that is, under a fixed and limited number of training samples, the classification accuracy of spectral data decreases when the dimensionality of spectral data increases (Herrmann et al., 2012). Therefore, it is necessary to develop effective low-dimensional asteroid spectral feature learning methods and find the latent discriminative information for different kinds of asteroid spectrum, which are very beneficial for the precise classification of asteroids.

It has developed rapidly in recent years for machine learning techniques in spectral data processing applications, such as classification and target detection (Dong et al., 2021; Guo et al., 2020a, 2020b; Luo et al., 2020; Zhang & Zhang, 2015). For example, the classic PCA has been applied to extract meaningful features from the observed spectral data without using the prior label information. PCA is also useful for asteroid and meteorite spectra analysis because many of the variables, that is, the reflectance at different wavelengths, are highly correlated (Demeo et al., 2009; Hotelling, 1933; Shui et al., 1995). Linear discriminant analysis

(LDA) can make full use of the label priors by concurrently minimizing the within-class scatter and maximizing the between-class scatter in a dimension-reduced subspace (Fisher, 1938). In addition to the above statistics-based methods, some geometry theory-based methods have also been proposed for the problem of data dimensionality reduction. For example, the locality preserving projections (LPPs) assume that neighboring samples are likely to have a similar label, and the affinity relationships among samples should be preserved in subspace learning (He & Niyogi, 2002). Locality preserving discriminant projections (LPDP) has been developed with locality and Fisher criteria, which can be seen as the combination of LDA and LPP (Gui et al., 2009; Zhang et al., 2019).

Based on the well-labeled asteroid spectral dataset described in Table 2.2, we aim to study the pattern characteristics of different categories of asteroids from the perspective of data-driven machine learning methodology, and develop efficient asteroid spectral features learning and classification method, as shown in Fig. 2.1. To be specific, it is assumed that not only the specified absorption bands, such as the 1 and 2 μm, but also all the spectral bands contain some useful diagnostic compositional information for asteroid category identification, and will contribute to the accurate classification of different

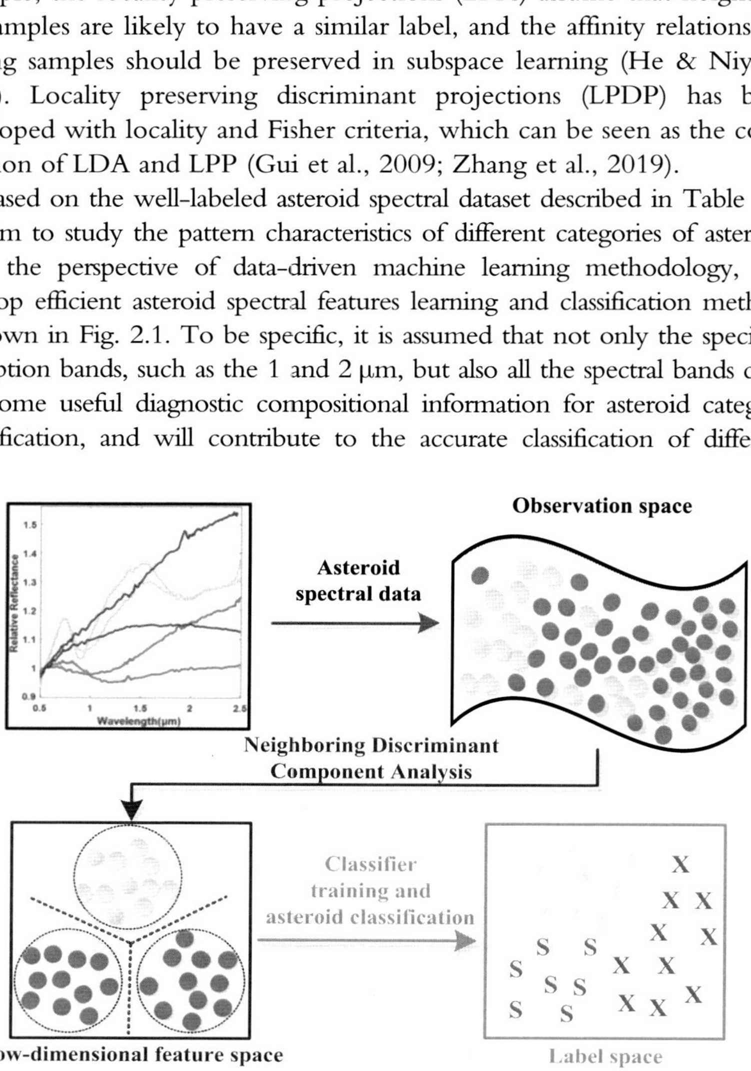

Figure 2.1 Overview of the data-driven asteroid feature learning and classification scheme.

kinds of asteroids. As a result, the spectral data spanning over the visible to near-infrared wavelengths (0.45–2.45 μm) are treated as a whole to discover the key category-related discriminate information within neighboring samples by excluding the outliers and abnormal samples for efficient asteroid spectral feature learning and classification. The main contributions of this chapter are summarized below.

1. Rather than empirically determine the spectral features via the presence or absence of specific spectral features to define new class boundaries and classify asteroids, this chapter introduces a novel supervised date-driven machine learning method named Neighboring discriminant components analysis (NDCA) model for discriminative asteroid spectral feature learning by simultaneously maximizing neighboring between-class scatter and data variances, and minimizing the neighboring within-class scatter to avoid the overfitting problem caused by outliers and enhance the generalization ability of the model.
2. With the neighboring discrimination, the proposed NDCA method has a higher tolerance to abnormal samples or outliers and can avoid overfitting problems. In addition, the proposed NDCA method transforms the data into a more separable subspace and the most discriminative structure information can be well discovered and preserved for different classes of asteroids with neighboring structure preservation and label prior guidance principles.
3. The performance of the proposed NDCA model is verified on the real-world asteroid spectral data spanning over the wavelength range 0.45–2.45 μm by combining with several baseline classifier models, including the nearest neighbor (NN), support vector machine (SVM), and extreme learning machine (ELM). In particular, the best results are obtained by ELM with a classification accuracy of about 95.19%.

The remainder of this chapter is arranged as follows. Section 2.2 introduces related works on subspace learning and machine learning classifier models. The proposed NDCA method is introduced in Section 2.3 in detail. Section 2.4 presents the experimental results and discussions. The conclusion is given in Section 2.5.

2.2 Related work

2.2.1 Notations used in this chapter

In this chapter, the observed asteroid visible and near-infrared spectroscopy dataset is denoted as $\mathbf{X} = [\mathbf{x}_1, \mathbf{x}_2, \ldots, \mathbf{x}_N] \in \Re^{D \times N}$, which is composed of N spectral samples with dimensionality D from C classes. N_i is the number of the samples in ith class. The label matrix for $\mathbf{X}$ is denoted as $\mathbf{T} = [t_1, t_2, \ldots, t_N] \in \Re^{C \times N}$ with t_i as the label vector for $\mathbf{x}_i$. The label of each sample in $\mathbf{X}$ is coded as a C-dimensional vector, and the jth entry of t_i is $+1$ with the remaining entities as 0, which indicates that sample $\mathbf{x}_i$ belongs to the jth category. The basic idea of linear low-dimensional feature learning, that is, dimension reduction, is to automatically learn a transformation matrix $\mathbf{P} = [\mathbf{p}_1, \mathbf{p}_2, \ldots, \mathbf{p}_N] \in \Re^{D \times d}$ with $d < D$, which can project the observed spectral data from the original high D-dimensional observation space into a lower d-dimensional feature subspace, and obtain the low-dimensional meaningful features $\mathbf{Y} \in \Re^{d \times N}$ of $\mathbf{X}$ via $\mathbf{Y} = \mathbf{P}^T\mathbf{X} = [\mathbf{y}_1, \mathbf{y}_2, \ldots, \mathbf{y}_N] \in \Re^{d \times N}$.

2.2.2 Low-dimensional feature learning for spectral data

In the process of low-dimensional feature learning, the key data knowledge and information, such as the discriminative data structures, should be enhanced and preserved. Meanwhile, the noise and redundant information should be removed and suppressed. PCA is a widely applied unsupervised statistical dimension reduction and feature learning algorithm, which focuses on maximizing the variance of the data with significant principal components. A formulation for PCA can be derived by solving the following least-squares problem.

$$\min_{\boldsymbol{P}} \left\| \mathbf{X} - \boldsymbol{P}\boldsymbol{P}^T\mathbf{X} \right\|_F^2 s.t. \boldsymbol{P}^T\boldsymbol{P} = \mathbf{I}_d \tag{2.1}$$

where $\| \cdot \|_F^2$ means the Frobenius norm of a matrix and $\mathbf{I}_d$ is an identity matrix with the size of d. Formula (2.1) is equivalent to maximizing the variance of the transformed data as follows.

$$\max_{\boldsymbol{P}} \mathrm{Tr}\left(\boldsymbol{P}^T\mathbf{X}\mathbf{X}^T\boldsymbol{P}\right) s.t. \boldsymbol{P}^T\boldsymbol{P} = \mathbf{I}_d \tag{2.2}$$

Unlike PCA, LDA is a supervised dimension reduction learning method and aims to maximize the separability between different classes and enhance the compactness within each class with the guidance of label information as below.

$$\min_{\boldsymbol{P}} \frac{\mathrm{Tr}\left(\boldsymbol{P}^T\boldsymbol{S}_{\mathrm{W}}\boldsymbol{P}\right)}{\mathrm{Tr}\left(\boldsymbol{P}^T\boldsymbol{S}_{\mathrm{B}}\boldsymbol{P}\right)} \tag{2.3}$$

where $\boldsymbol{S}_{\mathrm{W}}$ and $\boldsymbol{S}_{\mathrm{B}}$ are respectively the within-class and between-class scatter matrices of data, which are calculated in the following way.

$$\mathbf{S}_{\mathrm{W}} = \sum_{i=1}^{C} \sum_{j=1}^{N_i} \left(\mathbf{x}^{ij} - \boldsymbol{\mu}^{i}\right)\left(\mathbf{x}^{ij} - \boldsymbol{\mu}^{i}\right)^{T} \tag{2.4}$$

$$\boldsymbol{S}_{\mathrm{B}} = \sum_{i=1}^{C} N_i\left(\boldsymbol{\mu}^{i} - \boldsymbol{\mu}\right)\left(\boldsymbol{\mu}^{i} - \boldsymbol{\mu}\right)^{T} \tag{2.5}$$

where $\boldsymbol{x}^{ij}$ is the jth sample of the ith class, and $\boldsymbol{\mu}^{i}$ and $\boldsymbol{\mu}$ are the mean value of the samples in ith class and all the samples in $\mathbf{X}$, respectively.

2.2.3 Classifier models for spectral data classification

Classifier models, such as NN, SVM (Vapnik, 1998), and ELM (Huang et al., 2012; Zhang & Zhang, 2015), have been commonly utilized in the contexts of machine learning and pattern recognition to recognize or classify data points. To be specific, SVM aims to solve the following minimization problem with some inequality constraints according to the structural risk minimization principle.

$$\begin{gathered} \min_{\mathbf{w},\xi_i} \frac{1}{2}||\mathbf{w}||^2 + \eta\cdot\sum_{i=1}^{N}\xi_i \\ \text{s.t., } \xi_i \geq 0,\ y_i\left[\mathbf{w}^T\varphi(\mathbf{x}_i) + b\right] \geq 1 - \xi_i \end{gathered} \tag{2.6}$$

where $\varphi(\cdot)$ is a linear/nonlinear mapping function, $\mathbf{w}$ and b are the parameters of the classifier hyperplane. In general, the original problem (2.6) can be transformed into its dual formulation with equality constraint by using the Lagrange multiplier method. As a result, the following Lagrange function can be obtained.

$$\begin{gathered} L\left(\mathbf{w}, b, \xi_i; \gamma_i, \lambda_i\right) = \frac{1}{2}||\mathbf{w}||^2 + \eta\cdot\sum_{i=1}^{N}\xi_i \\ - \sum_{i=1}^{N} \gamma_i\left(y_i\left[\mathbf{w}^T\varphi(\mathbf{x}_i) + b\right] - 1 + \xi_i\right) - \sum_{i=1}^{N} \lambda_i\xi_i \end{gathered} \tag{2.7}$$

where $\gamma_i \geq 0$ and $\lambda_i \geq 0$ Lagrange multipliers. The solution can be given by the saddle point of Lagrange function (2.7) by solving

$$\max_{\gamma_i,\lambda_i} \min_{w,b,\xi_i} L\left(w, b, \xi_i;\gamma_i, \lambda_i\right) \tag{2.8}$$

Through calculating the partial derivatives of Lagrange function with respect to $\mathbf{w}$, b and ξ_i, the following equations will be obtained:

$$\begin{cases} \dfrac{\partial L(\mathbf{w}, b, \xi_i; \gamma_i, \lambda_i)}{\partial \mathbf{w}} = 0 \rightarrow \mathbf{w} = \sum_{i=1}^{N} \gamma_i y_i \varphi(\mathbf{x}_i) \\ \dfrac{\partial L(\mathbf{w}, b, \xi_i; \gamma_i, \lambda_i)}{\partial b} = 0 \rightarrow \quad \sum_{i=1}^{N} \gamma_i y_i = 0 \\ \dfrac{\partial L(\mathbf{w}, b, \xi_i; \gamma_i, \lambda_i)}{\partial \xi_i} = 0 \rightarrow \quad 0 \leq \gamma_i \leq \eta \end{cases} \tag{2.9}$$

Furthermore, the above equation can be rewritten in the following form:

$$\begin{aligned} &\max_\gamma \sum_i \gamma_i - \frac{1}{2} \sum_{i,j} y_i y_j \gamma_i \gamma_j \varphi(\mathbf{x}_i)^T \varphi(\mathbf{x}_j) \\ &\text{s.t.} \sum_{i=1}^{N} \gamma_i y_i = 0, 0 \leq \gamma_i \leq \eta \end{aligned} \tag{2.10}$$

By solving $\boldsymbol{\gamma}$ of the dual problem with quadratic programming, the goal of SVM is to construct the following decision function.

$$f(\boldsymbol{x}) = \text{sgn}\left(\sum_{i=1}^{M} \gamma_i y_i \mathbb{k}(\mathbf{x}_i, \mathbf{x}) + b\right) \tag{2.11}$$

where $\mathbb{k}(\cdot)$ is a kernel function. $\mathbb{k}(\mathbf{x}_i, \mathbf{x}) = \varphi(\mathbf{x}_i)^T \varphi(\mathbf{x}) = \mathbf{x}_i^T \mathbf{x}$ for the linear SVM and $\mathbb{k}(\mathbf{x}_i, \mathbf{x}) = \exp\left(- \left\| \mathbf{x}_i - \mathbf{x} \right\|^2 / \sigma^2 \right)$ for the RBF-SVM.

ELM is a newly developed machine learning paradigm for the generalized single hidden layer feedforward neural networks and has been widely studied and applied due to its some unique characteristics, that is, fast training speed, excellent generalization, and universal approximation/classification abilities.

The most noteworthy feature of ELM is that the weights between the input and the hidden layers are randomly generated without further adjustments. The objective function of ELM is formulated as

$$\min_\beta \frac{1}{2} ||\beta||_F^2 + \frac{\alpha}{2} \sum_{i=1}^{N} ||\xi_i||^2 \text{s.t.} h(\boldsymbol{x}_i)\beta = t_i - \xi_i, i = 1, 2 \ldots, N \Leftrightarrow \boldsymbol{H}\beta = \boldsymbol{T} - \xi \tag{2.12}$$

where $\boldsymbol{\beta} \in \Re^{L \times C}$ denotes the output weights connecting the hidden layer and the output layer. $\boldsymbol{\xi} = \left[\boldsymbol{\xi}_1, \boldsymbol{\xi}_2, \ldots, \boldsymbol{\xi}_N\right]^T \in \Re^{N \times C}$ is the prediction error matrix with respect to the training data. $\mathbf{H} \in \Re^{N \times L}$ is the hidden layer output matrix, and is computed in the following fashion:

$$\mathbf{H} = \begin{bmatrix} h\left(\mathbf{w}_1^T \mathbf{x}_1 + b_1\right) & \cdots & h\left(\mathbf{w}_1^T \mathbf{x}_1 + b_L\right) \\ \vdots & \ddots & \vdots \\ h\left(\mathbf{w}_1^T \mathbf{x}_N + b_1\right) & \cdots & h\left(\mathbf{w}_1^T x_N + b_L\right) \end{bmatrix} \tag{2.13}$$

where $h(\cdot)$ is the activation function in the hidden layer, such as the sigmoid function. $\mathbf{W} = [\mathbf{w}_1, \mathbf{w}_2, \ldots, \mathbf{w}_\mathrm{L}] \in \Re^{d \times L}$ and $\mathbf{b} = [\mathbf{b}_1, \mathbf{b}_2, \ldots, \mathbf{b}_\mathrm{L}]^T \in \Re^L$ denote the randomly generated input weights and bias, respectively. The output weight matrix $\boldsymbol{\beta}$ is used to transform the data from the L-dimensional hidden layer space into the C-dimensional high-level label space and is analytically calculated as in the following manner:

$$\boldsymbol{\beta}^* = \begin{cases} \left(\mathbf{H}^T\mathbf{H} + \frac{\mathbf{I}_{L \times L}}{\alpha}\right)^{-1} \mathbf{H}^T\mathbf{T}, & if \mathrm{N} \geq \mathrm{L} \\ \mathbf{H}^T\left(\mathbf{H}\mathbf{H}^T + \frac{\mathbf{I}_{N \times N}}{\alpha}\right)^{-1} \mathbf{T}, & if \mathrm{N} < \mathrm{L} \end{cases} \tag{2.14}$$

With the optimal output weight matrix $\boldsymbol{\beta}^*$ obtained, the predicted label for a new test sample $\mathbf{z}$ can be computed as follows.

$$\mathrm{label}(\mathbf{z}) = \mathbf{h}(\mathbf{z})\boldsymbol{\beta}^* \tag{2.15}$$

where $\mathbf{h}(\mathbf{z})$ is the hidden layer output for test sample $\mathbf{z}$. Since the output weights can be determined directly from (2.14), the parameters in ELM are never updated by iterative backpropagation, which makes ELM extremely efficient and flexible.

2.3 Neighboring discriminant component analysis: a data-driven machine learning model for asteroid spectra feature learning and classification

The remote observed asteroid spectral data usually contain noise and outliers, which will mix different categories of asteroids and make them inseparable. In addition, learning with outliers will easily cause overfitting problems, which will decrease the generalization ability of machine learning models for testing samples. Thus, the key problem is to distinguish the outliers and select the key valuable samples for the learning of low-dimensional feature subspace and preserve the discriminative data knowledge for different classes of asteroids. To this end, as shown in Fig. 2.2, the idea of neighboring learning is introduced to find a neighboring group of valuable samples from all the training samples as well as the samples in each class, and the outliers and noised samples will be excluded to enhance the generalization ability of the model. To achieve this goal, the neighboring between-class and within-class scatter matrices are needed to be calculated to characterize the neighboring and discriminative properties of observed asteroid spectra.

Neighboring between-class scatter matrix ($\boldsymbol{S}_{Nb}$) computation: Firstly, calculate the global centroid $\boldsymbol{m}_b = (1/N) \times \sum_{i=1}^{N} \mathbf{x}_i$ for all the samples in training dataset $\mathbf{X}$ and find $N_b = Rb \cdot N$ neighboring samples to $\boldsymbol{m}_b$ using between-class neighboring ratio $(0 < Rb < 1)$. Thus, $N_b = [N_{b1}, N_{b2}, \ldots, N_{bc}, \ldots, N_{bC}]$ global neighboring samples $\mathbf{X}_b$ can be

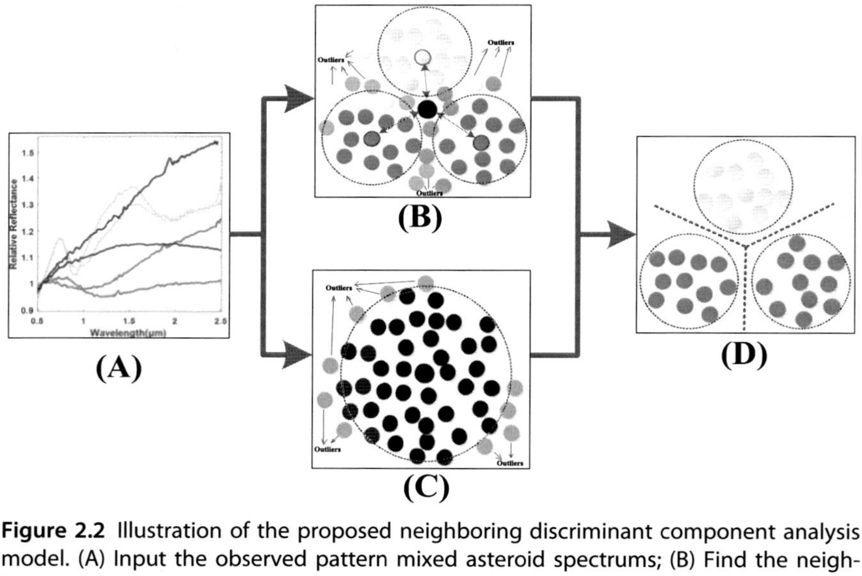

Figure 2.2 Illustration of the proposed neighboring discriminant component analysis model. (A) Input the observed pattern mixed asteroid spectrums; (B) Find the neighboring samples in each asteroid class to characterize the neighboring within-class and between-class properties of data; (C) Find the neighboring samples from all the samples for neighboring principal components to preserve the most valuable variances of data. With the basic principles of (B) and (C), a clearer class boundary can be derived to alleviate the overfitting problems caused by outliers and noised samples and enhance the neighboring and discriminative principal components of data for efficient spectral feature learning, as shown in (D).

obtained with N_{bc} as the number of the neighboring samples in the cth class for computing the neighboring between-class scatter matrix. Secondly, compute the local centroid $\boldsymbol{m}_{bc} = \left(1/N_{bc}\right) \times \sum_{j=1}^{N_{bc}} \mathbf{x}_{bcj}$ for the cth class, and $\mathbf{x}_{bcj}$ is the jth sample in the cth class of the neighboring samples $\mathbf{X}_b$. Finally, the neighboring between-class scatter matrix is calculated as follows.

$$\mathbf{S}_{Nb} = \sum_{c=1}^{C} N_{bc}(\boldsymbol{m}_{bc} - \boldsymbol{m}_b)(\boldsymbol{m}_{bc} - \boldsymbol{m}_b)^T \tag{2.16}$$

At the same time, the N_b global neighboring samples are used to calculate the covariance matrix as below.

$$\sum_{i=1}^{N_b} \mathbf{x}_{bi}(\mathbf{x}_{bi})^T \tag{2.17}$$

Neighboring within-class scatter matrix ($\boldsymbol{S}_{Nw}$) computation: Firstly, calculate the basic local centroid $\boldsymbol{m}_{wc} = \left(1/Nc\right) \times \sum_{i=1}^{Nc} \boldsymbol{x}_{ci}$, where $\boldsymbol{x}_{ci}$ is the ith sample in the cth class of $\mathbf{X}$, and then find the samples group containing $N_{wc} = Rw \cdot N_c$ neighboring samples to $\boldsymbol{m}_{wc}$ using within-class neighboring ratio Rw $(0 < Rw < 1)$ in the ith class. Secondly, refine the local centroid of each class using the samples in the obtained neighboring group of samples $\mathbf{X}_{wc}$. Finally, compute the neighboring within-class scatter matrix as follows.

$$\mathbf{S}_{Nw} = \sum_{c=1}^{C} \sum_{i=1}^{N_{wc}} (\mathbf{x}_{wci} - \boldsymbol{m}_{wc})(\mathbf{x}_{wci} - \boldsymbol{m}_{wc})^T \tag{2.18}$$

where $\boldsymbol{m}_{wc} = \left(1/N_{wc}\right) \times \sum_{i=1}^{N_{wc}} \mathbf{x}_{wci}$ is the refined centroid of each class based on the neighboring sample groups $\mathbf{X}_{wc}$, and $\mathbf{x}_{wci}$ is the ith samples of cth class samples in the neighboring group. By comprehensively consider (2.16), (2.17), and (2.18) in a dimension-reduced subspace, the following optimization problem is formulated.

$$\max_{\boldsymbol{P}} Tr\left(\boldsymbol{P}^T \mathbf{S}_{Nb} \boldsymbol{P}\right) + \gamma \mathrm{Tr}\left(\boldsymbol{P}^T \mathbf{X}_b \mathbf{X}_b^T \boldsymbol{P}\right) - \mu \mathrm{Tr}\left(\boldsymbol{P}^T \mathbf{S}_{Nw} \boldsymbol{P}\right) \tag{2.19}$$

where γ and μ are the tradeoff parameters for balancing the corresponding components in the objective function, from which one can see that in the subspace spanned by $\mathbf{P}$, the goals of neighboring between-class scatter maximization, within-class scatter minimization and neighboring principal components preservation can be simultaneously achieved. Accordingly, the side effects of outliers and noised samples will be suppressed to the largest extent. As a result, the global and local neighboring discriminative structures and principal components will be enhanced and preserved by using the neighboring learning mechanism. Furthermore, optimization problem (2.19) can be transformed to be the following one by introducing an equality constraint.

$$\max_{\boldsymbol{P}} \mathrm{Tr}\left(\boldsymbol{P}^T\mathbf{S}_{Nb}\boldsymbol{P}\right) \quad \text{s.t.} \quad \mu\mathrm{Tr}\left(\boldsymbol{P}^T\mathbf{S}_{Nw}\boldsymbol{P}\right) - \gamma\mathrm{Tr}\left(\boldsymbol{P}^T\mathbf{X}_b\mathbf{X}_b^T\boldsymbol{P}\right) = \varpi \tag{2.20}$$

where ϖ is a constant used to ensure a unique solution for model (2.9). The objective function for model (2.20) can be formulated as the following unconstrained one by introducing the Lagrange multiplier λ.

$$\mathscr{L}(\mathbf{P},\lambda) = \mathrm{Tr}\left(\boldsymbol{P}^T\mathbf{S}_{Nb}\boldsymbol{P}\right) - \lambda\left(\mu\mathrm{Tr}\left(\boldsymbol{P}^T\mathbf{S}_{Nw}\boldsymbol{P}\right) - \gamma\mathrm{Tr}\left(\boldsymbol{P}^T\mathbf{X}_b\mathbf{X}_b^T\boldsymbol{P}\right) - \varpi\right) \tag{2.21}$$

Then, the partial derivative of the objective function (2.21) with respect to **P** is calculated and set as zero, leading to the following equations:

$$\frac{\partial\mathscr{L}(\mathbf{P},\lambda)}{\partial\mathbf{P}} = \mathbf{S}_{Nb}\boldsymbol{P} - \lambda\left(\mu\mathbf{S}_{Nw}\boldsymbol{P} - \gamma\mathbf{X}_b\mathbf{X}_b^T\boldsymbol{P}\right) = \mathbf{0} \tag{2.22}$$

$$\mathbf{S}_{Nb}\boldsymbol{P} = \lambda\left(\mu\mathbf{S}_{Nw} - \gamma\mathbf{X}_b\mathbf{X}_b^T\right)\boldsymbol{P} \tag{2.23}$$

where the projection matrix $\mathbf{P} = [\mathbf{p}_1, \ \mathbf{p}_2, \ldots, \ \mathbf{p}_d]$ can be acquired and is composed of the eigenvectors corresponding to the first d largest eigenvalues λ_1, λ_2 ..., λ_d of the eigenvalue decomposition problem as below.

$$\left(\mu\mathbf{S}_{Nw} - \gamma\mathbf{X}_b\mathbf{X}_b^T\right)^{-1}\mathbf{S}_{Nb}\boldsymbol{p} = \lambda\boldsymbol{p} \tag{2.24}$$

Once the above optimal projection matrix **P** is calculated, the training data are projected into the subspace using **P** to acquire the low-dimensional discriminative feature of observed spectrums. Afterward, a classifier model is trained by the dimension-reduced training data. For testing, asteroid spectral sample with the unknown label is firstly transformed into the subspace using the optimal projection matrix **P** and then classified by the trained classifier model.

2.4 Experiments

2.4.1 Preprocessing for the asteroid spectral data

As shown in Table 2.3, a part of the samples described in Table 2.2 was used in the study. The data preprocessing procedures were firstly

Table 2.3 Description of asteroid spectral datasets used in the experiments.

Class	"A"	"C"	"D"	"K"	"L"	"Q"	"S"	"V"	"X"	Total
#Samples	6	45	16	16	22	8	199	17	32	361

performed to preliminarily reduce the influences of noise for ease of classification. Firstly, the original spectral data were filtered and smoothed using some data filtering method, such as the moving average filter. Secondly, the discrete spectrum measurements were fitted using the high-order polynomial method. Thirdly, the obtained fitted spectral curve within the spectral range from 0.45 to 2.45 μm was sampled with a certain step interval. Several examples for different categories of the original spectrums smoothed spectrums, and fitted spectrums are shown in Figs. 2.3–2.5, from which one can see that the abnormal noises in some spectral bands were suppressed to a certain extent. More examples for reflectance spectra characteristics of some representative asteroids from different categories used in this chapter can be found in Appendix A.

2.4.2 Experimental setup and results

As previously mentioned, the smoothed asteroid spectral curves were fitted using high-order polynomial, which was furthered sampled in the wavelength region from 0.45 to 2.45 μm with an increment step interval of 0.05 μm, obtaining 41 measurements for each asteroid spectrum. The data from different classes were then approximately equally divided into five folds as shown in Table 2.4 and Fig. 2.6. The five-fold cross-validation (CV) strategy was adopted for the performance evaluation of different methods. Specifically, random four folds of samples were used as the training set and the remaining one fold of samples were utilized for testing, and five experiments were thus carried out. A detailed description for the experiment settings can be found in Table 2.5, and the individual and average classification accuracy of different methods on the five experiments will be reported.

The proposed NDCA method was compared with several representative subspace learning methods, including PCA, LDA, LPP, and LPDP. Besides, the sampled raw asteroid without feature learning was also included for comparison. In addition, some baseline classifier models, such as the NN, SVM, and ELM, were adopted in the experiments for the classification of the asteroid features.

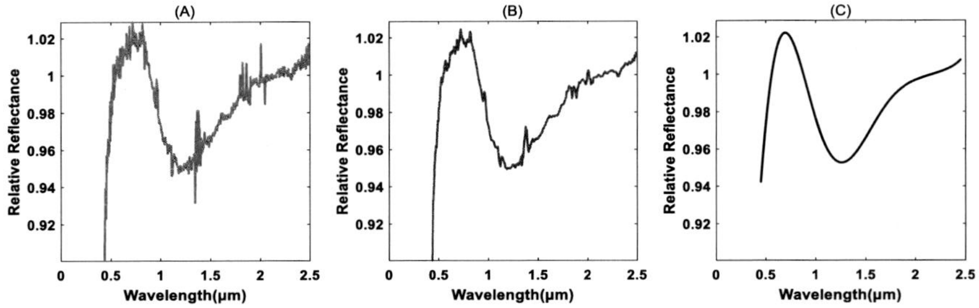

Figure 2.3 The spectral preprocessing for (1) *Ceres* with the category "C." (A) Original spectrums; (B) Smoothed spectrum; (C) Fitted spectrums spanning the wavelength range 0.45–2.45 μm.

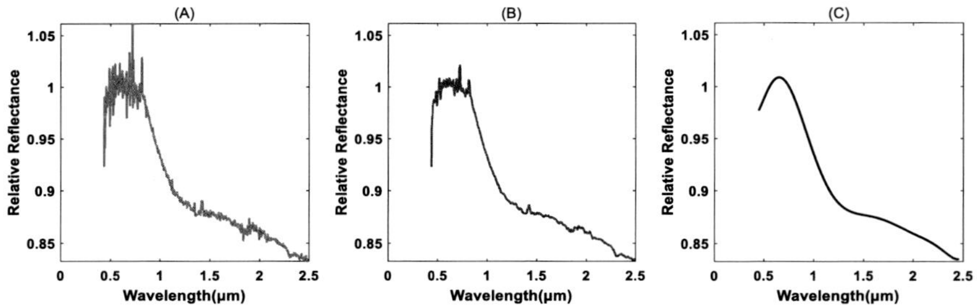

Figure 2.4 The spectral preprocessing for (2) *Pallas* with category "B." (A) Original spectrums; (B) Smoothed spectrums; (C) Fitted spectrums spanning the wavelength range 0.45–2.45 μm.

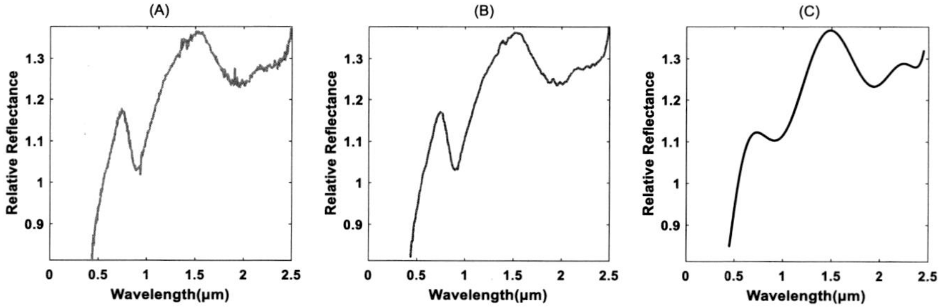

Figure 2.5 The spectral preprocessing for (5) *Astraea* with category "S." (A) Original spectrums; (B) Smoothed spectrums; (C) Fitted spectrums spanning the wavelength range 0.45–2.45 μm.

The classification accuracy of different dimension reduction methods achieved by using different baseline classifier models, that is, NN, SVM, and ELM, are reported in Tables 2.6–2.8, respectively. In addition, the performance of different dimension reduction methods in combination

Table 2.4 Experimental data partition of the five-folds.

Class	Fold 1	Fold 2	Fold 3	Fold 4	Fold 5	Total
"A"	1	1	1	1	2	6
"C"	9	9	9	9	9	45
"D"	3	3	3	4	3	16
"K"	4	3	3	3	3	16
"L"	4	4	4	5	5	22
"Q"	2	2	1	2	1	8
"S"	40	40	40	40	39	199
"V"	3	4	4	3	3	17
"X"	6	6	7	6	7	32
# samples	72	72	72	73	72	361

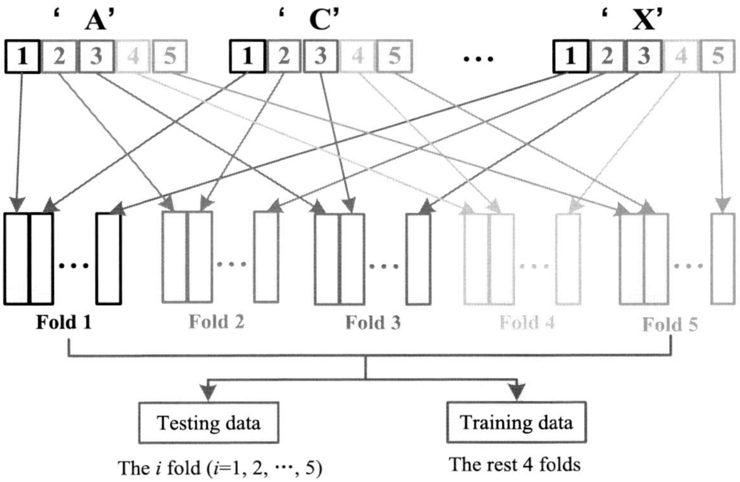

Figure 2.6 Five-fold cross verification scheme for asteroid spectral data.

with different classifier models under varying reduced dimension d (from 2 to 41 with an interval of 1) is shown in Figs. 2.7–2.9. The results show that the highest accuracy can be achieved when d is around 9 (the total number of the categories).

Furthermore, the two-dimensional scatter points for the first two dimensions acquired by different methods are visualized in Fig. 2.10 to further intuitively observe the low-dimensional feature learning

Table 2.5 Experiment settings with different fold partitions.

Experiments	Training dataset	Testing dataset
Exp 1	fold 1, fold 2, fold 3 and fold 4 (*289 samples in total*)	fold 5 (*72 samples in total*)
Exp 2	fold 1, fold 2, fold 3 and fold 5 (*288 samples in total*)	fold 4 (*73 samples in total*)
Exp 3	fold 1, fold 2, fold 4 and fold 5 (*289 samples in total*)	fold 3 (*72 samples in total*)
Exp 4	fold 1, fold 3, fold 4 and fold 5 (*289 samples in total*)	fold 2 (*72 samples in total*)
Exp 5	fold 2, fold 3, fold 4 and fold 5 (*289 samples in total*)	fold 1 (*72 samples in total*)

Table 2.6 Classification accuracy (%) of different dimension reduction algorithms using nearest neighbor.

Methods	Exp 1	Exp 2	Exp 3	Exp 4	Exp 5	Average
Raw	94.4444	84.9315	87.5000	93.0556	86.1111	89.2085
PCA	94.4444	84.9315	87.5000	93.0556	86.1111	89.2085
LDA	95.8333	90.4110	88.8889	97.2222	88.8889	92.2489
LPP	90.2778	87.6712	90.2778	91.6667	88.8889	89.7565
LPDP	95.8333	90.4110	91.6667	97.2222	88.8889	92.8044
NDCA	97.2222	89.0411	93.0556	98.6111	93.0556	94.1971

Table 2.7 The best classification accuracy (%) of different dimension reduction algorithms using support vector machine.

Methods	Exp 1	Exp 2	Exp 3	Exp 4	Exp 5	Average
Raw	94.4444	86.3014	93.0556	93.0556	91.6667	91.7047
PCA	94.4444	89.0411	93.0556	93.0556	91.6667	92.2527
LDA	94.4444	90.4110	88.8889	94.4444	91.6667	91.9711
LPP	97.2222	86.3014	90.2778	95.8333	94.4444	92.8158
LPDP	94.4444	90.4110	93.0556	94.4444	91.6667	92.8044
NDCA	94.4444	90.4110	94.4444	95.8333	93.0556	93.6377

performance. In contrast, the two-dimensional scatter points by comparing methods have serious data mixture effects between different classes, especially the "K," "L" and "Q" classes, which will lead to lower classification performance. From the figures, it can be observed that the

Table 2.8 Classification accuracy (%) of different dimension reduction algorithms using extreme learning machine.

Methods	Exp 1	Exp 2	Exp 3	Exp 4	Exp 5	Average
Raw	94.8611	89.3151	92.0833	95.6944	91.5278	92.6963
PCA	95.0000	90.4110	92.0833	96.5278	92.3611	93.2766
LDA	95.4167	94.5205	91.8056	97.2222	93.4722	94.4874
LPP	95.8333	90.4110	93.0556	98.6111	94.4444	94.4711
LPDP	95.9722	94.5205	92.6389	97.2222	93.3333	94.7374
NDCA	97.7778	91.7808	93.4722	97.2222	95.6944	95.1895

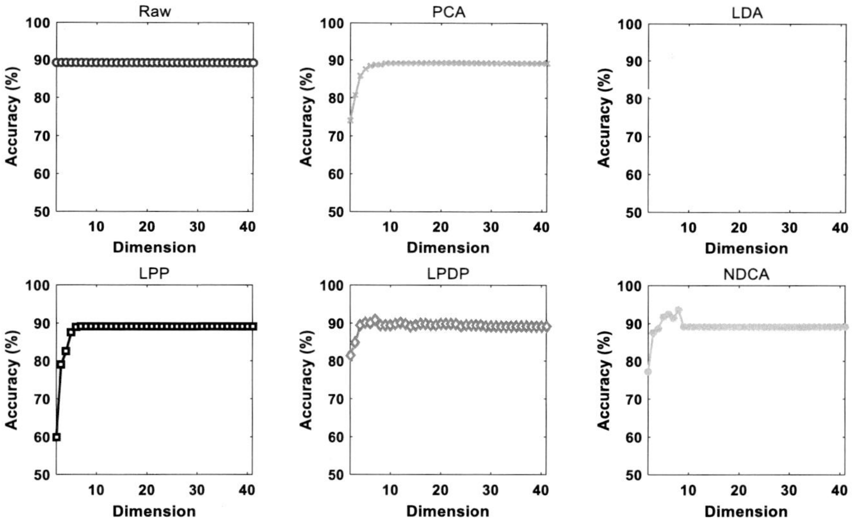

Figure 2.7 The performance of different dimension reduction methods under different reduced dimensions using nearest neighbor as the classifier.

two-dimensional scatter points derived by the proposed NDCA method show better within-class compactness and between-class separation characteristics with clearer category boundaries. With the proposed NDCA method, the spectral characteristic within each class can be better explored and the discriminant between different classes of asteroids can be well enhanced. By combining with off-the-shelf classifier models, the category boundaries between different classes of asteroids can be easily found and automatically defined, which will lead to promising generalization and classification performance.

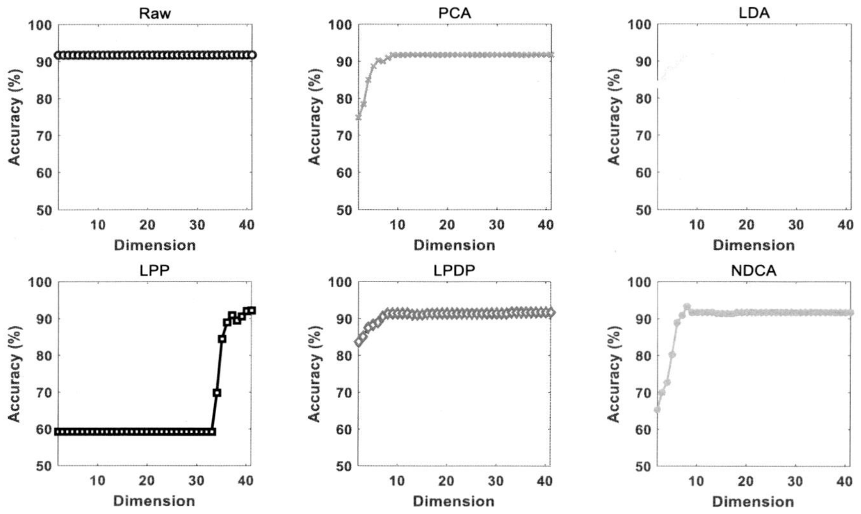

Figure 2.8 The performance of different dimension reduction methods under different reduced dimensions using support vector machine as the classifier.

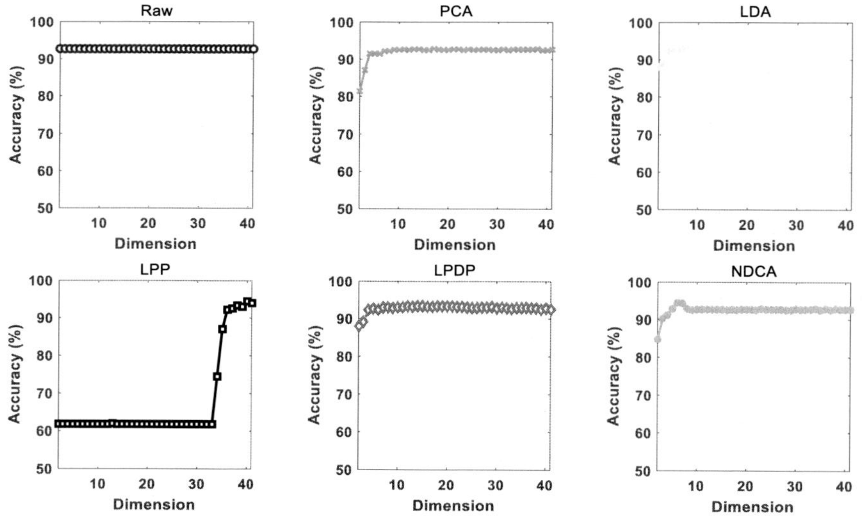

Figure 2.9 The performance of different dimension reduction methods under different reduced dimensions using extreme learning machine as the classifier.

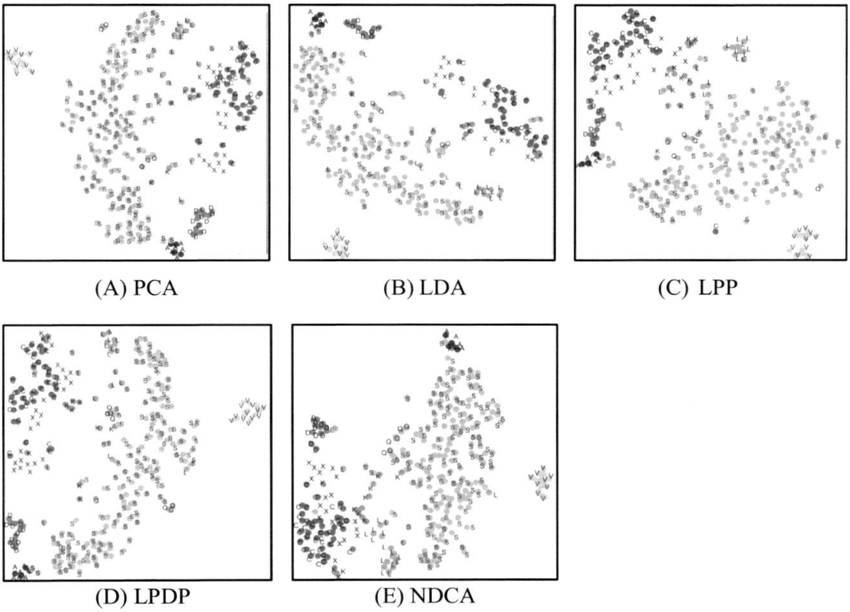

Figure 2.10 Two-dimensional visualization for the scatter points of the first two components acquired by different methods. By comparison, the proposed NDCA method shows better within-class compactness and between-class separation characteristics.

2.4.3 Analysis for neighboring discriminant component analysi parameters

Apart from the dimensionality of the derived subspace d, the proposed NDCA model has several key other parameters, including the between-class neighboring ratio Rb, the within-class neighboring ratio Rw, and the balance parameters γ and μ in the model formulation (2.9). Obviously, different parameter settings will result in fluctuating performances. Thus, parameter sensitivity analyses were needed to be conducted to show the classification performance variation with respect to these parameters. Specifically, the 4 parameters were divided into two groups, that is, (γ, μ), and (Rw, Rb). Among them, γ and μ were selected from the candidate parameter set $\{10^g, g = -4, -3, -2, -1, 0, 1, 2, 3, 4\}$, while Rw and Rb were selected from the candidate parameter set {0.5, 0.55, 0.6, 0.65, 0.7, 0.75, 0.8, 0.85, 0.9, 0.95, 1}. From the figures shown in Figs. 2.11–2.13, one can observe that

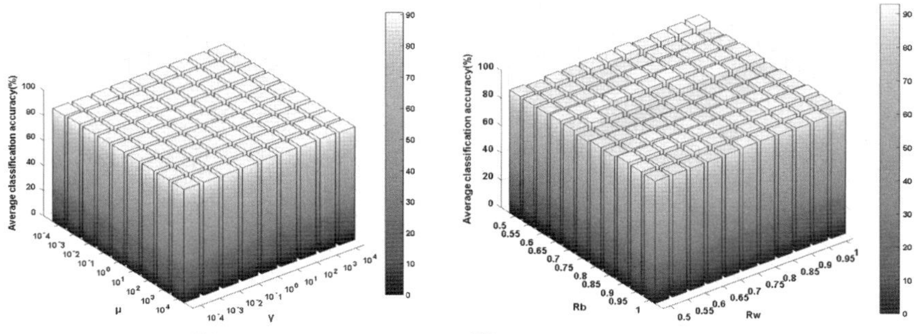

(A) (B)

Figure 2.11 Nearest neighbor plus neighboring discriminant component analysis performance under different combinations of parameters. (A) μ and γ (*best*, 90.60%; *worst*, 89.76%); (B) *Rb* and *Rw* (*best*, 92.81%; *worst*, 87.82%).

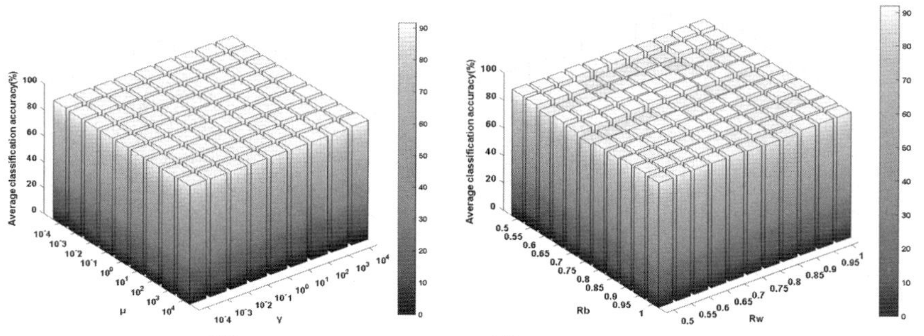

(A) (B)

Figure 2.12 Support vector machine plus neighboring discriminant component analysis performance under different combinations of parameters. (A) *Rb* and *Rw* (*best*, 91.9787%; *worst*, 87.2679%); (B) μ and γ (*best*, 91.7009%; *worst*, 90.8676%).

the average classification performance change surfaces in Figs. 2.11A–2.13A are smoother and more stable within a wide parameter setting range, which means the classification is insensitive to the settings of parameter pair (γ, μ). In contrast, the classification performance changes more acutely with the variations of different parameter pairs (*Rw*, *Rb*).

2.4.4 Analysis for extreme learning machine classifier parameters

The former experiments show that the proposed NDCA method can generally achieve the highest classification accuracy in combination with

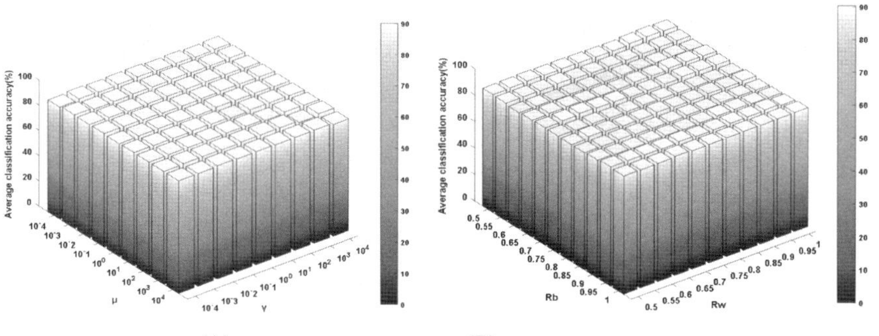

(A) (B)

Figure 2.13 Extreme learning machin plus neighboring discriminant component analysis performance under different combinations of parameters. (A) μ and γ (*best*, 90.0894%; *worst*, 88.4365%); (B) *Rb* and *Rw* (*best*, 90.3676%; *worst*, 87.2660%).

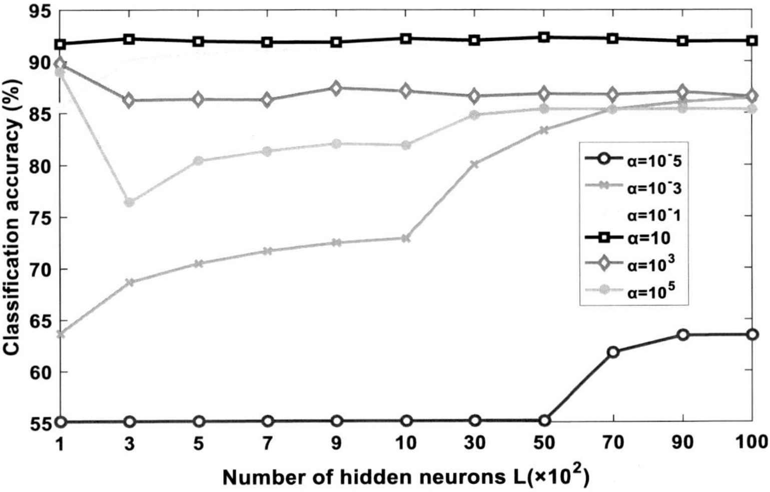

Figure 2.14 Classification performance variations of neighboring discriminant component analysis plus extreme learning machin under different settings of *L* and α.

ELM. As shown in formulation (2.15), ELM has two key hyperparameters, that is, the number of hidden neurons L and the balance parameter α. Fig. 2.14 shows that classification performance changes with different settings of L and α. In general, with the increase of the hidden neurons, the classification accuracy increases first and then tends to be stable. In the experiments, the number of hidden neurons in ELM is empirically set around 9000. As for the tradeoff parameter α, the classification accuracy first improves when α increases from 10^{-5} to 10 and then degrades when α decreases from 10 to 10^{-5}. In the experiments, α can be set around 10, by which promising performance can be expected. From the above experimental results, we observe that:

1. The effects of feature learning on asteroid spectrum classification. In the experimental results shown in Tables 2.4–2.6, the original raw observed spectra without feature learning were directly input into the classifier models, that is, NN, SVM, and ELM, for classification. The average classification performance achieved by NN, SVM, and ELM were 89.2085%, 91.7047%, and 92.6963%, respectively, which were generally the worst performance among all the comparing methods. Comparatively speaking, the classification performance yielded by the same classifier models after feature learning got some improvement. For example, PCA plus NN, SVM, and ELM can respectively achieve the improved classification accuracies of 92.2489%, 91.9711%, and 94.4874%. The results can verify the benefits of feature learning for the improvements of asteroid classification.
2. The advantages of the proposed NDCA method. In comparison with several representative low-dimensional feature learning methods, the proposed NDCA can generally yield better classification accuracy by combining with different classifiers. Specifically, NDCA plus NN, SVM, and ELM can yield the highest classification accuracies of 94.1971%, 93.6377%, and 95.1895%, respectively. The improvements are mainly due to the following two aspects. Firstly, the NDCA method is supervised and inherits the merits of existing methods, which can fully utilize the label knowledge for efficient discriminative feature learning. Secondly, the introduction of a neighboring learning methodology can significantly reduce the side effects of outliers and noised samples to avoid overfitting problems, which will enhance the robustness of the leaned features and finally improve the generalization ability and classification performance of the proposed model for testing.
3. The advantages of ELM. Three baseline classifier models, including NN, SVM, and ELM, were used in the experiments. In particular, the best results are obtained by NDCA plus ELM with a classification accuracy of about 95.19%. To our best knowledge, this work is the first attempt to apply ELM for the spectrum classification of asteroids and achieve the best performance, which can provide new application perspectives and insights for the ELM community.

4. Future work discussion. First, future work will consider employing feature selection methods to study asteroid spectral characteristics. Different from feature learning/extraction methods, which adopt the idea of data transformation, feature/band selection uses the idea of selection and aims to select a small subset of representative spectral bands to remove spectral redundancy while preserving the significant spectral knowledge. Since feature selection is performed in the original observation space, the specifically selected bands have clearer physical meanings with better interpretability. As a result, feature/band selection is an important technique for spectral dimensionality reduction and has room for further improvement. Second, the two-dimensional visualization figures for the scatter points of the first two components acquired by different methods showed that some classes of asteroid spectra with limited training samples are mixed and overlapped. One possible reason is that the numbers of training samples from different classes were unbalanced. For example, the number of samples for "S" class asteroid is 199, while the "A" class asteroid only has six samples. As a result, the sample imbalance problem will lead to learning bias and the generalization ability of the obtained model is thus restricted. Future work will consider incorporating the information of imbalanced class distribution such that the problem of imbalanced class distribution can be well handled.

2.5 Conclusion

This chapter has introduced a novel NDCA model for asteroid spectral feature learning and classification. The key motivation is to distinguish the outliers and noised samples and avoid overfitting problems in feature learning such that the classification performance can be improved. The goals are technically achieved by simultaneously maximizing the neighboring between-class scatter, minimizing the within-class scatter, and preserving the neighboring principal components. Experimental results on reflectance spectrum characteristics measured over the wavelength range 0.45–2.45 μm show the effectiveness of the proposed model by combining with different baseline classifier models, such as NN, SVM, and ELM, and the highest classification accuracy is achieved by using the ELM classifier, which also verifies the superiority of ELM for multiclass classification problem.

Acknowledgment

This work is supported by The Science and Technology Development Fund, Macau SAR (No. 0073/2019/A2). Tan Guo is also funded by The Macao Young Scholars Program under Grant AM2020008, The Natural Science Foundation of Chongqing under Grant cstc2020jcyj-msxmX0636, The Key Scientific and Technological Innovation Project for "Chengdu-Chongqing Double City Economic Circle" under grant KJCXZD2020025, The 2019 Out-standing Chinese and Foreign Youth Exchange Program of China Association of Science and Technology, and The National Key Research and Development Program of China under Grant 2019YFB2102001. Keping Yu is funded by The Japan Society for the Promotion of Science (JSPS) Grants-in-Aid for Scientific Research (KAKENHI) under Grant JP18K18044 and JP21K17736.

Appendix A Reflectance spectra characteristics for some representative asteroids from different categories are used in this chapter

(Figs. A2.1—A2.32).

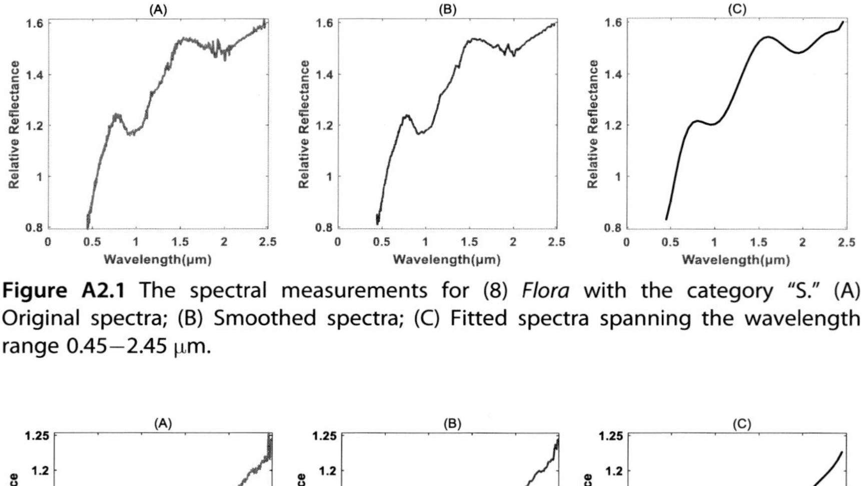

Figure A2.1 The spectral measurements for (8) *Flora* with the category "S." (A) Original spectra; (B) Smoothed spectra; (C) Fitted spectra spanning the wavelength range 0.45—2.45 μm.

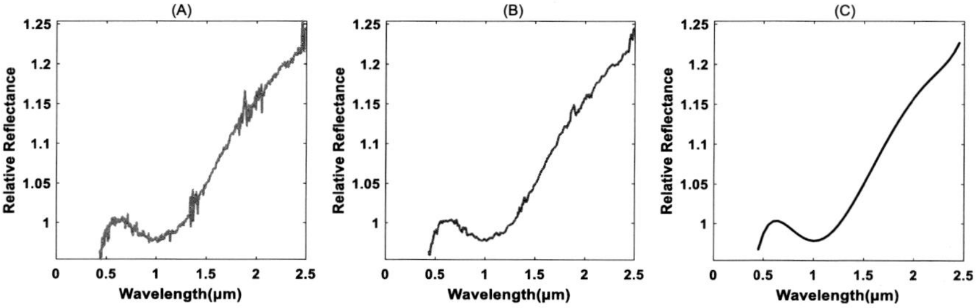

Figure A2.2 The spectral measurements for (10) *Hygiea* with the category "C." (A) Original spectra; (B) Smoothed spectra; (C) Fitted spectra spanning the wavelength range 0.45—2.45 μm.

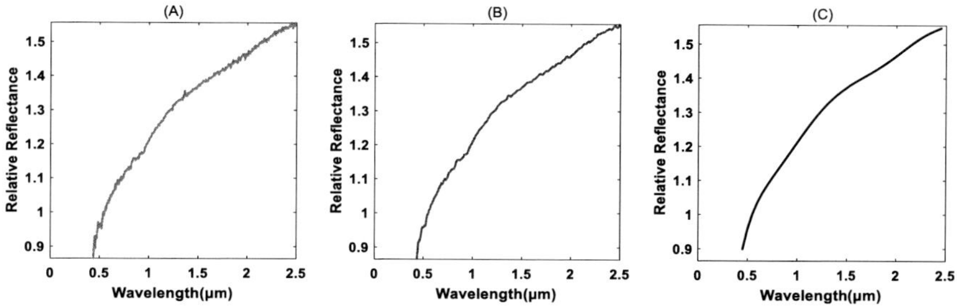

Figure A2.3 The spectral measurements for (16) *Psyche* with the category "X." (A) Original spectra; (B) Smoothed spectra; (C) Fitted spectra spanning the wavelength range 0.45–2.45 μm.

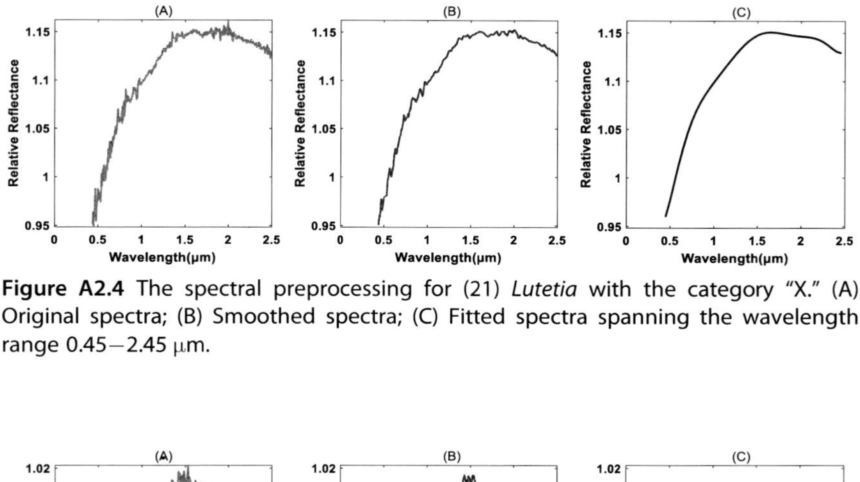

Figure A2.4 The spectral preprocessing for (21) *Lutetia* with the category "X." (A) Original spectra; (B) Smoothed spectra; (C) Fitted spectra spanning the wavelength range 0.45–2.45 μm.

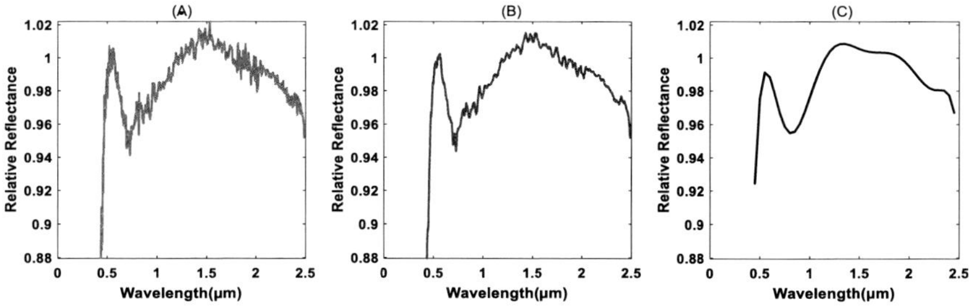

Figure A2.5 The spectral preprocessing for (49) *Pales* with the category "C." (A) Original spectra; (B) Smoothed spectra; (C) Fitted spectra spanning the wavelength range 0.45–2.45 μm.

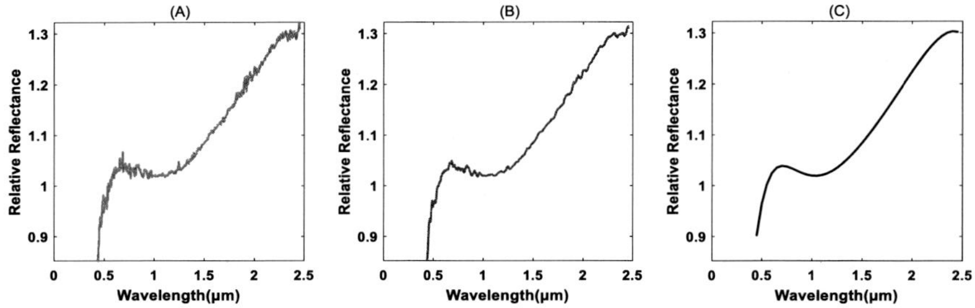

Figure A2.6 The spectral preprocessing for (52) *Europa* with the category "C." (A) Original spectra; (B) Smoothed spectra; (C) Fitted spectra spanning the wavelength range 0.45–2.45 μm.

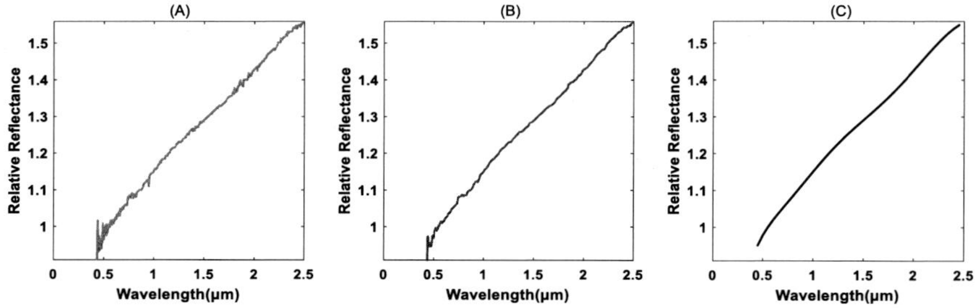

Figure A2.7 The spectral preprocessing for (87) *Sylvia* with the category "X." (A) Original spectra; (B) Smoothed spectra; (C) Fitted spectra spanning the wavelength range 0.45–2.45 μm.

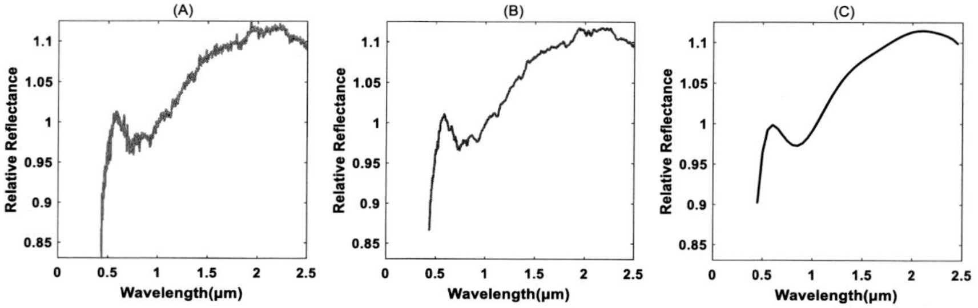

Figure A2.8 The spectral preprocessing for (130) *Elektra* with the category "C." (A) Original spectra; (B) Smoothed spectra; (C) Fitted spectra spanning the wavelength range 0.45–2.45 μm.

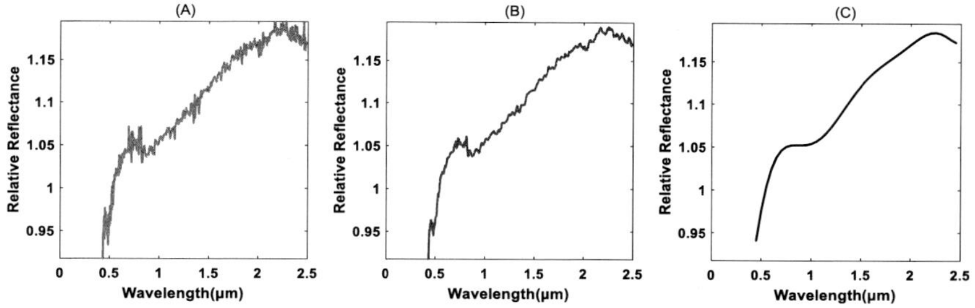

Figure A2.9 The spectral preprocessing for (132) *Aethra* with the category "X." (A) Original spectra; (B) Smoothed spectra; (C) Fitted spectra spanning the wavelength range 0.45–2.45 μm.

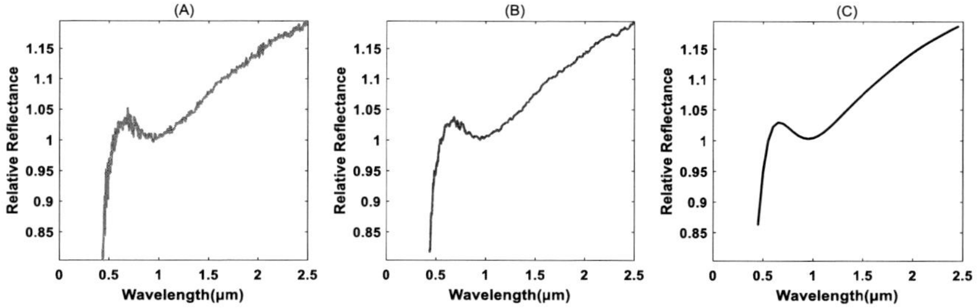

Figure A2.10 The spectral preprocessing for (175) *Andromache* with the category "C." (A) Original spectra; (B) Smoothed spectra; (C) Fitted spectra spanning the wavelength range 0.45–2.45 μm.

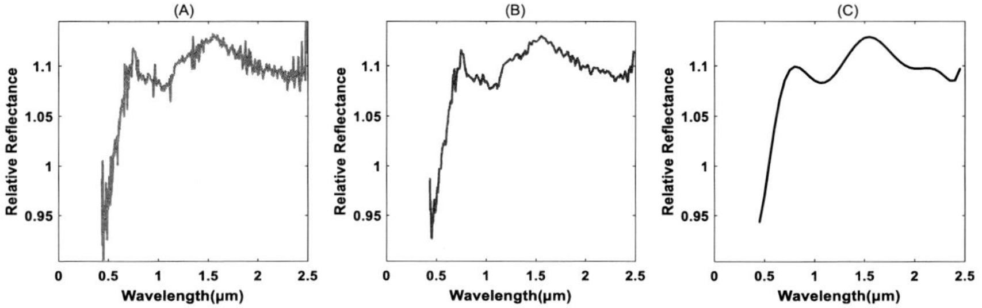

Figure A2.11 The spectral preprocessing for (181) *Eucharis* with the category "X." (A) Original spectra; (B) Smoothed spectra; (C) Fitted spectra spanning the wavelength range 0.45–2.45 μm.

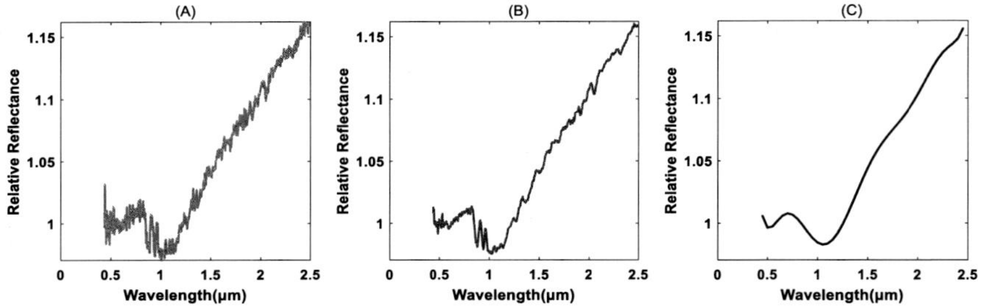

Figure A2.12 The spectral preprocessing for (210) *Isabella* with the category "C." (A) Original spectra; (B) Smoothed spectra; (C) Fitted spectra spanning the wavelength range 0.45 to 2.45 μm.

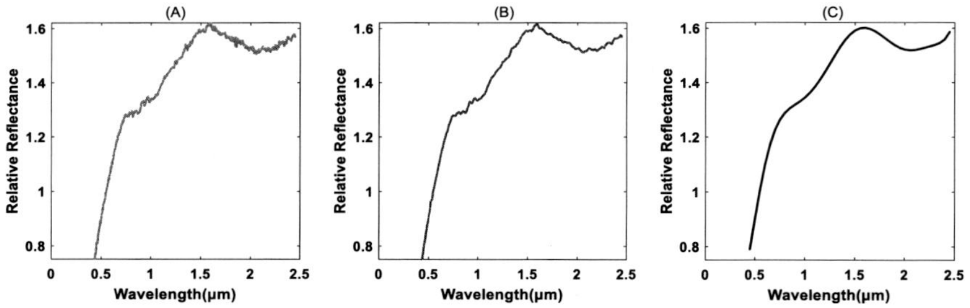

Figure A2.13 The spectral preprocessing for (234) *Barbara* with the category "L." (A) Original spectra; (B) Smoothed spectra; (C) Fitted spectra spanning the wavelength range 0.45–2.45 μm.

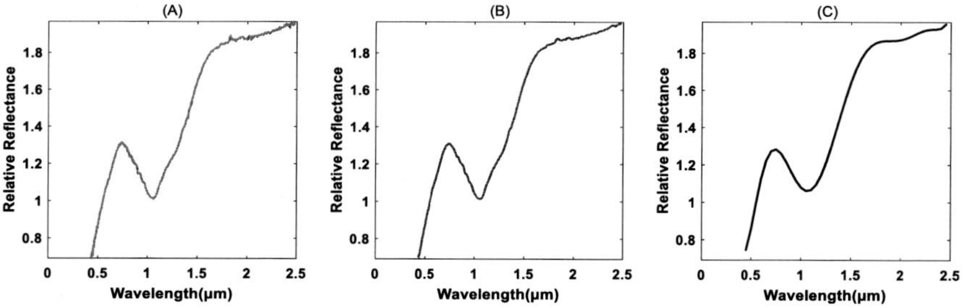

Figure A2.14 The spectral preprocessing for (246) *Asporina* with the category "A." (A) Original spectra; (B) Smoothed spectra; (C) Fitted spectra spanning the wavelength range 0.45–2.45 μm.

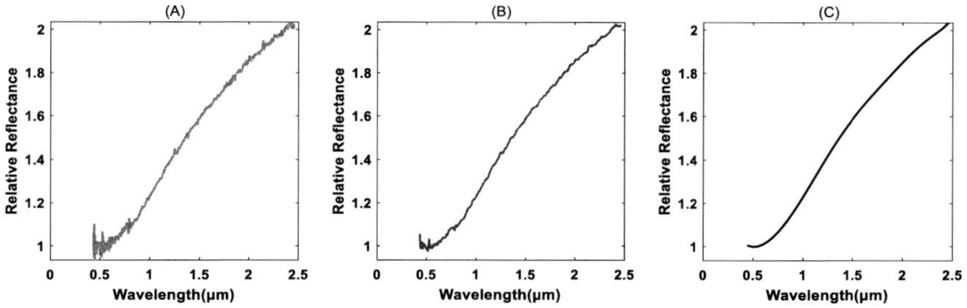

Figure A2.15 The spectral preprocessing for (279) *Thule* with the category "D." (A) Original spectra; (B) Smoothed spectra; (C) Fitted spectra spanning the wavelength range 0.45–2.45 μm.

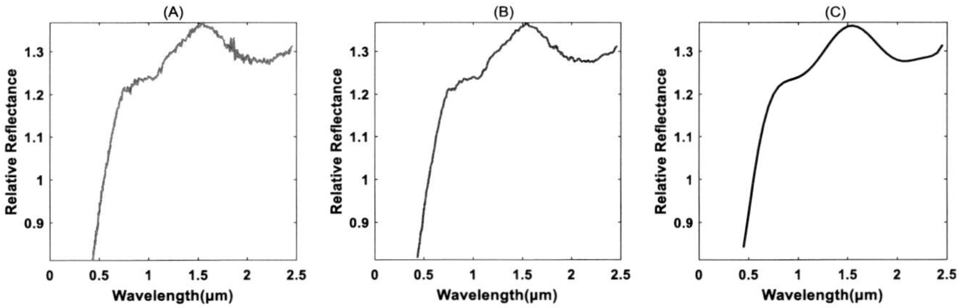

Figure A2.16 The spectral preprocessing for (387) *Aquitania* with the category "L." (A) Original spectra; (B) Smoothed spectra; (C) Fitted spectra spanning the wavelength range 0.45–2.45 μm.

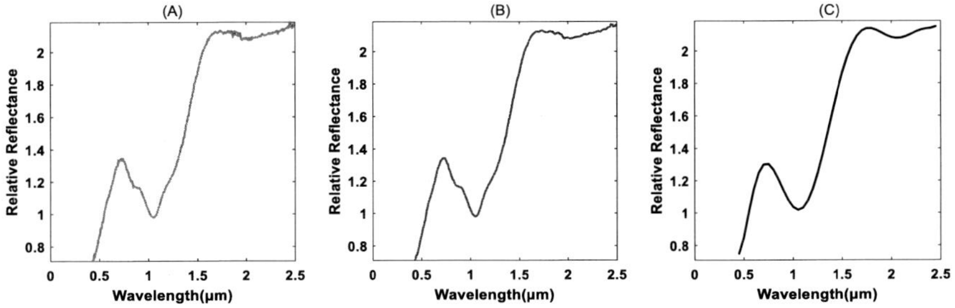

Figure A2.17 The spectral preprocessing for (446) *Aeternitas* with the category "A." (A) Original spectra; (B) Smoothed spectra; (C) Fitted spectra spanning the wavelength range 0.45–2.45 μm.

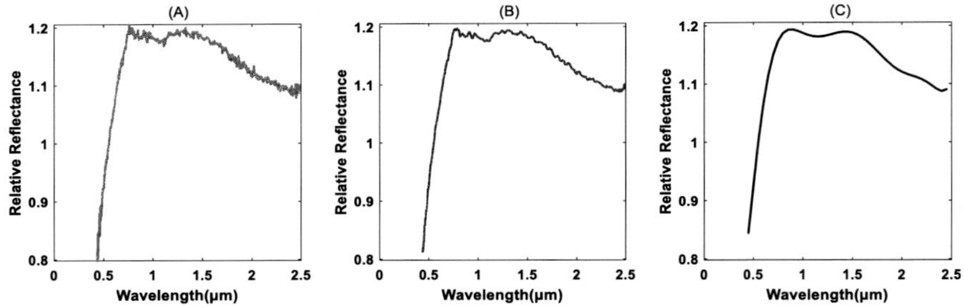

Figure A2.18 The spectral preprocessing for (824) *Anastasia* with the category "L." (A) Original spectra; (B) Smoothed spectra; (C) Fitted spectra spanning the wavelength range 0.45–2.45 μm.

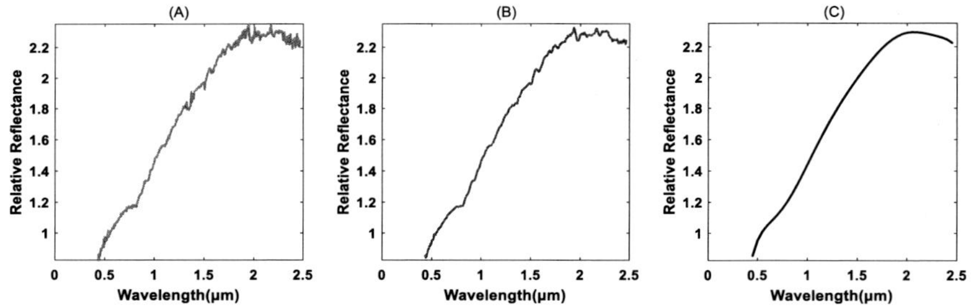

Figure A2.19 The spectral preprocessing for (908) *Buda* with the category "D." (A) Original spectra; (B) Smoothed spectra; (C) Fitted spectra spanning the wavelength range 0.45–2.45 μm.

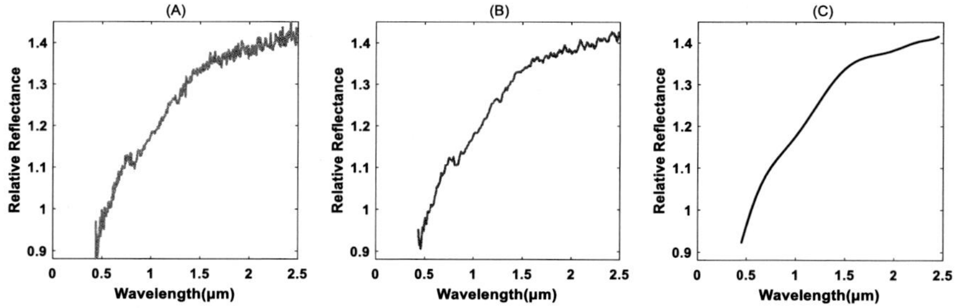

Figure A2.20 The spectral preprocessing for (1094) *Siberia* with the category "X." (A) Original spectra; (B) Smoothed spectra; (C) Fitted spectra spanning the wavelength range 0.45–2.45 μm.

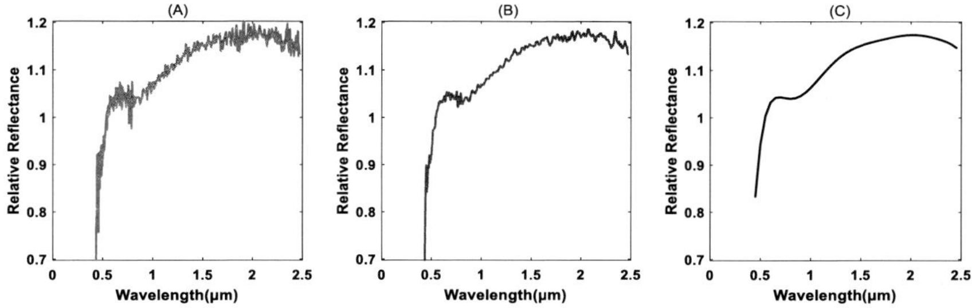

Figure A2.21 The spectral preprocessing for (1300) *Marcelle* with the category "C." (A) Original spectra; (B) Smoothed spectra; (C) Fitted spectra spanning the wavelength range 0.45–2.45 μm.

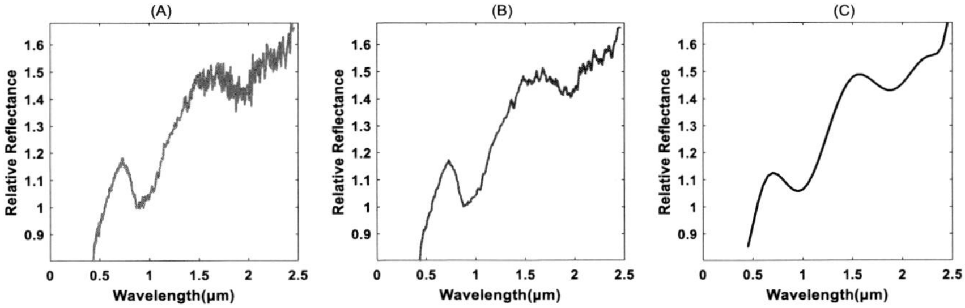

Figure A2.22 The spectral preprocessing for (1807) *Slovakia* with the category "S." (A) Original spectra; (B) Smoothed spectra; (C) Fitted spectra spanning the wavelength range 0.45–2.45 μm.

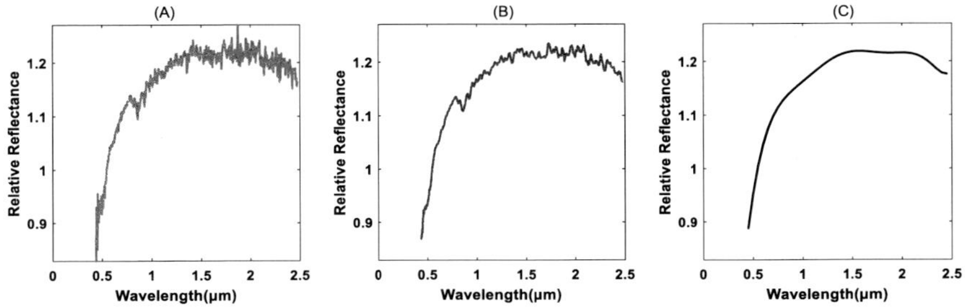

Figure A2.23 The spectral preprocessing for (2035) *Stearns* with the category "X." (A) Original spectra; (B) Smoothed spectra; (C) Fitted spectra spanning the wavelength range 0.45 to 2.45 μm.

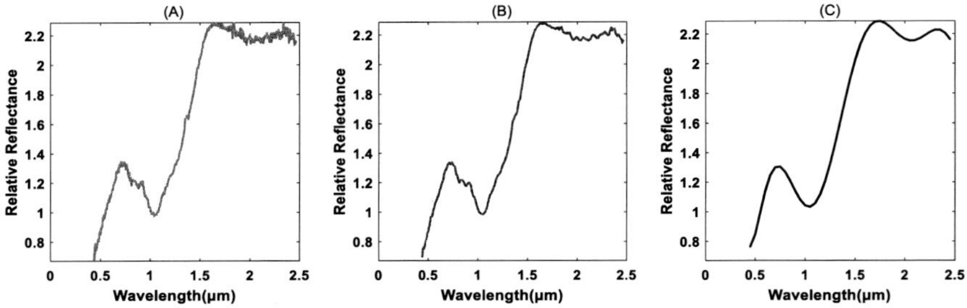

Figure A2.4 The spectral preprocessing for (2501) *Lohja* with the category "A." (A) Original spectra; (B) Smoothed spectra; (C) Fitted spectra spanning the wavelength range 0.45–2.45 μm.

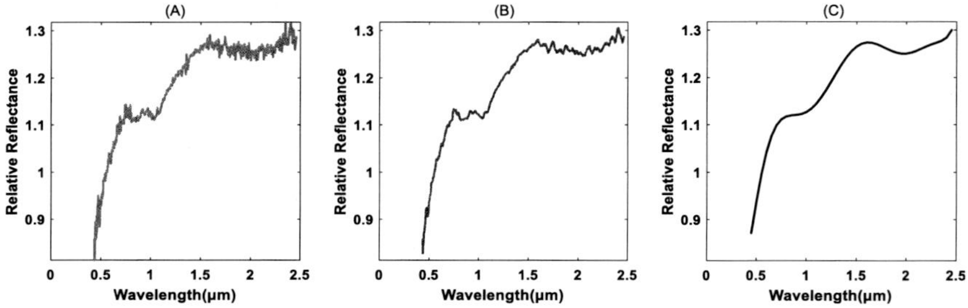

Figure A2.25 The spectral preprocessing for (2957) *Tatsuo* with the category "K." (A) Original spectra; (B) Smoothed spectra; (C) Fitted spectra spanning the wavelength range 0.45–2.45 μm.

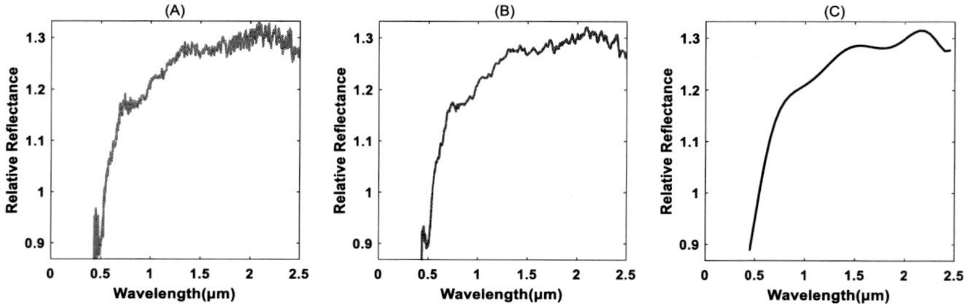

Figure A2.26 The spectral preprocessing for (3103) *Eger* with the category "X." (A) Original spectra; (B) Smoothed spectra; (C) Fitted spectra spanning the wavelength range 0.45–2.45 μm.

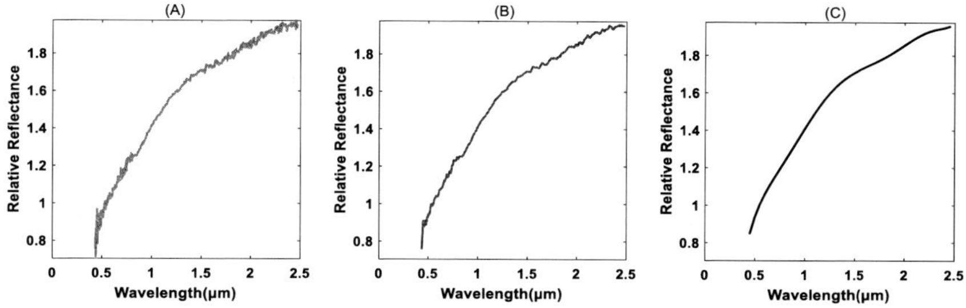

Figure A2.27 The spectral measurements for (3248) *Farinella* with the category "D." (A) Original spectra; (B) Smoothed spectra; (C) Fitted spectra spanning the wavelength range 0.45–2.45 μm.

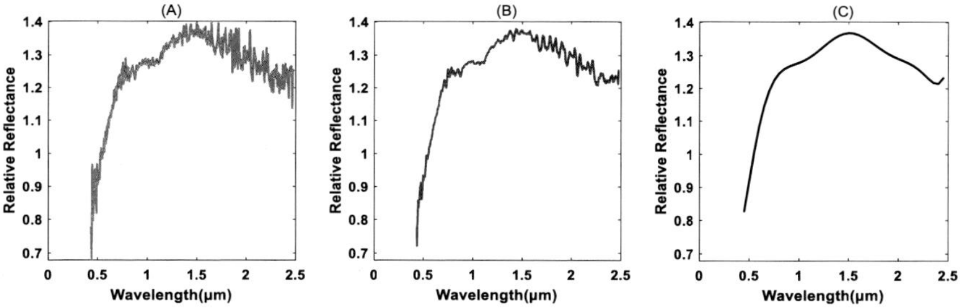

Figure A2.28 The spectral measurements for (3734) *Waland* with the category "L." (A) Original spectra; (B) Smoothed spectra; (C) Fitted spectra spanning the wavelength range 0.45–2.45 μm.

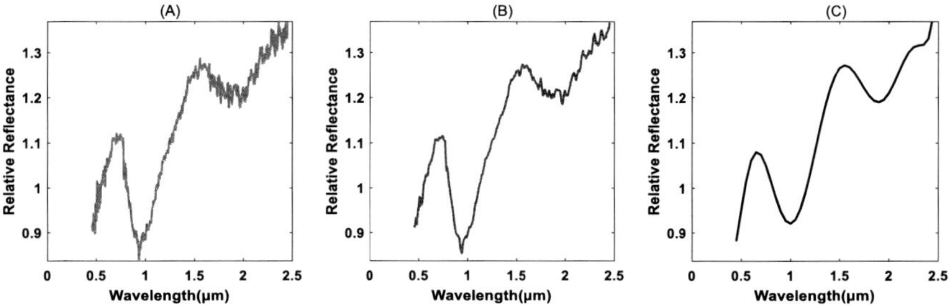

Figure A2.29 The spectral measurements for (3753) *Cruithne* with the category "Q." (A) Original spectra; (B) Smoothed spectra; (C) Fitted spectra spanning the wavelength range 0.45–2.45 μm.

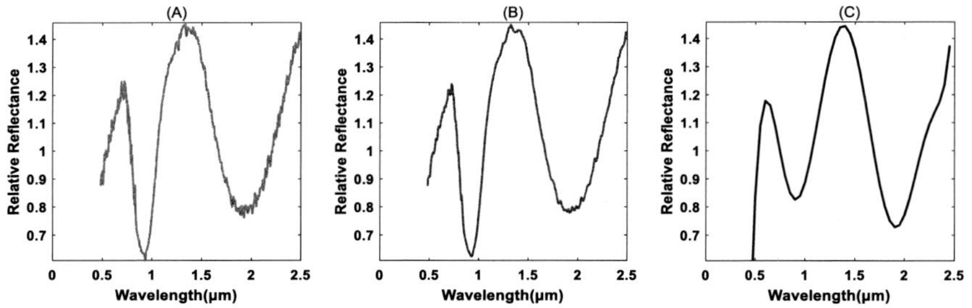

Figure A2.30 The spectral measurements for (4055) *Magellan* with the category "V." (A) Original spectra; (B) Smoothed spectra; (C) Fitted spectra spanning the wavelength range 0.45–2.45 μm.

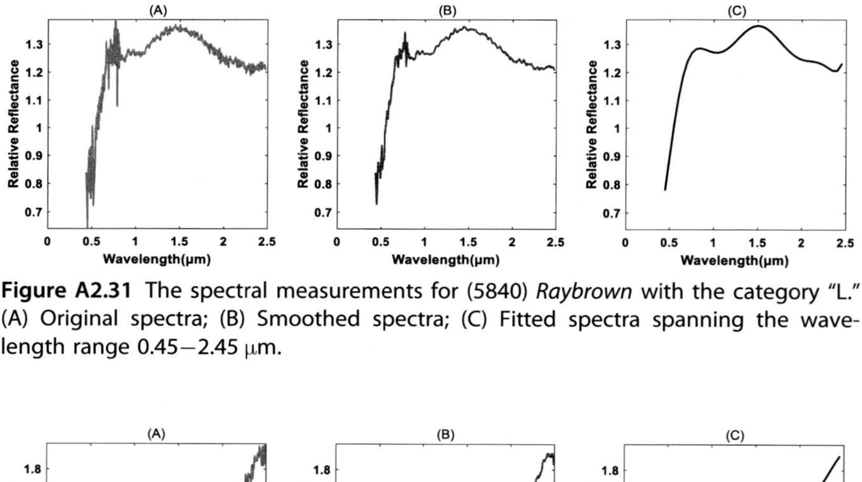

Figure A2.31 The spectral measurements for (5840) *Raybrown* with the category "L." (A) Original spectra; (B) Smoothed spectra; (C) Fitted spectra spanning the wavelength range 0.45–2.45 μm.

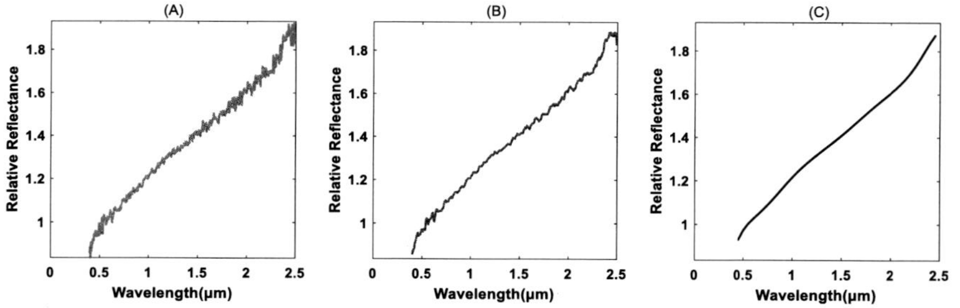

Figure A2.32 The spectral measurements for (17274) *2000 LC16* with the category "D." (A) Original spectra; (B) Smoothed spectra; (C) Fitted spectra spanning the wavelength range 0.45–2.45 μm.

References

Binzel, R. P., Harris, A. W., Bus, S. J., & Burbine, T. H. (2001). Spectral properties of near-Earth objects: Palomar and IRTF results for 48 objects including spacecraft targets (9969) Braille and (10302) 1989 ML. *Icarus*, *151*, 139–149.

Burbine, T. H., & Binzel, R. P. (2002). Small main-belt asteroid spectroscopic survey in the near-infrared. *Icarus*, *159*, 468–499.

Bus, S.J. (1999). *Compositional structure in the asteroid belt: Results of a spectroscopic survey*. Ph. D. thesis, Massachusetts Institute of Technology.

Bus, S. J., & Binzel, R. P. (2002). Phase II of the small main-belt asteroid spectroscopic survey: A feature-based taxonomy. *Icarus*, *158*(1), 106–145.

Carry, B. (2012). Density of asteroids. *Planetary & Space Science*, *73*(1), 98–118.

Demeo, F. E., Binzel, R. P., Slivan, S. M., et al. (2009). An extension of the bus asteroid taxonomy into the near-infrared. *Icarus*, *202*(1), 160–180.

Dong, Y., Liang, T., Zhang, Y., & Du, B. (2021). Spectral-spatial weighted kernel manifold embedded distribution alignment for remote sensing image classification. *IEEE Transactions on Cybernetics*, *51*(6), 3185–3197.

Dorsky, L. I. (2001). Trends in instrument systems for deep space exploration. *IEEE Aerospace & Electronics Systems Magazine*, *16*(12), 3–12.

Fisher, R. A. (1938). The statistical utilization of multiple measurements. *Annals of Human Genetics*, *8*(4), 376–386.

Gaffey, M. J., Burbine, T. H., & Binzel, R. P. (1993). Asteroid spectroscopy: Progress and perspectives. *Meteoritics*, *28*, 161–187.

Gaffey, M. J., Burbine, T. H., & Binzel, R. P. (2012). Asteroid spectroscopy-progress and perspectives. *Meteoritics*, *28*(2), 161–187.

Gui, J., Wang, C., & Zhu, L. (2009). Locality preserving discriminant projections, in *Int. Conf. Intelligent Computing* (pp. 566–572). Icic, Ulsan, SouthKorea, September 16–19.

Guo, T., Luo, F., Zhang, L., Tan, X., Liu, J., & Zhou, X. (2020a). Target detection in hyperspectral imagery via sparse and dense hybrid representation. *IEEE Geoscience and Remote Sensing Letters*, *17*(4), 716–720.

Guo, T., Luo, F., Zhang, L., Zhang, B., Tan, X., & Zhou, X. (2020b). Learning structurally incoherent background and target dictionaries for hyperspectral target detection. *IEEE Journal of Selected Topics in Applied Earth Observations and Remote Sensing*, *13*, 3521–3533.

Guo, T., Lu, X.-P., Zhang, Y.-X., & Yu, K. (2021). Neighboring discriminant component analysis for asteroid spectrum classification. *Remote Sensing*, *13*(16), 3306.

He, X., & Niyogi, P. (2002). Locality preserving projections (lpp). *NIPS*, *16*(1), 186–197.

Herrmann, F. J., Friedlander, M. P., & Yilmaz, O. (2012). Fighting the curse of dimensionality: Compressive sensing in exploration seismology. *IEEE Signal Processing Magazine*, *29*(3), 88–100.

Hotelling, H. H. (1933). Analysis of complex statistical variables into principal components. *British Journal of Educational Psychology*, *24*(6), 417–520.

Howell, E. S., et al. (1994). Classification of asteroid spectra using a neural network. *Journal of Geophysical Research*, *99*(E5), 10847-10,865.

Huang, G. B., Zhou, H., Ding, X., et al. (2012). Extreme learning machine for regression and multiclass classification. *IEEE Transactions on Systems Man & Cybernetics Part B*, *42* (2), 513–529.

Imani, M., & Ghassemian, H. (2014). Band clustering-based feature extraction for classification of hyperspectral images using limited training samples. *IEEE Geoscience and Remote Sensing Letters*, *11*(8), 1325–1329.

Keil, K. (2000). Thermal alteration of asteroids: Evidence from meteorites. *Planetary and Space Science*, *48*(10), 887–903.
Kerr, R. A. (2004). Planetary science: Dirty old ice ball found at Saturn. *Science (New York, N.Y.)*, *304*(5678), 1727–1727.
Liou, J.-C. (2006). Risks in space from orbiting debris. *Science (New York, N.Y.)*, *311* (5759).
Lu, X. P., & Jewitt, D. (2019). Dependence of lightcurves on phase angle and asteroid Shape. *The Astronomical Journal*, *158*(6), 220.
Luo, F., Zhang, L., Zhou, X., Guo, T., Cheng, Y., & Yin, T. (2020). Sparse-adaptive hypergraph discriminant analysis for hyperspectral image classification. *IEEE Geoscience and Remote Sensing Letters*, *17*(6), 1082–1086.
Sun, W., & Du, Q. (2019). Hyperspectral band selection: A review. *IEEE Geoscience and Remote Sensing Magazine*, 7(2), 118–139.
Taskin, G., Kaya, H., & Bruzzone, L. (2017). Feature selection based on high dimensional model representation for hyperspectral images. *IEEE Transactions on Image Processing: A Publication of the IEEE Signal Processing Society*, *26*(6), 2918–2928.
Tholen, D. J. (1984). *Asteroid taxonomy from cluster analysis of photometry*. University of Arizona, Ph.D. thesis.
Vapnik, V. (1998). *Statistical learning theory*. New York: John Wiley.
Vilas, F., & Mcfadden, L. A. (1992). CCD reflectance spectra of selected asteroids: I. Presentation and data analysis considerations. *Icarus*, *100*(1), 85–94.
Wood, X. H. J., & Kuiper, G. P. (1963). Photometric studies of asteroids. *The Astrophysical Journal*, *137*, 1279–1285.
Wu, W. R., Liu, W. W., Qiao, D., et al. (2012). Investigation on the development of deep space exploration. *Science China Technological Sciences*, *04*, 1086–1091.
Xu, S. (1994). *CCD photometry and spectroscopy of small main-belt asteroids*. Ph.D. thesis, Massachusetts Institute of Technology.
Xu, S., Binzel, R. P., Burbine, T. H., et al. (1995). Small main-belt asteroid spectroscopic survey: Initial results. *Icarus*, *115*(1), 1–35.
Xu, S., Binzel, R. P., Burbine, T. H., & Bus, S. J. (1993). Small main-belt asteroid spectroscopic survey. *Bulletin of the American Astronomical Society*, *25*, 1135.
Zellner, B., Tholen, D. J., & Tedesco, E. F. (1985). The eight-color asteroid survey: Results for 589 minor planets. *Icarus*, *61*(3), 355–416.
Zhang, L., Zhang, L., Tao, D., Huang, X., & Du, B. (2014). Hyperspectral remote sensing image subpixel target detection based on supervised metric learning. *IEEE Transactions on Geoscience and Remote Sensing*, *52*(8), 4955–4965.
Zhang, L., Wang, X., Huang, G., Liu, T., & Tan, X. (2019). Taste recognition in E-tongue using local discriminant preservation projection. *IEEE Transactions on Cybernetics*, *49*(3), 947–960.
Zhang, L., & Zhang, D. (2015). Domain adaptation extreme learning machines for drift compensation in e-nose systems. *IEEE Transactions on Instrumentation Measurement*, *64* (7), 1790–1801.
Zhang, Y., Jiang, J., & Zhang, G. (2021). Compression of remotely sensed astronomical image using wavelet-based compressed sensing in deep space exploration. *Remote Sensing*, *13*(2), 288.

CHAPTER 3

Recognition of target spacecraft based on shape features

Na Dong[1], Xinyu Liu[1], Donghui Li[1], Andrew W.H. Ip[2] and K.L. Yung[3]
[1]School of Electrical and Information Engineering, Tianjin University, Tianjin, P.R. China
[2]Department of Industrial and Systems Engineering, The Hong Kong Polytechnic University, Hong Kong SAR, P.R. China
[3]Department of Industrial and Systems Engineering, The Hong Kong Polytechnic University, Hong Kong, P.R. China

3.1 Introduction

3.1.1 Background

In recent decades, the aerospace industry has flourished (Weiss, Leung, & Yung, 2010). Thousands of artificial satellites are running in various types of orbits. For all kinds of space missions in the past few decades, many spacecraft failed to enter into the specific tracks properly or became invalid and obsolete in orbit. Geosynchronous orbital resources are limited, that is, the number of geostationary satellites that can be accommodated is limited, therefore, the orbital value is extremely high. Because of the failure of on-orbit devices, the depletion of fuel or the expiration of satellite life, some satellites failed, but they still occupy a valuable orbital position. Therefore, it is of great significance to provide on-orbit services for these failed (noncooperative) satellites, such as refueling, component upgrading, or preventing derailment (Wang, 2016). Several countries have already carried out or are planning on-orbit service tasks (Jia & Tian, 2012).

On-orbit servicing technology has wide applications and commonly includes repairing, upgrading, refueling, and reorbiting on-orbit spacecraft (Xu, Liang, & Li, 2011). The tethered space robot (TSR) system (Huang, Wang, & Meng, 2015; Wang, Wu, & Ip, 2008), as a new type of space robot, uses space tethers to overcome these shortcomings and is helpful for capturing noncooperative target (NCT). The TSR includes a robot platform, a space tether, and an operating robot, as shown in Fig. 3.1.

The National Academy of Sciences, the Chinese Space Research Board, and the Aerospace Engineering Board defined NCT in the Hubble Space

IoT and Spacecraft Informatics
DOI: https://doi.org/10.1016/B978-0-12-821051-2.00001-5

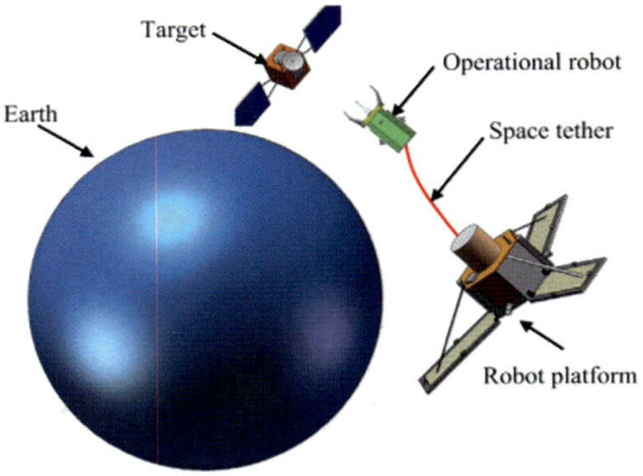

Figure 3.1 Tethered space robot system (Huang et al., 2015).

Telescope maintenance program evaluation report as follows (Liang, Du, & Li, 2012): A NCT refers to a space target that has no communication response device or active identification sensor, and other spacecraft cannot identify and locate through communication signal feedback. NCT has the following characteristics (Du, Liang, & Xu, 2011): (1) there is no special interface installed to capture docking; (2) there is no suitable reflector or sensor for measurement. (3) the movement of the target satellite cannot be controlled. Because NCT cannot provide effective cooperative identification information to the tracker, it is difficult to perform space operations such as mechanical arm grasping and autonomous rendezvous and docking.

Under conditions without cooperation information, making full use of the natural structural features and surrounding features of NCT on photos to recognize target spacecraft is the key technology for space NCT on-orbit service. Compared with microwave radar, laser radar and other sensors, the vision navigation system has the advantages of high accuracy, low power consumption, low cost and so on. It has become the main detection method for space missions of short distances. Target recognition can be acquired using the two methods, template-based or feature-based ones, and in this paper, are the emphases for research.

Since the key structures of artificial targets are mostly composed of cubes and cylindrical modules, it is worth considering a method to recognize artificial targets based on simple shape features imaged by optical sensors. The common modules on the NCT are usually used to operate the ROI of the robot (Sansone, Branz, & Francesconi, 2014), including the bolt, docking ring, apogee rocket engine injection, solar panel span, etc. separated from the satellite and rocket, as shown in Fig. 3.2. A proper

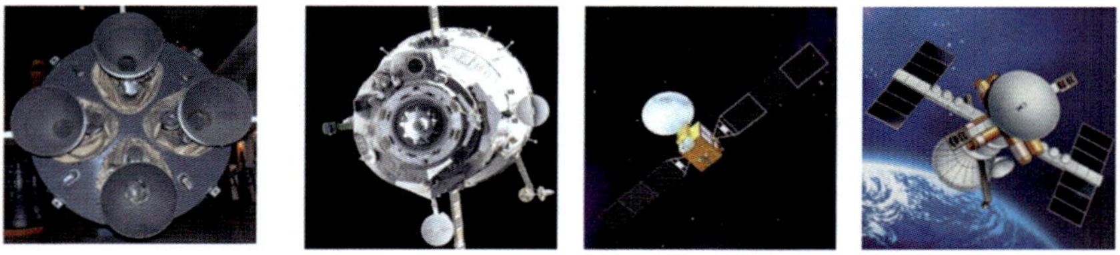

(A) nozzle (B) docking ring (C) span of solar panels (D) antenna

Figure 3.2 Common detection components on noncooperative targets referred from (Sansone et al., 2014). (A) nozzle, (B) docking ring, (C) span of solar panels, (D) antenna.

grasping structure should be necessary equipment for a spacecraft and ensure the generalization of the grasping system with a simple and reliable shape (rectangle, triangle, circle, etc.) In addition, another method based on template matching has been widely used in aerospace target recognition and tracking (Bao, Cai, & Qi, 2016; Cai, Zhu, & Wu, 2014; Opromolla, Fasano, & Rufino, 2015; Wang et al., 2008; Yoo, Hwang, & Kim, 2014). The templates can be selected as a subimage such as a module in NCT or an object that always exists in its surroundings. In view of the above analysis, when the charged coupled device (CCD) cameras on operating robots communicate with a computer vision system, which consists of advanced image detection algorithms, TSR can accurately and quickly capture NCT.

3.1.2 Related works

Object localization based on computer vision has been widely used for grasping NCT. For example, a new template matching algorithm was proposed to solve the problem of NCT recognition in the process of close-range rendezvous (Cai, Huang, & Zhang, 2015). Chen, Huang, and Cai (2016) introduced a method of target location, which uses the features of gradient histogram and support vector machine to predict the target area before extracting the target area, to achieve target grabbing. This method can significantly reduce the search range of the target and run fast. Liu, Xie, and Wang (2016) showed a practical detection method based on an ellipse fitting for the adapter ring on NCT. Here, we first adopt the circle-detection method to recognize circular modules on NCT. Selecting the circular modules on NCT, such as the circular bolts, docking ring, or motor injector as grasping structure, how to accurately detect and identify NCT is viewed as a circle detection problem. Then template matching method is employed to recognize other modules. To recognize multiple objects at the same time, multicircle detection and multitemplate matching (MTM) are the main work of this paper.

The circle detection problem has been widely studied, and most detection methods can be divided into two categories: technologies based on Hough transform and the least-square fitting method (Pan, Chu, & Saragih, 2011). The Hough transform has been widely used and improved in circle detection (Djekoune, Messaoudi, & Amara, 2017; Dong, Wu, & Ip, 2012a,b; Luo, Sun, & Bu, 2016; Marco, Cazzato, & Leo, 2015; Zhang, Wiklund, & Andersson, 2016), for example, applying the improved Hough transform technology to the development of the iris automatic detection system (Djekoune et al., 2017). The least-square fitting method is also applied to circle detection. Zelniker et al. used the convolution-based least-square fitting method to estimate the parameters of a circular object (Zelniker & Clarkson, 2006). Frosio and Borghese (2008) used the prior knowledge of foreground and background statistics to estimate the probability of circular object parameters. However, the contradiction between the computation cost and accuracy for the methods based on Hough transform still exists, and the methods based on least-squares have low robustness.

The template matching method started from the idea of searching the specified image part in the related image (Gonzalez & Woods, 2002). The matching function is realized by using correlation. The most commonly used correlation laws are the Hamming distance sum (HDS), normalized cross-correlation (NCC), absolute difference sum (SAD), square difference sum (SSD), and distance transformation. According to the above introduction, a template matching algorithm is worth considering to solve this difficult problem. Template matching technology has been widely used in target recognition and tracking. For example, the MTM method is applied to perform recognition in the natural environment (Bao et al., 2016). An effective automatic detection method is proposed for PCB component inspection using MTM technology (Wang et al., 2008). Since most template matching methods are time-consuming, they cannot be used for many real-time applications, Cai et al. (2014) used coarse-to-fine search strategies to improve matching efficiency and proposed partial calculation elimination schemes to make the search process faster. Yoo et al. (2014) proposed a template matching method based on histograms, which can deal with the large-scale difference between the target image and template image.

The matching degree is usually determined by evaluating the NCC value. NCC is always an effective and simple similarity measurement method in feature matching. The basic idea of template matching is to make the template traverse all pixels in the captured image and compare the similarity. To recognize NCT, the idea of MTM is the focus of this work.

In essence, multiobject detection, including multicircle detection and MTM, can be viewed as an optimization process of the multipeak function (Dong et al., 2012a,b). Multipeak optimization is used to find all the best values within the search space and has been widely researched by scholars. A number of intelligent algorithms based on kinds of means have been used to solve the multipeak optimization problem (Deng, Yao, & Du, 2012; Dong, Wu, & Ip, 2014; Jia & Tian, 2012; Luo et al., 2016; Xiao-Jun & Wang, 2011). A "niche" artificial bee colony (ABC) algorithm was proposed since niche technology can maintain swarm diversity and avoid the algorithm from converging only to the global optimal solution (Xiao-Jun & Wang, 2011). For the shortcomings of insufficient search ability and low optimization accuracy, Deng et al. (2012) proposed an improved artificial fish swarm hybrid algorithm (AFSHA) to solve the multipeak optimization problem. In order to improve the swarm diversity of the invasive weed optimization (IWO) algorithm, a niche weed optimization algorithm (NIWO) was proposed (Jia & Tian, 2012), which divides weed swarms according to the Euclidean distance among individuals in the swarm and adopts adaptive niche strategy to determine the number of categories. However, most multipeak optimization methods still suffer from the defects of lower accuracy and longer running time. Thus, the concept of species is proposed to realize multipeak optimization (Deng et al., 2012; Dong et al., 2014; Jia & Tian, 2012; Luo et al., 2016; Xiao-Jun & Wang, 2011).

The ABC algorithm was proposed by the Turkish scholar Karaboga (2005), and the basic idea is that the bee colony collaborates to take honey through division of labor and information exchange. ABC has the advantages of greater optimization speed, few control parameters, higher searching accuracy, stronger robustness and more simple operation (Qin, Cheng, & Li, 2014; Zhang, Zheng, & Wang, 2011). Wang (2013) pointed out that the quality of the solution solved by ABC algorithm is relatively good compared with the Genetic Algorithm (GA), Differential Evolution (DE) algorithm and Particle Swarm Optimization (PSO) algorithm. Such characteristics have encouraged the use of ABC to solve different kinds of engineering problems, such as signal processing (Chen, Lü, & Wang, 2014), flow shop scheduling (Li & Pan, 2015), structural inverse analysis (Kang, Li, & Xu, 2009), clustering (Ozturk, Hancer, & Karaboga, 2015) vehicle path planning (Li, Chiong, & Gong, 2014) and electromagnetism (Kitagawa & Takeshita, 2012).

In view of the above analysis, our aim is to propose a species-based artificial bee colony (SABC) algorithm for multipeak optimization and further to solve the multiobject detection for NCT. The rest of this chapter is organized as follows: ABC algorithm is briefly introduced in Section 3.2. Section 3.3 gives a detailed description of the novel algorithm SABC and verifies its optimized performance for multipeak functions. Section 3.4 applies SABC to circle detection and illustrates the method to evaluate the circular integrity and accuracy. In Section 3.5, we provide contrast experiments on circular modules of NCT in various environments to verify the proposed circle-detection method. Similarly, SABC is employed by MTM and further applied to recognition of NCT in Section 3.6. Section 3.7 concludes the work.

3.2 Artificial bee colony algorithm

The basic model of bee colony to realize collective intelligence includes four elements: food source, employed bees, surrounded bees and scouts, and two basic types of behavior: bee recruitment and food source abandonment (Karaboga, 2005; Zhang et al., 2011).

The algorithm starts from a randomly generated initial swarm. Then half of the individuals with better fitness values begin searching, and a competitive survival strategy is applied to reserve individuals, called "employed bees search." The other half with worse fitness values are treated as the onlooker bees and scouts. Each onlooker bee uses the "roulette wheel selection" method to select a good individual, and greedily searches round it to form another half of a new swarm. This process is called "onlooker bees search." The individuals generated by employed bees and onlooker bees form a new swarm, meanwhile, "scouts" are introduced to the new swarm to avoid loss of swarm diversity. After the above mentioned processes, the swarm completes the update. The algorithm approaches the optimal solution through iterating and calculating continuously, preserving good individuals and giving up inferior individuals.

A nonlinear minimization problem is taken as an example to describe the operation process of the ABC algorithm in detail. The problem of minimizing the nonlinear function can be expressed as $\min f(X)$, $X^L \leq X \leq X^U$, where X^U and X^L are the upper and lower bounds, respectively of the variable $X = (X_1, X_2, ..., X_n)$, and X is the n -dimension vector. To solve a nonlinear minimization problem by ABC algorithm, the first step is to generate the initial swarm containing NP individuals within the range of X. Assuming maximum cycles number of the algorithm is *maxCycle*, the i-th individual in the t-th iteration swarm can be expressed as $x_i^t = (x_i^t(1), ..., x_i^t(n))$, where $i = 1, 2, ..., NP$.

3.3 Species-based artificial bee colony algorithm

Multimodal optimization is used to find all the optimal values in the search space. This optimization problem has been widely examined by scholars. Many similar algorithms have been proposed (Deng et al., 2012; Dong et al., 2014; Jia & Tian, 2012; Luo et al., 2016; Xiao-Jun & Wang, 2011). Inspired by (Dong et al., 2014), the concept of species is introduced into the ABC algorithm, and a SABC algorithm is proposed to solve the multimodal optimization problem.

3.3.1 Species

The introduction of the population into swarm intelligence optimization algorithm is considered to be a feasible method to solve multimodal optimization problems (Dong et al., 2014; Luo et al., 2016). The species technique aims to generate multiple species in parallel by dividing the large swarm and searching synchronously for multiple optima. The core of the SABC algorithm is the notion of the species. A species can be defined as a group of individuals with similar characteristics. This similarity can be determined by the Euclidean distance. The shorter the Euclidean distance between the two individuals, the greater their similarity is. A distance parameter γ_s is introduced to denote the distance between the species center and the edge. The center of a species, also named its seed, is usually defined as the individual with the highest fitness value. All individuals within γ_s from their seeds are considered to belong to the same species.

Steps of dividing the swarm into species can be summarized as follows:

1. Sort all individuals by fitness value (descending).
2. Find the best individual whose fitness value is the highest as the first seed.
3. Each individual is assigned to a different species based on Euclidean distance, and the seed of each species is determined.
4. Repeat steps c until all individuals are divided.

Fig. 3.3 illustrates the iterative process of dividing species.

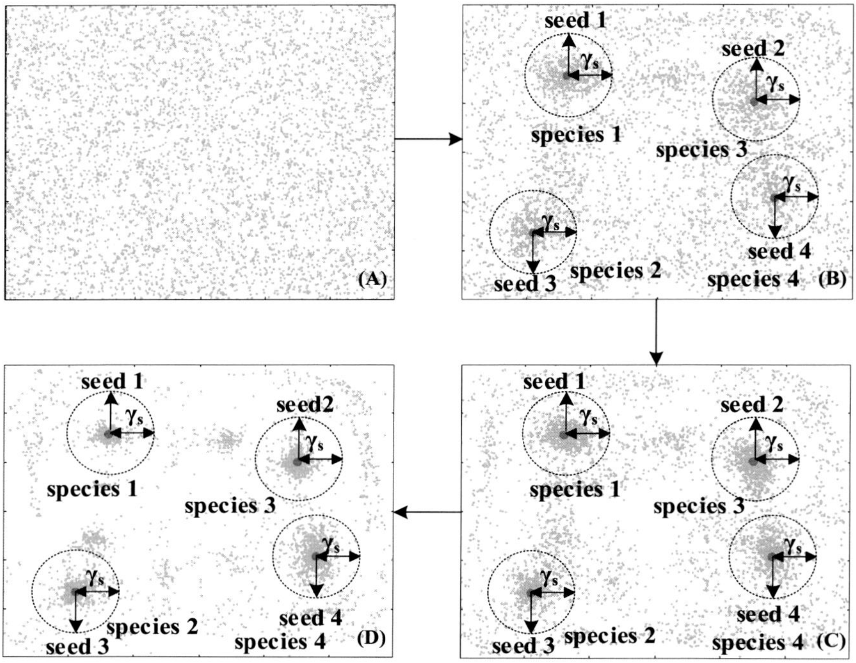

Figure 3.3 Iterative process of dividing species. (A) The initial distribution of swarm. (B) The distribution of species after several searches. (C) The process of converging. (D) The final distribution of swarm.

Taking the optimization objective function with four global optima, for example, the swarm distributions in iterative order are shown in Fig. 3.3. Fig. 3.3A is the swarm distribution after initialing, and it is evenly distributed. Then all individuals are classified into four species and searches within each species several times. Fig. 3.3B shows the updated swarm, and the four species and their seeds are indicated. After several iterations, as seen in Fig. 3.3C, the swarm converges to the four best values gradually and the species are more obvious. The final distribution is as Fig. 3.3D, and all individuals are seen around the four optima. In this process, the bee colony evolves gradually with iteration, and finally, the algorithm locates all the optima successfully.

3.3.2 Species-based artificial bee colony algorithm

Here, we propose a new multipeak optimization algorithm SABC by incorporating the concept of species into ABC. The novelty of SABC lies in that it searches according to ABC mechanism within species instead of the whole swarm. In SABC, after initializing and ranking the fitness values of all individuals, according to the Euclid distance described in Section 3.3.1, the population is divided into several species, and each species searches independently as the ABC mechanism.

Formula (3.1) shows the initialization, where X^U and X^L are upper bound and lower bound respectively, *NP* and *D* are the number and dimension of individuals, and *Foods* is the position vector that stores all individual positions. In addition, a vector *Bas* is initialized to record the number of consecutive stays of each individual at the same location

$$Foods = X^L + (X^U - X^L)\cdot rand(NP, D) \tag{3.1}$$

Referring to Fig. 3.4, the principle of SABC is illustrated in detail. As seen from Fig. 3.4A, the individuals are distributed evenly in the search space after initialing. Then all individuals are divided into some species and the seeds are determined. The locations of individuals within each species are stored in $sFoods\{q\}$ separately, and the corresponding flags are stored in $sBas\{q\}$, where the parameter q is used as an index to distinguish the species. Fig. 3.4B illustrates how to divide the swarm into species, where the discrete points denote individuals. Here, the number of species is assumed as three, but in fact, the number and scale of the species are also related to γ_s and the range of search space. The locations of the seeds are also pointed out in Fig. 3.4B, and we can observe that each species is composed of the individuals within the globe whose center is the seed and the radius is γ_s. After determining the species, the individuals within species are divided into three the kinds of bees to start the search.

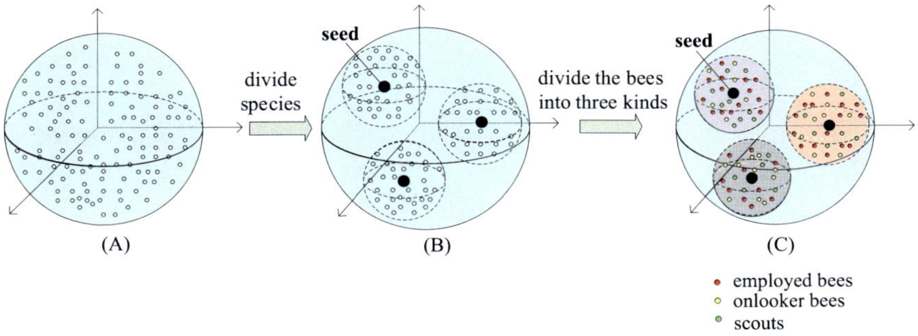

Figure 3.4 Principle of species-based artificial bee colony. (A) The initial distribution of swarm. (B) The way of dividing into species. (C) The way of dividing bees into sepecies.

As shown in Fig. 3.4C, three different colors are used to show three kinds of bees. The method of determining three kinds of bees is described in Section 3.2.

Employed bee: within species, every employed bee will find another food source by formula (3.2).

$$sol\{q\}(i,j) = sFoods\{q\}(i,j) + (sFoods\{q\}(i,j)\text{-}sFoods\{q\}(neibour,j))\cdot(rand\text{-}0.5)\cdot 2 \quad (3.2)$$

Among them, i and j respectively represent the index and random dimension of randomly selected honeybees from the current species and *neibour* represent the other randomly selected honeybees. $sol\{q\}(i,j)$ and $sFoods\{q\}(i,j)$ represent new and current sources of food, respectively. Unlike ABC, because SABC searches within species, all formulas are added to `q'. Then compare the fitness value of $sol\{q\}(i,j)$ and $sFoods\{q\}(i,j)$, and use the greedy selection method to keep better fitness values. If the employed bee still stays in the current source, the flag vector should be added. This is SABC's "employed bees search."

Then, the probability vector $P\{q\}$ is obtained using formula (3.3) for employed bees in every species, which is proportional to honey in the food source.

$$P\{q\} = (0.9\cdot sFitness\{q\}/max(sFitness\{q\}) + 0.1 \quad (3.3)$$

Onlooker bee: each onlooker bee chooses a good source of food $sFoods\{q\}(i,j)$ from the renewal Employed bees using the "roulette selection" method. From formula (3.3), it can be seen that the larger the fitness is, the more likely the food source is to be selected, which makes the population continue to converge. Then the onlooker bee's search nearby $sFoods\{q\}(i,j)$ for a new source of food $sol\{q\}(t,j)$, such as formula (3.2). In the end, it is well preserved as a new spectator. This is SABC's "onlooker bee search."

Scout: after all the hired bees and the onlookers' search, if any bee i stays in the same place several times *Limit*, that is to say $sBas\{q\}(i) > Limit$, the current food source will be abandoned, and the bees will become scouts. Scouts will randomly search for a new food source in the search space as formula (3.1), so that the diversity of the group is not lost. This is SABC's "Scout search."

In one cycle, all species independently search once according to the above process. Finally, both the location $sFoods\{q\}$ and flag $sBas\{q\}$ of all species are restored respectively in *Foods* and *Bas* to prepare for the next iteration.

According to the principle, the implementation of SABC can be summarized by the following steps:

Step 1. Initialize the following parameters: the number of individuals *NP*, the maximum of keeping in the same food source continuously *Limit*, maximum cycles *maxCycle*, the current cycle iter, the number of scouts *SearchNumber* and the radius of species γ_s. The location and flag vector of the swarm are named as *Foods* and *Bas*, respectively.

Step 2. Divide the species and determine the seeds. Rank all individuals in descending order and divide them into species according to steps in Section 3.3.1. Individuals belonging to species q are stored $sFoods\{q\}$ and their flag vector is called as $sBas\{q\}$. For species q, the following steps *a–e* are executed:

1. Within the species q, individuals are classified into three kinds of bees according to the fitness values, and the flag vector $sBas\{q\}$ is used to record how many times the bees remain in the current location continuously.
2. Every employed bee i searches for a new food source. If the new bee has a greater fitness value than the current food source, the current location of the food source is replaced, and $sBas\{q\}(i)$ is reset as zero, otherwise, update the flag vector as $sBas\{q\}(i) = sBas\{q\}(i) + 1$.
3. Calculate the probability vector $P\{q\}$ according to the fitness of employed bees. Every onlooker bee chooses an employed bee with probability $P\{q\}$ and searches around it to generate a new food source, then reserves the better one.
4. If the number of certainly employed bees or onlooker bees stays in the same location continuously, more than *Limit*, the bee will abandon the current location and become a scout to search a new food source.
5. If the species q achieves the above search process, the updated location sFoods$\{q\}$ and the flag vector $sBas\{q\}$ are restored into the swarm.

Step 3. Two stop criteria are employed: either all optima have been found or the number of cycles *iter* attains an upper limit *maxCycle.* Output locations and fitness corresponding to seeds of all species if a termination criterion is met.

The procedure of SABC is exhibited in Fig. 3.5.

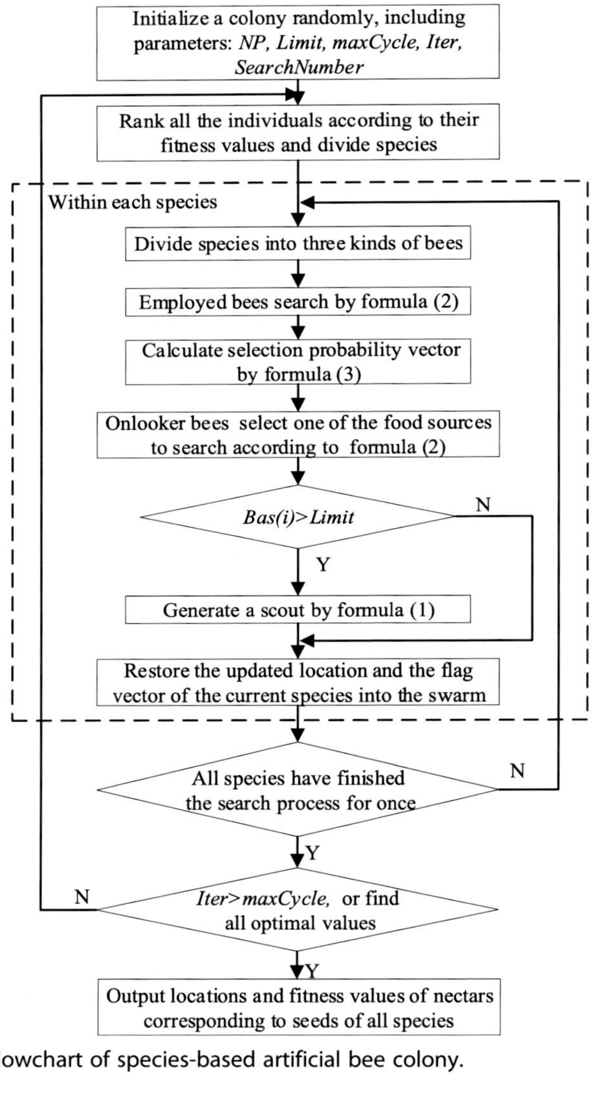

Figure 3.5 Flowchart of species-based artificial bee colony.

3.3.3 Benchmark test

Five benchmark functions are used to test the performance of the newly proposed method. Table 3.1 shows the definition, range of variables, and comment of the test functions. All above benchmarks are maximization of objectives. An Intel (R) Core(TM) i5-6500 CPU @ 3.20 GHz 8.00 G

Table 3.1 Benchmark test functions.

Function	Range	Comment
$F1(x) = \sin^6(5\pi x)$	[0,1]	Five global optima with equal heights
$F2(x) = \exp\left(-2\log(2)\cdot\left(\frac{x-0.1}{0.8}\right)^2\right)\cdot\sin^6(5px)$	[0,1]	One global optimum and four local optima
$F3(x) = \sin^6(5\pi(x^{\frac{3}{4}} - 0.05))$	[0,1]	Five global optima unevenly spaced
$F4(x, y) = 200 - (x^2 + y - 11)^2 - (x + y^2 - 7)^2$	[−6,6]	Four global optima with equal heights
$F5(x, y) = \left(1 - 2y - \frac{\sin(4\pi y - x)}{20}\right)^2 + \left(y - \frac{\sin(2\pi x)}{2}\right)^2$	[−10,10]	Five global optima with equal heights

RAM computer is used to execute all experiments under MATLAB programming.

For enhancement of the analysis during different algorithms, the concept of species is also introduced to GA and PSO to compare with SABC (Dong et al., 2014). The experimental parameters are fixed as $maxCycle = 2000$, $NP = 60$, $Limit = 20$. The inertia weight ω and the acceleration coefficient c_1, c_2 of SPSO are set as 2 and 0.85, respectively. Crossover probability and mutation probability of SGA are fixed as 0.75 and 0.1, respectively. Other parameters are the same as SABC. Each algorithm is tested 50 times, and the related indicators are recorded in Table 3.2 to show the optimization performance.

As can be seen from Table 3.1, $F1$, $F2$, and $F3$ are univariate functions, while $F4$ and $F5$ are bivariate functions.$F1$, $F4$, and $F5$ all have global optima with equal heights, while global optimal points of $F2$ and $F3$ are unevenly distributed. It can be observed from Table 3.2, SPSO and SABC can locate all the optima every time for all functions, while SGA has 96% and 94% success rate for $F2$ and $F3$, which means SGA has 2 and 3 misdetections out of 50 trials. Considering the execution-time needed in these three algorithms, SABC works faster for each function. Then, comparing the optimization accuracy, optima solved by SABC are the most stable and closest to the theoretical values.

Table 3.2 Benchmark test results.

Function	Measurement	γ_s	SGA	SPSO	SABC
F1	Accuracy	0.1	− 7.92 e-05 ± 2.58 e-04	0 ± 0	0 ± 0
	Time (s)		0.5851	0.2252	0.0622
	Success rate[a] (%)		100	100	100
F2	Accuracy	0.1	− 3.60 e-04 ± 3.09 e-05	− 1.56 e-05 ± 2.02 e-04	− 7.23 e-06 ± 1.17 e-06
	Time (s)		0.6265	0.0726	0.0385
	Success rate[a] (%)		96	100	100
F3	Accuracy	0.1	− 4.58 e-05 ± 6.39 e-05	0 ± 0	0 ± 0
	Time (s)		0.6646	0.1171	0.0693
	Success rate[a] (%)		94	100	100
F4	Accuracy	3	− 0.0763 ± 7.94 e-03	− 2.39 e-07 ± 8.94 e-07	− 1.80 e-09 ± 1.00 e-09
	Time (s)		0.6375	0.6654	0.2155
	Success rate[a] (%)		100	100	100
F5	Accuracy	0.15	0.0815 ± 4.79 e-03	7.62 e-03 ± 6.37 e-04	5.05 e-04 ± 5.45 e-05
	Time (s)		0.6931	0.3614	0.1503
	Success rate[a] (%)		100	100	100

[a]The success rate represents the percentage of finding all the optima successfully in 50 trials.

After experiments and comparison, it can be concluded that the proposed algorithm SABC has an excellent performance in multipeak optimization problems.

3.4 The application of species-based artificial bee colony in circle detection

In this part, multicircle detection is considered a multipeak optimization problem. We design a multicircle detection method based on SABC. Using the three-edged point positioning method (Dong et al., 2012a,b) to represent the circle effectively reduces the search space and eliminates the infeasible position.

3.4.1 Representation of the circle

In this representation, after the edge is extracted, all the edge points in the image are stored as the index of their relative positions in the edge array V. Three edge points are stored in the form of the serial number corresponding to their coordinates V_i, V_j, V_k to determine a circle passing through them. Each circle is represented by a center (x_0, y_0) and radius. Through the coordinates of the three points on the edge graph, the candidate circle can be determined by the following formula:

$$(x\text{-}x_0)^2 + (y\text{-}y_0)^2 = r^2 \tag{3.4}$$

$$x_0 = \frac{\begin{vmatrix} x_j^2 + y_j^2 - (x_i^2 + y_i^2) & 2(y_j - y_i) \\ x_k^2 + y_k^2 - (x_i^2 + y_i^2) & 2(y_k - y_i) \end{vmatrix}}{4\left((x_j - x_i)(y_k - y_i) - (x_k - x_i)(y_j - y_i)\right)} \tag{3.5}$$

$$y_0 = \frac{\begin{vmatrix} 2(x_j - x_i) & x_j^2 + y_j^2 - (x_i^2 + y_i^2) \\ 2(x_k - x_i) & x_k^2 + y_k^2 - (x_i^2 + y_i^2) \end{vmatrix}}{4\left((x_j - x_i)(y_k - y_i) - (x_k - x_i)(y_j - y_i)\right)} \tag{3.6}$$

Therefore, the parameters of a circle can be represented by the serial numbers of three edge points i, j, k :

$$[x_0, y_0, r] = T(i, j, k) \tag{3.7}$$

Among them, T is the transformation of formulas (3.4)–(3.6).

The test set of points is $S = \{s_1, s_2, \ldots, s_{N_s}\}$, where N_s is the number of test points on the circular edge, and the test set of points S is evenly sampled from the circular edge. Each point s_i is a two-dimensional vector whose coordinates (x_i, y_i) are obtained according to the following formula:

$$\begin{cases} x_i = x_0 + r \cdot \cos \dfrac{2\pi i}{N_s} \\ y_i = y_0 + r \cdot \sin \dfrac{2\pi i}{N_s} \end{cases} \tag{3.8}$$

3.4.2 Assessment of circular accuracy

Each circle corresponds to an individual in the swarm, and the algorithm finds the optimal solution, that is, the circle is detected. In general, fitness functions are used to evaluate individuals. Here, for the evaluation of accuracy, the fitness function represents the existence of the measured point on the edge of the actual circle, which is defined as

$$F(C) = \left(\sum\nolimits_{i-0}^{N_s-1} E(x_i, y_i)\right)/N_s \tag{3.9}$$

where $E(x_i, y_i)$ is the gray value of the image coordinates (x_i, y_i), so $F(C) \in \begin{bmatrix} 0 & 255 \end{bmatrix}$. The fitness value reflects the degree of overlap between the candidate circle and the circle existing in the actual image. So, larger $F(C)$ means better accuracy. For clarity, a parameter P_C is defined to describe the percentage of coincidence with the actual one.

$$P_c = N_t / N_s \tag{3.10}$$

Here, N_t is the actual number of test points in the edge image, so $P_C \in \begin{bmatrix} 0 & 1 \end{bmatrix}$. The bigger P_C is, the better the accuracy. Under the guidance of the fitness function, the SABC algorithm is used to evolve the set of candidate circles corresponding to all seeds, so that multiple optimal candidate circles can fit into the actual circles. If the corresponding individual fitness meets the set value (180) and the center x_0, y_0 and radius r of the circle are determined, the circle is considered to be correctly detected. Then we calculate the difference between the theoretical value and the actual circle $\Delta x, \Delta y, \Delta r$ (Cai, Huang, & Chen, 2016) to describe the detection accuracy.

The complete process of the proposed multicircle detection algorithm based on SABC is shown in Fig. 3.6.

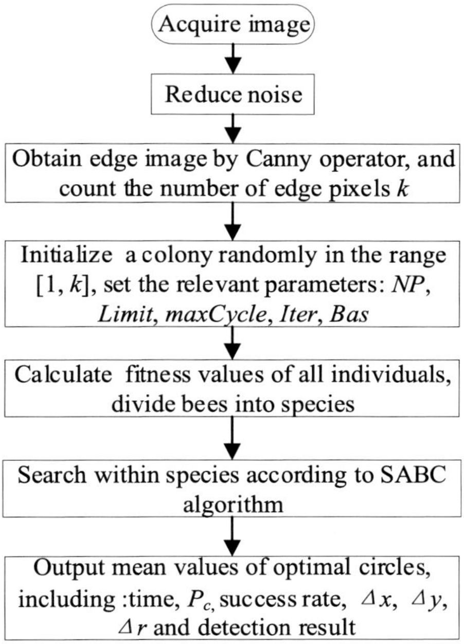

Figure 3.6 Flowchart of the circle detection method based on species-based artificial bee colony.

It is worth mentioning that the number of detected circles is a preset value, and the algorithm searches until all circles have been correctly detected or the cycle reaches *maxCycle*.

3.5 The application of species-based artificial bee colony in multicircle detection

In this section, two test images are firstly introduced for experimental tests to verify SABC's performance in circle detection, then the method is further applied to detect circular modules on NCT.

3.5.1 Test experiments on drawn sketches

Firstly, the performance of the proposed multicircle detection method is tested by using two test images of randomly distributed circles. The distance between two individuals in a species is defined as the distance between the centers of two individuals. The number of circles in each test image and the distance between circles is known, and the species radius γ_s

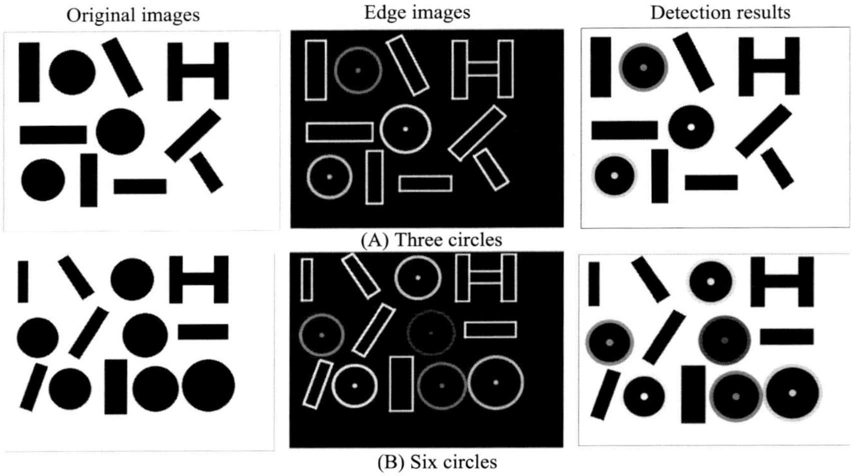

Figure 3.7 Test images and the detection results. (A) Three circles. (B) Six circles.

is set to a value less than the distance between the nearest two circles. Fig. 3.7 shows a test image, a detection result in an edge map, a detection result in an original image, and a detected circle and its center marked with the same color.

The algorithm parameters for the two images are set as follows: the size of the population is 500, the species radius are 170 and 100, respectively. The inertia weight of SPSO is adjusted between 0.2 and 0.9, and the acceleration coefficient is set to 2. The crossover probability of SGA is 0.75, and the mutation probability is 0.1.

Each algorithm runs independently 50 times. The running results including the average execution time, average value, and standard deviation of P_C, the success rate and $(\Delta x, \Delta y)/\Delta r$ are given to show the performance of SABC in Table 3.3.

As can be seen from the running results, for Fig. 3.7, only the SABC algorithm can locate all circles every time, while SGA and the SPSO algorithm have several failures. Specifically, for Fig. 3.7B, the success rate of SGA and the SPSO algorithm only can reach 82% and 80%, which means that there are 9 and 10 times failure out of 50 experiments in locating all circles. As for execution time, the new algorithm SABC works fastest and the time consumption is stable, while the SPSO and SGA take too long. Furthermore, P_C of SABC can reach 96% and 93%, which is higher than that of SPSO and SGA and indicates a higher coinciding rate and detection accuracy.

Table 3.3 The running results on the two test images out of 50 runs.

Image	Method	SGA	SPSO	SABC
Fig. 3.7A	Average time(s)	3.2966	2.6328	1.8633
	P_C (A. v and S. d)	0.9120 ± 0.0503	0.9285 ± 0.0363	0.9637 ± 4.56 e-04
	Success rate[a] (%)	88	90	100
	$(\Delta x, \Delta y)/\Delta r$	(0.32,0.68)/0.46	(0.22, 0.39)/0.34	(0.11, 0.35)/0.25
Fig. 3.7B	Average time(s)	3.9778	3.1350	2.0709
	P_C (A. v and S. d)	0.9076 ± 4.32 e-02	0.9183 ± 1.05 e-04	0.9391 ± 2.69 e-05
	Success rate[a] (%)	92	94	100
	$(\Delta x, \Delta y)/\Delta r$	(1.01, 0.52)/0.53	(0.42, 0.24)/0.42	(0.39, 0.19)/0.31

[a]The success rate represents the percentage of finding all circles successfully in 50 trials.

From the results, SABC can locate multiple circles with a high success rate and accuracy as well as less computational time. All the above demonstrates the good performance of SABC in multicircle detection.

3.5.2 Detection for circular modules on noncooperative targets

In this part, we report the results of SABC application in NCT and compare it with SGA and SPSO. In addition, to fully verify our method, we also use another method GHC (Cai et al., 2016) based on gradient region growth and Euclidean distance histogram distribution as a comparison, and has been proved to be superior to many other methods. As a classical circle detection method, the random Hough transform (RHT) (Xu & Oja, 1993) is also introduced for comparison.

It is difficult to collect photos of real rockets, spacecraft or space stations, due to confidentiality and cost. Therefore, an ideal method is to use photos of real spacecraft models. In our experiment, the models of the spacecraft Shenzhou 8 and the space station Tiangong 1 from China are used, as shown in Fig. 3.8 (Cai et al., 2016). As shown in Fig. 3.8, there are several circular parts on the model, such as hatch, docking ring, engine nozzle, radar antenna, etc.

As shown in Fig. 3.9A–D, four circular modules were used to test the performance of the five methods. (A) is the circular antenna. (B) is the image of docking ring. (C) and (D) show the motor injectors. After running all methods on four images, the detected results are shown in Fig. 3.9. Among the four images, (A) has a single circle, others have multiple circles, on which the multiple-circle recognition capability can be tested. It is observed that, for the simple image (A), the five methods all can detect the

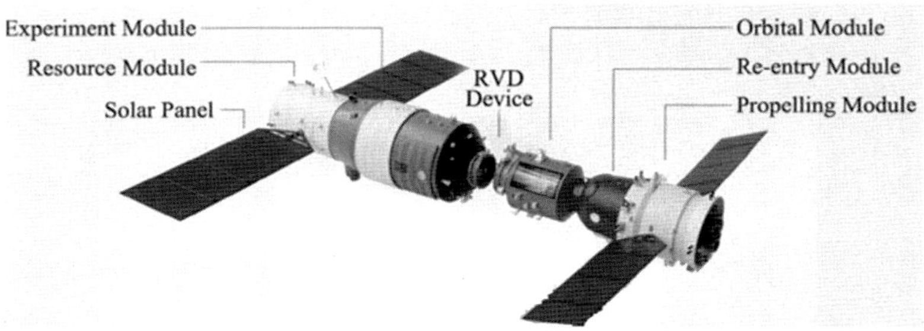

Figure 3.8 The models of spacecraft "Shenzhou 8" and space station "Tiangong" referred from (Cai et al., 2016).

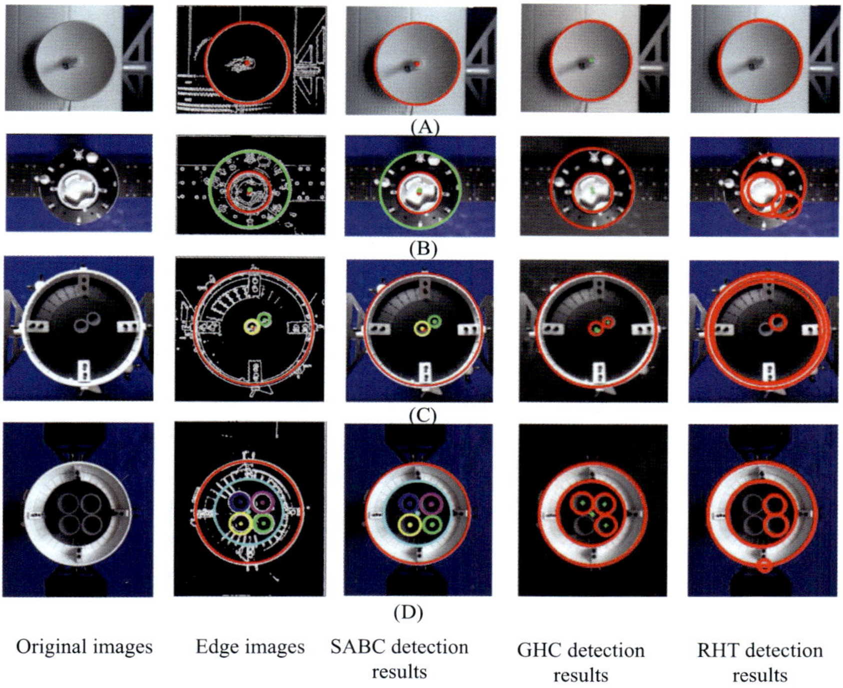

Figure 3.9 Detection results for circular modules on noncooperative targets. (A) circular antenna, (B) docking ring, (C) motor injector, (D) motor injector.

circle successfully. With the number of circles increasing in (B) and (C), detection by the first four methods stills were correct, while the RHT began misdetection and missed detection. Specifically, RHT detected false circle in (B) and missed the correct circles in (B) and (C). Then in (D), GHC was also defeated by the three methods based on optimization algorithms due to a missed detected circle. So, as far as the correctness of the detection results, the SGA, SPSO, and SABC work better than GHC and RHT. Moreover, the experiments (B), (C) and (D) also reflect the good performance of our method in multiple-circle recognition.

Table 3.4 shows the comparison results of SGA, SPSO, and SABC for the above four images. From the table, the data indicated that the method based on SABC performs best in execution-time, accuracy, success rate, deviation with theoretical values, which once again confirmed the previous results.

The contrast between SABC, GHC, and RHT is shown in Table 3.5, where the average time of our method was much less than that of other

Table 3.4 The running results of the three algorithms in Fig. 3.10.

Image	Algorithm	SGA	SPSO	SABC
Fig. 3.9A	Average time(s)	2.0754	1.2046	0.9523
	P_C (A. v and S. d)	0.8427 ± 0.7376	0.8893 ± 0.4823	0.9273 ± 6.28 e-04
	Success rate[a] (%)	92	96	100
	$(\Delta x, \Delta y)/\Delta r$	(0.36,0.49)/0.57	(0.32, 0.27)/0.48	(0.11, 0.19)/0.35
Fig. 3.9B	Average time(s)	2.9658	2.0134	1.1731
	P_C (A. v and S. d)	0.8936 ± 6.85 e-02	0.9073 ± 6.42 e-04	0.9264 ± 5.53 e-05
	Success rate[a] (%)	94	94	100
	$(\Delta x, \Delta y)/\Delta r$	(0.57, 0.61)/0.63	(0.50, 0.54)/0.73	(0.49, 0.35)/0.41
Fig. 3.9C	Average time(s)	4.0738	3.1692	2.3764
	P_C (A. v and S. d)	0.8274 ± 4.74 e-02	0.8536 ± 2.85 e-04	0.9163 ± 7.23 e-04
	Success rate[a] (%)	90	92	100
	$(\Delta x, \Delta y)/\Delta r$	(0.61, 0.52)/0.68	(0.73, 0.44), 0.62	(0.58, 0.39), 0.43
Fig. 3.9D	Average time(s)	4.3854	3.9435	2.6852
	P_C (A. v and S. d)	0.8037 ± 2.89 e-02	0.8463 ± 3.64 e-04	0.8846 ± 6.93 e-04
	Success rate[a] (%)	92	90	100
	$(\Delta x, \Delta y)/\Delta r$	(1.25, 0.63),0.76	(0.97, 0.56),0.82	(0.82, 0.46),0.63

[a]The success rate represents the percentage of finding all circles successfully in 50 trials.

Table 3.5 The running results by species-based artificial bee colony compared with GHC and random Hough transform.

Image	Algorithm	SABC	GHC	RHT
Fig. 3.9A	Average time (s)	0.9523	1.56	1.09
	$(\Delta x, \Delta y)/\Delta r$	(0.11, 0.19)/ 0.35	(0.36,0.47)/ 0.52	(0.25, 0.34)/ 0.37
Fig. 3.9B	Average time (s)	1.1731	1.80	2.42
	$(\Delta x, \Delta y)/\Delta r$	(0.49, 0.35)/ 0.41	(0.65,0.58)/ 0.61	NAN
Fig. 3.9C	Average time (s)	2.3764	3.7	0.91
	$(\Delta x, \Delta y)/\Delta r$	(0.58, 0.39), 0.43	(0.72,0.63)/ 0.72	NAN
Fig. 3.9D	Average time (s)	2.6852	2.74	3.53
	$(\Delta x, \Delta y)/\Delta r$	(0.82, 0.46), 0.63	NAN	NAN

methods. Further, the detection deviation $(\Delta x, \Delta y), \Delta r$ by SABC was much smaller.

3.5.3 Detection performance with noise

The first step of this method is to denoise the image. Further experiments are carried out to evaluate the fault tolerance when the original image is directly affected by noise. In the test, we added three kinds of noise to the same image separately. Noise with different parameters was added to the original image, and then the method was used for detection.

Fig. 3.10 shows two original images: a circular component with two circles. With the noise parameters gradually increasing from the default value, the detection results are shown in Fig. 3.10. As observed in Fig. 3.10, our method acquired a good detection result when using the default parameter settings. Then the noise intensity was gradually increasing. When the images were seriously interfered with due to noise, the results were not ideal. Specifically, the method failed to find the small circle in Fig. 3.10 when the noise was large enough. Focusing on the misdetections, we can see that discovering the undetected circles are indeed too blurry to recognize through human vision, so the misdetections by our method can be accepted.

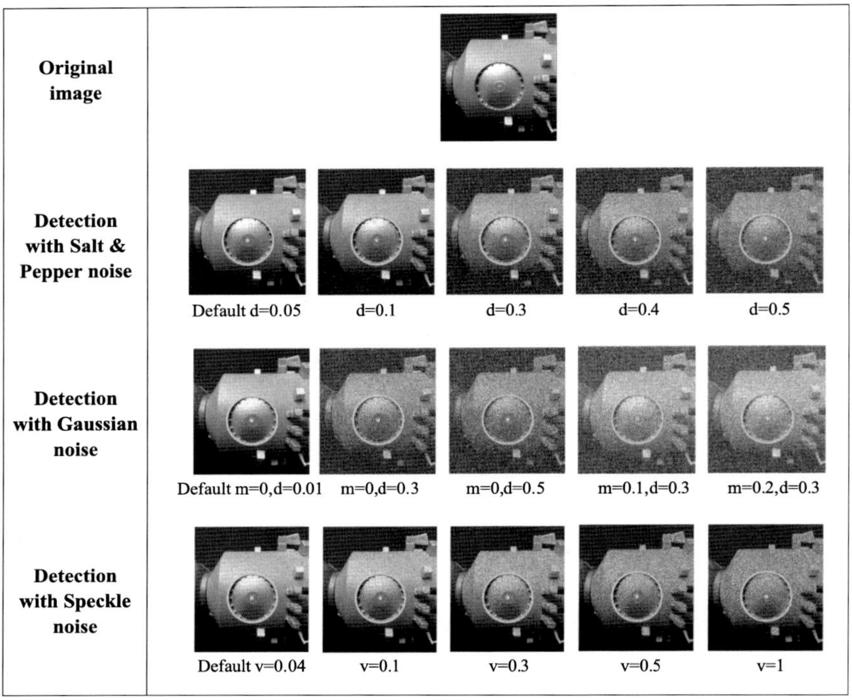

Figure 3.10 Detection results with noise.

The results show that the method has a good detection effect under the default parameters. Gaussian noise, speckle noise, and salt pepper noise do not affect the detection process and have strong robustness and stability for medium noise.

3.5.4 Detection performance under different light intensity

Actually, the images photographed under weak light maybe not clear. To test the detection performance in weak light, Fig. 3.9B and D were dimmed to compare with the detection results in normal conditions. Fig. 3.11 shows the detection results with the gradually weakened light intensity.

From the edge maps, we find that the extracted edges become less when the light weakens, which increases the difficulty of detection. Observing the results, all circles were detected except for only one misdetection in the darkest image of Fig. 3.11A, in which two small circles

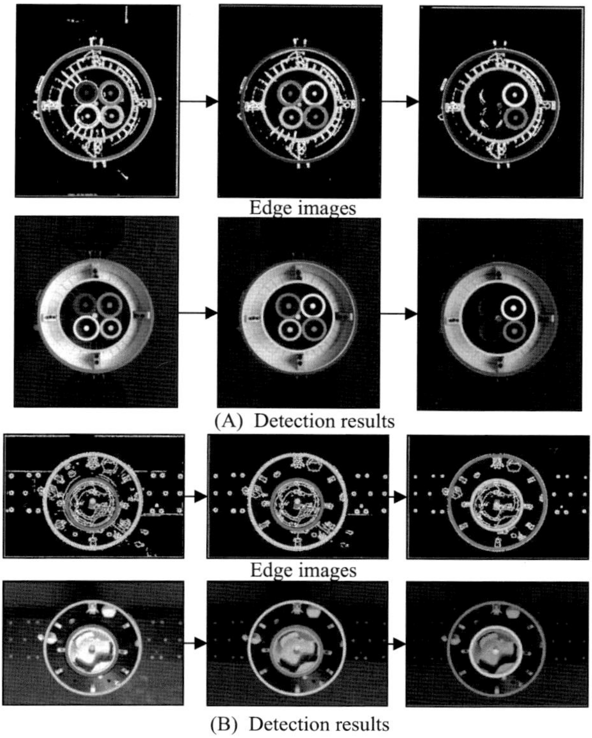

Figure 3.11 Detection results with gradually weakened brightness. (A) Detection results. (B) Detection results.

were undetected, while detection results in Fig. 3.11B were all correct. In fact, the missing circle is almost unseen, which makes the failure of our method understandable.

In all the above, we obtain satisfactory results by our method in most cases. Thus, we conclude that our method is robust for detection when the light intensity weakens.

3.5.5 Detection performance during continuous flight

In the above experiments, images are some static circular components. Although our method performed well on these images, it is worth considering the detection performance during continuous flight. Hence, other experiments were designed to test the performance during continuous flight. In reality, due to the influence of motion, the circle may deform and its size may change dramatically. Consider the problem of circular deformation, the detected circle is perfect no matter how the actual circle deforms, which means the coincidence degree between both lowers when serious deformation. In our method, when their coincidence degree meets the preset value, the circle is considered to be correctly detected. Therefore, to ensure that the circle is correctly detected, it is a feasible method to reduce the set value when the circle is slightly deformed. However, if the deformation is severe, our method is also ineffective. Furthermore, our method limits the radius for the detected circle. So during continuous flight, expanding the range of radius also is helpful for successful detection.

Here, photos taken during the continuous flight of spacecraft were collected from Apollo 9 Magazine for experiments. Detection results on two spacecraft from four continuous views are shown in Fig. 3.12. From the images, we find that the spacecraft gradually become larger when approaching, and the circles are not perfect, which transform into a shape more like an ellipse. Despite this, our method always finds the two circles

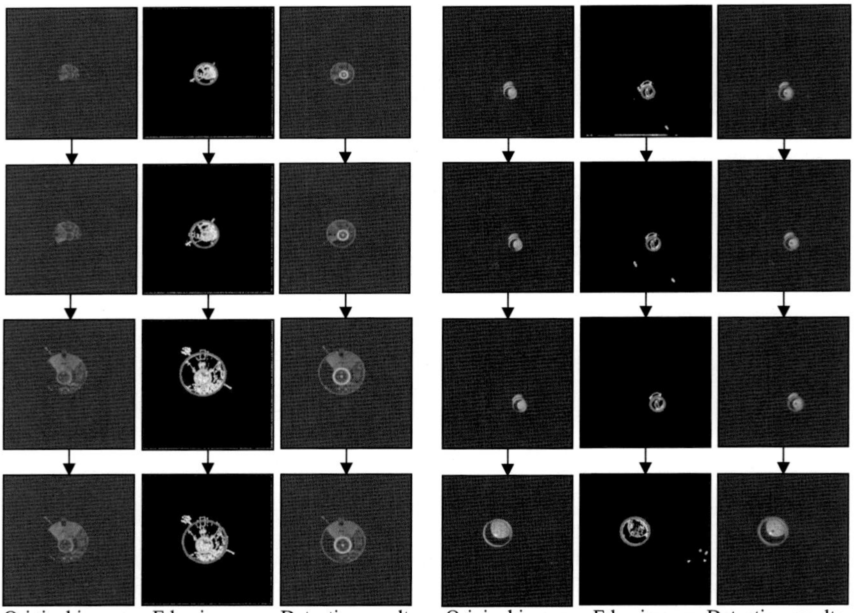

Original images Edge images Detection results
(A)

Original images Edge images Detection results
(B)

Figure 3.12 Detection results during continuous flights. (A) Detection for spacecraft with two circular components. (B) Detection for spacecraft with one circular components.

in Fig. 3.12A and one circle in Fig. 3.12B successfully. The slight transformation does not affect the detection results.

It can be seen that the proposed method is capable of dealing with slightly deformed and gradually approaching circles. From all the above experiments, it is sufficient to prove that our method can achieve excellent performance during the continuous flight of spacecraft.

3.6 The application of species-based artificial bee colony in multitemplate matching

To enrich multiobject detection, SABC was further introduced into MTM. Similar to multicircle detection, here, we employed SABC to search multiple templates on the edge images at one time. The MTM problem is formulated as in Dong et al. (2012a,b).

Here, the impact of the SABC algorithm was also evaluated using the open-source data set from Chang'e 3 space mission (Chinese Academy of Sciences, 2018). In December 2013, China's Chang'e-3 space mission was successfully launched, and the Yutu spacecraft landed on the surface of the moon. However, because of some technical problems with the solar control panel, the rover could go to the preset destination. In the limited driving range, the rover had been operated for over three years on the moon and thousands of photos were obtained in this mission. Using the improved TSR in future would help operate and repair the rover and overcome the problem of such NCT recognition and control.

We used several images from the Chang'e-3 lander to test our new algorithm. The results provide a variety of insights for the development of the camera pointing system (CPS) installed on the lander that was used to take images of the moon and lunar rover in the Chang'e space mission. The CPS developed by Hong Kong Polytechnic University in early 2013 was used to capture images of the moon and the rover's movement (Chinese Academy of Sciences, 2018; Polyu, 2017a,b,c). It can collect images, locate and navigate 360 degrees. In the future, TSR embedded with ABC algorithm can be applied for the operation and maintenance of vehicles or equipment in Chang'e mission. We test the proposed method and evaluate its feasibility in the future Chang'e space mission. Based on our previous CPS experience, the proposed SABC algorithm can enhance the operation and maintenance services of future TSR systems.

3.6.1 Multitemplate matching by species-based artificial bee colony

MTM based on the proposed algorithm SABC was applied on three images obtained from Chang'e 3; two rovers and one lander. To be clear, we recorded the template position detected in the edge images through MATLAB, and then marked the corresponding detection results on the original images, as shown in Fig. 3.13. Both detected templates were 50×50 pixel in size, and the detected areas were marked in the same color as the templates. For comparison, SPSO-based MTM algorithm and SGA-based MTM algorithm (Dong et al., 2012a,b) were also introduced. The performance of the three algorithms is shown below.

As shown in Table 3.6, the proposed SABC outperformed the other two template matching methods. In regard to the running time, the SABC-based template matching needed a shorter time than SGA and SPSO. Meanwhile, among the three images, templates were correctly recognized and positioned in different frames. In Fig. 3.14A, the targets were correctly identified and matched by SABC, only failing twice in fifty trials. In general, the correct recognition rate was more than 96%, which was quite acceptable.

As shown in Table 3.6, the proposed SABC is superior to the other two template matching methods. In running time, SABC-based template matching takes less time than SGA and SPSO. At the same time, in the three images, the template is correctly identified and located in different frames. In the 50 experiments of Fig. 3.14B, SABC only identified and matched the target wrongly twice. Generally speaking, the correct recognition rate is more than 96%, which is acceptable.

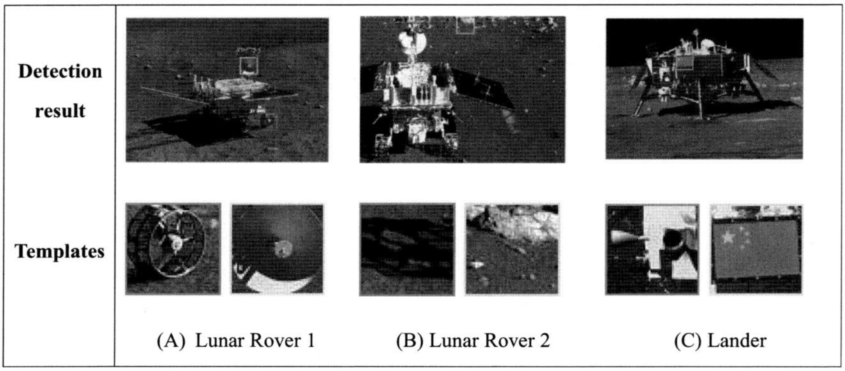

Figure 3.13 Multitemplate matching results in the Chang'e missions. (A) Detection for Lunar Rover 1. (B) Detection for Lunar Rover 2. (C) Detection for Lander.

Table 3.6 Matching performance of species-based artificial bee colony, SPSO and SGA.

Image	Algorithm	SABC	SPSO	SGA
Fig. 3.13A	Maximum execution time (s)	0.6825	0.9672	1.3974
	Maximum execution time (s)	2.2408	3.0895	3.0137
	Average time(s)	1.1498	1.4973	1.8804
	Success rate[a] (%)	100	98	96
Fig. 3.13B	Maximum execution time (s)	0.7317	0.9738	1.0368
	Maximum execution time (s)	3.0325	3.2870	3.4693
	Average time(s)	1.3684	1.6505	2.0737
	Success rate[a] (%)	96	92	90
Fig. 3.13C	Maximum execution time (s)	0.3026	0.5964	1.0568
	Maximum execution time (s)	0.8286	1.9685	2.4502
	Average time(s)	0.4194	1.0875	1.7653
	Success rate[a] (%)	100	96	94

[a]The success rate represents the percentage of finding all circles successfully in 50 trials.

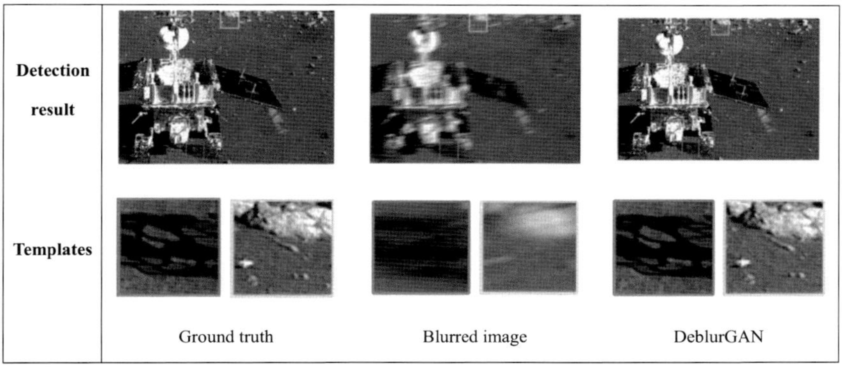

(A) Detection for blurred Lunar Rover

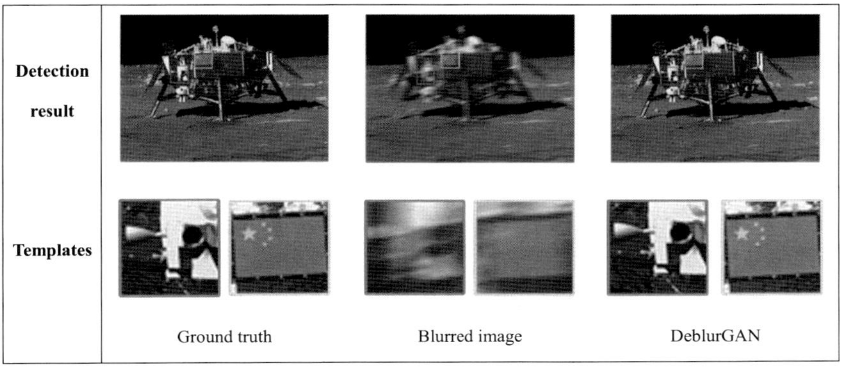

(B) Detection for blurred Lander

Figure 3.14 Multitemplate matching results for blurred images. (A) Detection for blurred Lunar Rover. (B) Detection for blurred Lander.

3.6.2 Multitemplate matching for blurred images

Considering that the captured image may be blurred during the movement of the target, here the blurred images were introduced to verify the MTM based on SABC in the blurred images and then the defuzzified images were detected again. Fig. 3.14 gives the results, and the two templates are found in the area with the same color frame as the original images.

As depicted in Fig. 3.14, although the templates were very blurred, the matching results all were correct. The average time consumption for blurred images was, respectively 1.8 and 1.2 s, a little longer than for the clear images. The success rate for the two blurred images can achieve a level higher than 96%. Thus, conclusions can be drawn that our MTM is also applicable for moving targets.

3.6.3 Multitemplate matching for images with noises

Experiments were further conducted to show whether the detection performance was affected by noise. To simulate the actual situation, two images that added three kinds of noises were used for the experiment, while the templates had no noise added. Fig. 3.15 shows the matching results by MTM based on SABC.

As shown in Fig. 3.15, the original images have been noise added, while the templates were the same as in Fig. 3.15. Despite this, we can see that the template images were matched relatively well. During the experiment process, the noise parameters all were default values, and we find that mismatching will occur when the noise is big enough. Moreover, despite of the added noise, the time needed is still around 1–2 s, not much increase, and the success rate stayed more than 90%. So it can be concluded that our method is robust to the general noise.

3.7 Conclusions

With exploration and use by human beings in outer space, there are more and more kinds of discarded spacecraft being left in space. However, the orbital resources are limited, and invalid satellites have become obstacles to current and future space activities. Therefore, the implementation of on-orbit servicing for space debris is essential. Such nonspecific identifiers and failure to communicate are often called NCTs. Accurate and fast NCT recognition is the key technology of TSRs in orbit service.

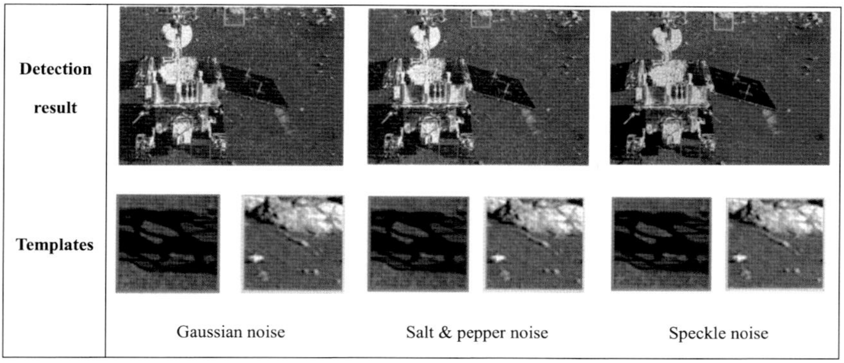

(A) Detection for Lunar Rover with noises

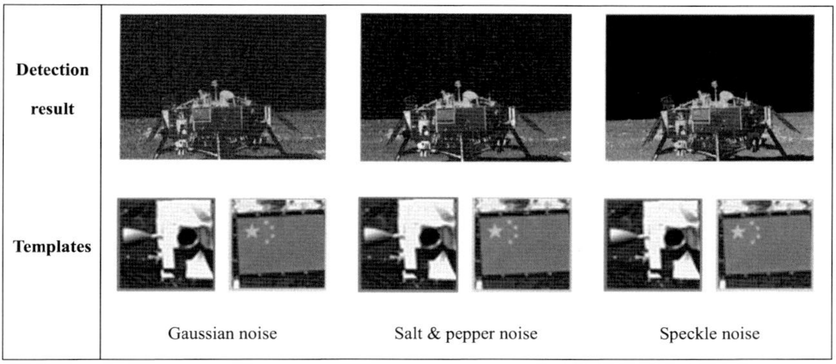

(B) Detection for Lander with noise

Figure 3.15 Multitemplate matching results for images with noises. (A) Detection for Lunar Rover with noises. (B) Detection for Lander with noise.

For such a target, the recognition method based on the intelligent computer vision system has advantages. Here, the multiobject detection method of NCT based on ABC algorithm is proposed.

In this work, a new algorithm, SABC, is proposed for solving multimodal optimization problems for the first time. Its novelty lies in the simultaneous generation of multiple species in parallel search as the ABC algorithm, so that all multiple optimal solutions can be obtained. The updated mechanism and iterative process of SABC are introduced in detail. Differing from the ABC algorithm, which can only find one optimal solution in one iteration, the SABC algorithm can find one or more optimal solutions in one optimization cycle. To evaluate the performance of the algorithm, experiments were carried out on five benchmark functions. Compared with SGA and SPSO, the results showed that the new

SABC algorithm can solve the multimodal optimization problem with high accuracy and a good success rate in a short time.

Furthermore, the multiobject detection problem is regarded as a multi-peak optimization process. SABC is applied to multicircle detection and MTM. The overall process of circle detection can be divided into three main steps: reducing noise, extracting the edge of the original image, and then using the combination of three edge points as candidate circles to search the circles in the whole edge map. We define a function to assess the accuracy of candidate circles on the actual circle. Under the guidance of this function, we use the ABC algorithm to evolve candidate circles of each species, so that multiple circles can match with the actual circles at the same time. Likewise, SABC is also used for MTM, so multiple templates can be detected at the same time. We then propose a multiobject detection method, including multicircle detection and MTM. As an effective detection method, it is used for TSR identification of modules installed on the NCT in the process of cross acquisition. For simulation of different environments, a large number of images of modules on NCT have experimented. The experimental results showed that the multiobject detection method proposed in this paper was superior to other related methods in speed and accuracy, and has strong robustness to light intensity and noise.

The purpose of this study is to demonstrate that the proposed SABC can be used as an attractive method to successfully solve multimodal optimization problems by combining the concept of species. On this basis, a multiobject detection method was proposed, and experiments were carried out on the Shenzhou 8 and Chang'e 3 space missions to verify the good applicability of this method to NCT. We expanded our methods to detect more spacecraft targets and performed more space missions, to improve the operation and maintenance services using the TSR system (Liu, Li, Dong, Yung, & Ip, 2019).

References

Bao, G., Cai, S., Qi, L., et al. (2016). Multi-template matching algorithm for cucumber recognition in natural environment. *Computers & Electronics in Agriculture*, *127*, 754–762.

Cai, H., Zhu, F., Wu, Q., et al. (2014). A new template matching method based on contour information. *International symposium on optoelectronic technology & application image processing & pattern recognition*, 9301:930109-930109-7.

Cai, J., Huang, P., Chen, L., et al. (2016). An efficient circle detector not relying on edge detection. *Advances in Space Research*, *57*(11), 2359–2375.

Cai, J., Huang, P., Zhang, B., et al. (2015). A TSR visual servoing system based on a novel dynamic template matching method. *Sensors*, *15*(12), 32152–32167.

Chen, L., Huang, P., Cai, J., et al. (2016). A Non-cooperative target grasping position prediction model for tethered space robot. *Aerospace Science & Technology*, *58*, 571–581.
Chen, Y., Lü, S. X., Wang, M. J., et al. (2014). A blind source separation method for chaotic signals based on artificial bee colony algorithm. *Acta Physica Sinica*, *63*(23), 0-0.
Chinese Academy of Sciences. China national space administration, the science and application center for moon and deepspace exploration. Chang'e 3 data: Rover Panoramic camera. Available at: http://planetary.s3.amazonaws.com/data/change3/pcam.html. Accessed 14.04.18.
Deng, T., Yao, H., & Du, J. (2012). Improved artificial fish swarm hybrid algorithm for multipeak function optimization. *Journal of Computer Applications*, *32*(10), 2904–2906. Available from https://doi.org/10.3724/SP.J.1087.2012.02904.
Djekoune, A. O., Messaoudi, K., & Amara, K. (2017). Incremental circle hough transform: An improved method for circle detection. *Optik - International Journal for Light and Electron Optics*, *133*, 17–31.
Dong, N., Wu, C. H., Ip, W. H., et al. (2012a). An opposition-based chaotic GA/PSO hybrid algorithm and its application in circle detection. *Computers & Mathematics with Applications*, *64*(6), 1886–1902.
Dong, N., Wu, C. H., Ip, W. H., et al. (2012b). Chaotic species based particle swarm optimization algorithms and its application in PCB components detection. *Expert Systems with Applications*, *39*(16), 12501–12511.
Dong, N., Wu, C. H., Ip, W. H., et al. (2014). Species-based chaotic hybrid optimizing algorithm and its application in image detection. *Applied Artificial Intelligence*, *28*(7), 647–674.
Du, X., Liang, B., Xu, W., et al. (2011). Pose measurement of large noncooperative satellite based on collaborative cameras. *Acta Astronautica*, *68*(11–12), 2047–2065.
Frosio, I., & Borghese, N. A. (2008). Real-time accurate circle fitting with occlusions. *Pattern Recognition*, *41*(3), 1041–1055.
Gonzalez, R. C., & Woods, R. E. (2002). *Digital image processing* (2nd (ed.)). Upper Saddle River, NJ, USA: Prentice Hall.
Huang, P., Wang, D., Meng, Z., et al. (2015). Adaptive postcapture backstepping control for tumbling tethered space robot–target combination. *Journal of Guidance Control & Dynamics*, *39*(1), 1–7.
Jia, P., & Tian, X. (2012). Improved invasive weed optimization based on adaptive niche algorithm. *Journal of Shanghai Dianji University*.
Kang, F., Li, J., & Xu, Q. (2009). Structural inverse analysis by hybrid simplex artificial bee colony algorithms. *Computers & Structures*, *87*(13–14), 861–870.
Karaboga, D. (2005). *An idea based on honey bee swarm for numerical optimization*.
Kitagawa, W., & Takeshita, T. (2012). Optimum design of electromagnetic solenoid by using Artificial Bee Colony (ABC) algorithm. In *Xxth international conference on electrical machines* (pp. 1393–1398). IEEE.
Li, B., Chiong, R., & Gong, L. G. (2014). Search-evasion path planning for submarines using the Artificial Bee Colony algorithm. *Evolutionary Computation. IEEE*, 528–535.
Li, J. Q., & Pan, Q. K. (2015). *Solving the large-scale hybrid flow shop scheduling problem with limited buffers by a hybrid artificial bee colony algorithm*. Elsevier Science Inc.
Liang, B., Du, X. D., Li, C., et al. (2012). Research progress of on-orbit service for space robots non-cooperative spacecraft. *Robot*, *34*(2), 242–256.
Liu, X. Y., Li, D. H., Dong, N., Yung, K. L., & Ip, W. H. (2019). Non-cooperative target detection of spacecraft objects based on artificial bee colony algorithm. *IEEE Intelligent Systems*, *34*(4), 3–15.
Liu, Y., Xie, Z., Wang, B., et al. (2016). A practical detection of noncooperative satellite based on ellipse fitting. In *IEEE international conference on mechatronics and automation* (pp. 1541–1546), IEEE.

Luo, W., Sun, J., Bu, C., et al. (2016). Species-based particle swarm optimizer enhanced by memory for dynamic optimization. *Applied Soft Computing*, *47*, 130–140.
Marco, T. D., Cazzato, D., Leo, M., et al. (2015). Randomized circle detection with isophotes curvature analysis. *Pattern Recognition*, *48*(2), 411–421.
Opromolla, R., Fasano, G., Rufino, G., et al. (2015). A model-based 3D template matching technique for pose acquisition of an uncooperative space object. *Sensors*, *15*(3), 6360–6382.
Ozturk, C., Hancer, E., & Karaboga, D. (2015). Dynamic clustering with improved binary artificial bee colony algorithm. *Applied Soft Computing*, *28*, 69–80.
Pan, L., Chu, W. S., Saragih, J. M., et al. (2011). Fast and robust circular object detection with probabilistic pairwise voting. *IEEE Signal Processing Letters*, *18*(11), 639–642.
Polyu.edu.hk. life on mars? *Award-wining PolyU device could dig up the answer*. Available at: http://www.polyu.edu.hk/openingminds/en/story.php?sid = 3. Accessed 10.11.17.
Polyu.edu.hk. *PolyU 70th anniversary*. Available at: https://www.polyu.edu.hk/cpa/70thanniversary/memories_achievement10.html. Accessed 03.11.17.
Polyu.edu.hk. *The Hong Kong Polytechnic University - The space exploration journey*. Available at: https://www.polyu.edu.hk/web/filemanager/en/content155/1960/Appendix_Milest_ones_SpaceProjects_.pdf. Accessed 24.11.17.
Qin, Q. D., Cheng, S., & Li, L. (2014). A review on artificial bee colony algorithm. *Journal of Intelligent Systems*, *9*(2), 127–135.
Sansone, F., Branz, F., Francesconi, A., et al. (2014). 2D Close-range navigation sensor for miniature cooperative spacecraft. *Aerospace & Electronic Systems IEEE Transactions on*, *50*(1), 160–169.
Wang, D. Z., Wu, C. H., Ip, A., et al. (2008). *Fast multi-template matching using a particle swarm optimization algorithm for PCB inspection*, *4974*(4), 365–370.
Wang, J. Y. (2013). Research and application of artificial bee colony algorithm. *Journal of Harbin Engineering University*.
Wang, X. H. (2016). Space-in-orbit service technology and its development status and trends. *Satellite and Network*, *3*, 70–76.
Weiss, P., Leung, W., & Yung, K. L. (2010). Feasibility study for near-earth-object tracking by a piggybacked micro-satellite with penetrators. *Planetary & Space Science*, *58*(6), 913–919.
Xiao-Jun, B. I., & Wang, Y. J. (2011). Niche artificial bee colony algorithm for multipeak function optimization. *Systems Engineering & Electronics*, *33*(11), 2564–2568.
Xu, L., & Oja, E. (1993). Randomized hough transform (RHT): Basic mechanisms, algorithms, and computational complexities. *Cvgip Image Understanding*, *57*(2), 131–154.
Xu, W., Liang, B., Li, B., et al. (2011). A universal on-orbit servicing system used in the geostationary orbit. *Advances in Space Research*, *48*(1), 95–119.
Yoo, J., Hwang, S. S., Kim, S. D., et al. (2014). Scale-invariant template matching using histogram of dominant gradients. *Pattern Recognition*, *47*(9), 3006–3018.
Zelniker, E. E., & Clarkson, I. V. L. (2006). Maximum-likelihood estimation of circle parameters via convolution. *IEEE Transactions on Image Processing A Publication of the IEEE Signal Processing Society*, *15*(4), 865–876.
Zhang, C. Q., Zheng, J. G., & Wang, X. (2011). A review of research on bee colony algorithm. *Application Research of Computers*, *28*(9), 3201–3205. Available from https://doi.org/10.3969/j.issn.1001-3695.2011.09.001.
Zhang, H., Wiklund, K., & Andersson, M. (2016). A fast and robust circle detection method using isosceles triangles sampling. *Pattern Recognition*, *54*, 218–228.

CHAPTER 4

Internet of Things, a vision of digital twins and case studies

Aparna Murthy[1,*], Muhammad Irshad[2,*], Sohail M. Noman[3,*], Xilang Tang[4], Bin Hu[5], Song Chen[6] and Ghadeer Khader[7]
[1]EIT, PEO, Toronto, ON, Canada
[2]Department of Electronic and Information Engineering, The Hong Kong Polytechnic University, Hong Kong, China
[3]Shantou University Medical College, Shantou, Guangdong, P.R. China
[4]Air Force Engineering University, Xi'an, P.R. China
[5]Changsha Normal University, Changsha, P.R. China
[6]Tianjin University of Science and Technology, Tianjin, P.R. China
[7]Diligent Trust Inc., IT Solutions, Toronto, Canada

4.1 Introduction to internet of things

Internet of things (IoT) is a term that is collectively used for seamless connectivity of devices that talk using the backbone called "internet" (Muzammal, Murugesan, & Jhanjhi, 2020). Modern-day appliances and gadgets that are connected to the internet, such as refrigerators, lighting, digital assistants such as Siri/Alexa, sensors for measuring, logging of temperature, and humidity, etc., are blurring the boundary of Wi-Fi enabled devices (Sohail et al., 2019). Collectively, those who use the internet framework have been classified as "IoT" (Sarkar, Patel, & Dave, 2020). Billions of embedded internet-enabled sensors from various domains worldwide churn a rich set of data that are used for analytics, track operations (Irshad, Liu, Arshad, et al., 2019), and cut back on manual processes. Hence the goal of IoT systems is to seamlessly connect devices with ease of use and provide business solutions based on consumer preferences (Pradhan et al., 2021). The most common use of IoT is seen in mobile clouds and interface/communication of one or more mobile devices (Huo, Hameed, Haq, Noman, & Chohan, 2020; Irshad et al., 2018; Patnaik & Popentiu-Vladicescu, 2019; Sadiq, Hameed, Abdullah, & Noman, 2020; Shah et al., 2021; Verma, Kawamoto, Fadlullah, Nishiyama, & Kato, 2017). IoT and its various examples are meant to understand the plethora of applications that it can potentially support. The applications from popular telecommunication to medical use are understood in this paper. With the everyday growth of IoT devices that are

* Equal contribution.

IoT and Spacecraft Informatics
DOI: https://doi.org/10.1016/B978-0-12-821051-2.00010-6

connected to the network, so are the data churned out by them. Many popular machine learning (ML) algorithms are used to analyze the data.

Consumer IoT refers to wearables, smart home devices, etc., that are marketed to the consumer directly. Thus, IoT enables better experience, increasing efficiency along with improving health and safety for the consumers (Thibaud, Chi, Zhou, & Piramuthu, 2018). Increasing efficiency includes in the field of energy-conserving, efficiency in agricultural produce for indoor management of climatic conditions, inventory management, and so on. Similarly, health and safety comprise disaster warning, caregiving, and environmental excellence to mention a few. The advancement of IoT and cloud computing has triggered many traditional applications and paved the way for smarter homes and cities. IoT and cloud-based management systems aim at confronting existing problems and providing realistic solutions by deploying sensors, collecting the sensory data that are incessantly transferred to a cloud server for storage and secondary analysis. Through a web-based human-machine interface (HMI), admins can control remotely the sensors that are used in a variety of applications (Song, Kang, Lee, & Kim, 2019). ML techniques help the process and analysis based on feature extraction (Asif, Arshad, Shakir, Noman, & Rehman, 2020; Irshad, Liu, Wang, and Khalil, 2019; Rizwan et al., 2020). Powerful computing resources such as GPU are evolving in the cloud environment and neural networks. Anomaly detection tasks take full advantage of IoT data collection in any circumstances.

In the rest of the chapter, we have discussed the major components of IoT and the idea of the digital twin (DT). The concept of the DT is a computer-based entity that takes the real-world object and its variables to generate outputs or simulate the system. The PO or system being simulated uses the data to develop a mathematical model that represents the object in digital space. DT construction is done in such a way that the input is received from sensors that gather data from the real world. This enables the twin to simulate the PO and offers intuitions into potential problems. The twin design is based on a prototype of its PO and delivers feedback as the product refines. Thus, a twin serves as a prototype even before the physical product is designed or built. A DT is found in some business sectors such as automotive, healthcare, and manufacturing.

4.2 Components of internet of things

The major components that comprise an IoT system include sensors or devices, connectivity, processing of the data, and user interface (UI).

4.2.1 Sensor/devices

Sensors are basic elements that gather data and are linked to the internet where each of the devices has an Internet Protocol (IP) address. The devices range from autonomous vehicles to simple sensors that monitor temperature/humidity in buildings. The role of sensors or devices is to collect data from the environment they are in. As mentioned, it could range from temperature values to complex information such as a video. The complexity of devices is not that they are standalone, but bundled together. As an example, consider a smartphone that has several sensors such as accelerometers, cameras, and a global positioning system (GPS) (Yu et al., 2020). Each of these sensors provides data for secondary usage and processing.

Basically, any device that can gather information from the physical world and send it back to the IoT ecosystem is a part of the IoT structure. Sensors can monitor a range of factors such as temperature, pressure in an industrial setup, the status of critical parts, vital signs of patients, an endless list of parameters that need to be logged. Generic devices such as Raspberry Pi and Arduino embedded systems allow us to build custom IoT endpoints.

4.2.2 Connectivity

The sensor and associated devices can be connected to the cloud using various networks from satellite, cellular infrastructure, Wi-Fi, Bluetooth, and local area networks (LAN). Hence, the communication between IoT devices is by using a variety of standard communication protocols. For example, Wi-Fi or Bluetooth low energy (BLE) is specialized for IoT stream. To improve the speed and bandwidth, the 5 G cellular network will benefit IoT (Lei, Cai, & Hua, 2021).

There is a lot of data coming in large quantities that have given rise to new technology, such as edge computing. Edge computing is a new paradigm for distributed devices where the data and power of computation are brought closer to the place where they are most needed (Hsieh, Chen, & Benslimane, 2018). Information is not processed by the filtered cloud, rather the cloud moves closer to the consumer, hence eliminating lag-time and improving the bandwidth. These machines or systems process the data and send only relevant information back to the centralized base for analysis. As an example, a network of surveillance camera systems, with each camera outputting 3 Mbps at 30 frames per second (fps) each frame of size 100 Kb, bombards the security operations center by simultaneously streaming the video. Edge computing can analyze the incoming video and alert the operations when a device detects, for example, movement. The next thing is you can also avoid the data traveling to a centralized system and rather direct it to the cloud. Thus, the elastic nature of cloud computing is more beneficial when data come intermittently. Many of the clouds from Google/Amazon provide IoT services too. Choice of connectivity boils down to IoT application and the end goal of getting data to the cloud.

4.2.3 Data processing

With the data in the cloud, the next step would be processing it using the software. To make data useful, they need to be collected in order, processed, filtered, and analyzed; each can be done in a particular way. We know that zettabytes (1021) of data are guided thru each edge gateway and sent for processing (Vandebroek, 2016); it should not overwhelm the system. Hence, IoT is devised for collecting information from the real world through sensors so that agile decisions are made in real-time. Thus, big data analytics is used to analyze the production data at the highest level, connecting enterprise software and cloud data. IoT-generated data vary in structure and are real-time. These large amounts of data need processing, analysis, and classification for decision-making. There is a necessity to develop techniques to convert the raw data to a usable form and knowledge. Hence, there is a need to identify and deal with data. IoT providers offer ML and artificial intelligence (AI) capabilities to analyze the data collected. For example, ARM is working on low-power chips that offer AI capabilities on IoT endpoints.

4.2.4 User interface

Data should not only be cleaned and formatted in a way but should also be used to alert end-users. As an example, the temperature in cold storage needs to be alerted. A UI will have an interface that allows end-users to proactively check. Hence, depending on the IoT applications, users may also react to the system inputs. For example, an alert to the phone is sent, then input from the user such as adjusting the parameters needs to be reflected, for example, control or regulating the temperature of the cold storage (Mohsin & Yellampalli, 2017). Some actions are automated rather than waiting for the user's input and the temperature can be regulated based on the predefined rules. In the home automation systems, rather than alerting the homeowner about an intruder, the IoT system can automatically notify the authorities.

4.3 Digital twin

DT has flourished in the manufacturing industry and moved to IoT and Cyber-Physical Systems (CPS) (Yi & Park, 2015). The DT notions have been extended to various aspects that can be adapted to different domains and usages. The definition of DT is that it is a thorough software description of an individual object. The things it includes are properties, behaviors, and conditions through data and models. It is a realistic model that represents and reflects its physical twin and remains as a virtual part of its lifecycle. Consolidated DT (cDT) has well-known and common properties and comprises functionality and mechanisms from various technical areas. DT has multiple roles to play in the field of IoT — it is implemented and identified as an approach for creating IoT applications. Also, it is correlated with the capability of sensing IoT technologies. IoT sensors are a part of DT and are used to predict the various outcomes based on the input. With additional data analytics and software, DT can optimize IoT deployment for increased efficiency and provide information about how they should operate when deployed in the real world. DT is aimed at extending over the lifecycle of a product, facilitating the design, production along with the virtual representation of a product, as shown in Fig. 4.1.

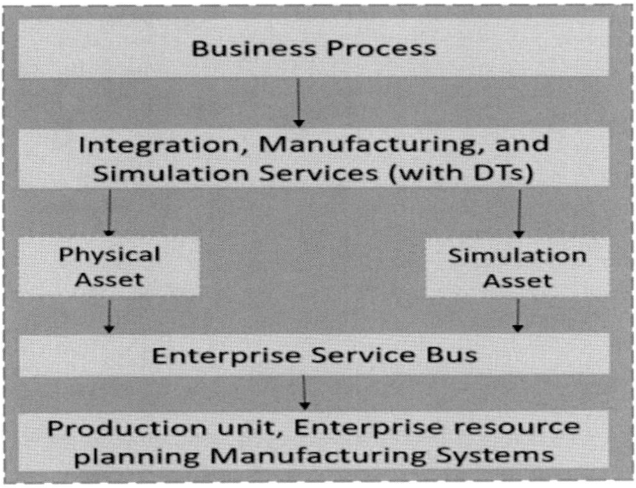

Figure 4.1 Typical architecture of industrial internet of things.

The physical and virtual components are linked to each other and can be used to construct, experiment, and assess the physical attributes of the real-world object at any stage (Minerva, Lee, & Crespi, 2020). The benefits of a DT are its capacity to test, support simulation, and consider different conditions. Usage of DT reduces the cost of prototyping and minimizes the expense of growth, testing, and physical wastages. Replicating the mechanisms of a full product compared to genuine analysis in a facility saves energy consumption and material cost. DT allows the prospect of functionally distributing virtual systems over various processing environments. The clear benefit of virtual systems is that they can have a modular structure that can be simulated in various environments. Having a modular approach to the detection of errors and acquiring new information would determine how to improve the entire system or even a subpart by forecasting the performance under stressful circumstances. This takes advantage of edge, fog, and cloud computing facilities to support the processing constraints of the DT.

With DT simulation, a high-risk case of a physical entity during its malfunctioning that leads to stressful situations can be avoided. Also, new functions, characteristics of the system, and features can be added with the least overheads, before the actual production. Virtualization of a PO aims at creating software using the application programming interface (API) and programming models. With no longer proprietary interfaces, specific advancements tend to divide the applicability of the solutions in terms of IoT.

The layers as seen are contributory to the identification of the basic services that give modular structure at different levels. These layers represent different sources as related to the enterprise systems that are a part of the manufacturing cycle. This integration supports a seamless flow of data and dispatching of well-informed knowledge to all the other components. Similarly, the changing of services that is possible to control the services and be managed is also formulated. Thus, servicing layer provides the competence to manage DT and to simulate futuristic behaviors. The topmost layer, also called the business layer, defines the business processes and logistics related to production. Hence a DT support is characterized by:

1. The flow of data between the manufacturing and management.
2. Exploiting big data analytics for processes and choices or decision-making.

A DT concept is useful in current systems such as Enterprise Resource Planning (ERP) and manufacturing systems that control the production unit. The information flow is related to the Enterprise Service Bus that gives a level of transparency and lets the design and functioning of the system run smoothly.

The product lifecycle management (PLM) of a DT is illustrated in Fig. 4.2 with each phase analogous to the phase of product development and design. The interconnectivity of operations is illustrated in four stages. The first phase is the design phase when the product is formulated and designed. The next phase is the production phase in which the manufactured goods are fabricated. The development, simulation, and testing tools are considered during this phase. In the next phase, namely the operations, is meant for the usage of the product and AI/ML models and algorithms are used. The final iterative stage is the disposal phase when the product is at the end of the lifecycle. The product is terminated rather than taken out of production in this phase.

DT has an important role in manufacturing in terms of software entity and lifecycle. Its usage in various phases of the product lifecycle continually helps in optimizing the product at each step. Hence, a DT relates to different items all in a descriptor, like a software entity, making it a metasystem representing the behavior of a product. An obvious role for IoT is resources in the production and operational phases. The sensors of the IoT can be used to build products and later used to measure the behavior and performance of products. As data are collected from the customers, factories, PLM operations become more accurate. Since the data are realtime and analyzed, the outcome is to ensure higher quality products with the best possible performance and predict the product modifications with the least overheads. Thus, DT provides a unique asset to the production line and enables a quick fix. Companies have implemented DT that eliminates unplanned downtime, reducing costs, and improving quality. In the lifecycle, a DT must be able to coordinate with revised engineering models and processes during all the phases, namely design, production, and operations.

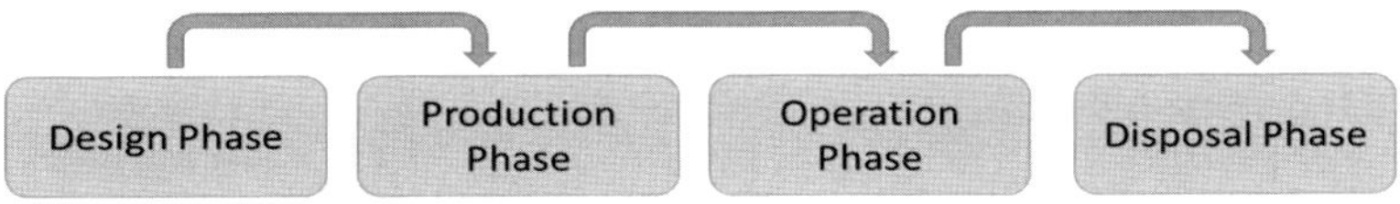

Figure 4.2 Various phases of the digital twin.

4.4 Digital twin description in internet of things context

The characteristic of a DT can be described as a set of features and principles that are formulated using the initial properties for identification (Jacoby & Usländer, 2020). DT must have an unequivocal identity, the relationship between physical and logical objects is 1-to-1 cardinality. This relation can also be 1-to-N in certain cases where different models exist. In this situation, each PO must have an exclusive identifier to the logical object representing it in both time and space. Given the definition by ITU: "An intrinsic depiction of an entity that is described at an appropriate level of abstraction in terms of its attributes and functions," a model specifies functionality and services based on object behavior and its interface. For logical entities, terms such as clone, duplicate, or other terms derived from physical entities or even software terminologies are used. There is also a term, cDT, that is used to fuse various specifications, technologies, and definitions that comprise the components of a DT. cDT defines perpetual entanglement between a PO and its software [logical object (LO)] counterpart. The software is executed in the virtual space and the nature of PO in the real world.

The objective of cDT is that it is meant to define appropriate characteristics and behavior of the PO and it is considered using three parameters:

1. Similarity—ability to reproduce logically.
2. Randomness—probability of arriving at differing features from the original.
3. Contextualization—functional context of the objects

The major attributes, status, events, actions, and other information of a PO are software-based LO, and vice-versa. In the IoT domain, these properties refer to the measurable quantities of the PO. Many such examples are not quantifiable with only variables and features. Modeling in such a case allows us to determine the aspects of IoT technologies that can be quantified as PO. Thus, reflection is a property that represents the application objectives. Thus, a PO is fully defined if all the principles are mapped to represent an LO.

4.5 Multiagent system architecture

IoT aims at connecting industrial technologies with one another so that it promotes data generated and shared amongst these holdings. The purpose of multiagent system (MAS) architecture is for cooperative learning and presenting its findings is for a predictive problem. Interaction with the POs is by analyzing through their data, resulting in controlling of the decision-making algorithms (Akram et al., 2021). The logical objects or DTs are an asset that analyses the data coming from itself. To enable cooperation between similar assets the distance among them is calculated and the data coming in are selected. A decentralized architecture lets complex tasks be divided into subtasks; such smaller tasks can be allocated to the best fit. It gives advantages like robustness, dexterity and reduced data delays.

4.5.1 Dynamic real-life environment

A multiagent system helps to adapt to a real-life dynamic environment. Single-agent learning does not give an optimal solution as against a MAS. A useful way to obtain partnership is to integrate social networking of IoT (SIoT). Such SIoTs have assets like sharing of data and collaboration to produce an optimal solution. Multiagent collaborative learning can be implemented using algorithms such as swarm intelligence and other social algorithms, one such example is the "Stochastic Search" which is among reward-based understanding. Other methods that combine collaboration between agents are swarm intelligence, such as bees or ant colonies optimization algorithms. An asset fleet is a cluster of supplies under consideration. Cooperation among assets or resources within a fleet increases their sensitivity, allowing events to be broadcast among friends. Thus, improving the accuracy of core algorithms by making a deeper dataset accessible for training and forecasting tasks.

4.5.2 Collaborative learning

A devastating failure can occur to a small subset of assets when a fleet is considered for which it is better to convey the reason for failure to other assets in the fleet. If this is not done, then an event that is informed would be unknown to the machine. Thus, the algorithm's response will fail to predict. Paths corresponding to a newly scheduled event that can be pooled among related assets and other agents can be informed or made aware of such events in a collaborative MAS. To make efficient communication between interasset efficient, data that are shared should be made effective rather than being verbose and bulky. To accomplish this, assets keep sharing prespecified parameters regularly under normal conditions.

So as soon as an asset comes across a new event, the data related to that framework the trajectory to that event are shared as a new training dataset notifying other assets. Consequently, DTs too are informed about the event that caused a failure. Apart from making the system vigorous, collaborative learning makes a system more agile and resourceful. Thus, the benefits of such learning in applications such as job-sequencing where there are agents that maintain scheduling and cooperative information produce an optimal activity level job schedule. MAS has three layers:

1. Virtual Asset (VA)—software components that ensure that data starting from the asset are driven to the agents in a reliable format and at fixed periods.
2. DTs—that run algorithms over the asset-data replicating their operational prerequisite for human intervention.
3. Social Platform—that is accommodated on a central server or a cloud platform. Mostly the communication to/from a manager and with the external world happens via this platform.

First layer: Comprises virtual assets and is distinctive from DT, which is the heterogeneous kind of real-world industrial asset fleets. For example, a production service may have a milling machine, lathe, and a packaging machine to name a few. These may have come from different manufactures and have different specs. A VA is a software that corresponds to each asset and is accountable for normalizing asset data before it reaches the DT. The data from VA have three features, namely machine identifier, Features, and Events. The machine identifier helps in the specific identity of an asset fleet. It involves asset information, location, and operator/features are sensor-generated while events can be messages /warnings.

Second layer: Instrumentation and heightened digitization along with heterogeneity in manufacturing systems have made the design of DT a challenge. With moderately homogenous asset fleets, they may vary and so do the data. Hence, addressing this is through generic DT. These DTs are efficient when working with a variety of assets, unlike specific DTs that are found in a typical industrial environment. A generic DT can be tailor-made to every type of asset and industrial problem. Fig. 4.3 shows the design of the data flow into the DT. As seen, there are two sources, namely a related asset and the social platform. The data collected are stored in the repository for processing through analytical algorithms. Depending on the kind of asset, the correlated DT will run specific kinds of algorithms. The choice of algorithm that is needed is handled by the asset manager and communicated by the social platform. A manager

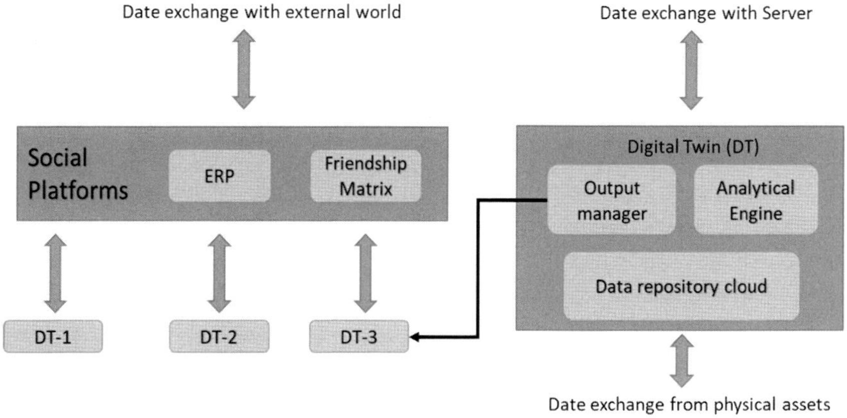

Figure 4.3 Block diagram of multiagent system architecture.

supervises the stream of data rolling out of the DT and is responsible for data sharing and cooperation with other asset fleets. The DT of the analytical engine may engage in tasks such as health management, optimization of the performance, and other features particular to the asset. The diagnostics and prognostics are conveyed to all of the assets, for example, load management and path resolving is done for transportation asset. Algorithms that support centralized clustering functioned by the social platform can be executed in the analytics engine. A flexible hierarchy is allowed depending on the computational capabilities of the agent. For example, when operating, DT can process data from itself and from the collaborating assets. This automized hierarchy can be prevented at any time based on the request of the social platform.

Third Layer:Presented in a single or many servers in the cloud is the social program that acts as an entryway for human interface and DTs, enabling asset-to-asset exchange. The internal groups are running algorithms that are implemented using data given by the whole network and by collecting important data in the repository. Analyzing data from several assets, the enterprise-level decision-making is made by the algorithms. The effectiveness of collaboration is maintained when assets are prioritized. ML algorithms are used for decision-making and they are trained over the large dataset, thus increasing the accuracy of ML algorithms. To efficiently manage this architecture, a matrix of similarities or distance between assets is maintained called the "friendship matrix." As the system operates, the similarities are periodically updated and stored in the social platform. The metric of similarity is based on the statistics such as data, machine type, etc.; this makes the platform the best informed to calculate the similarity. Typically, this can be carried out by enterprise-level k-means clustering. For each asset, its group of collaborative assets is given by N-closest resources in the friendship matrix. Using these "friends," data are shared and collaborative learning is carried out. The data collected by an asset from a friend may be weighted in the algorithms running in the analytical engine of the DT based on the similarity estimate.

4.6 The mathematical construct of a typical digital twin

The ability of DT to reflect the properties of PO is mathematically modeled. Fig. 4.4. shows more about the mapping functionality of PO and LO. Let the PO be described as a set of variables given as in following Eq. (4.1):

$$\mathrm{X} = \{x_1, x_2, \quad \ldots \quad x_n\} \tag{4.1}$$

where $\mathfrak{R}$, $\forall\ x_i \in \mathfrak{R}$ in a multidimensional space

The reflection property is shown in Eq. (4.2).

$$\exists \mathrm{f(X)} = X' \tag{4.2}$$

Here X' is a general logical object as described in Eq. (4.3).

$$X' = \{x_1{}', x_2{}', \quad \ldots x_n{}'\} \tag{4.3}$$

The X' is real for all $ix_i \equiv x_i{}'$. The function f(X) is an equality function that transforms X to X; the combined set from X to X' is injective, as described in Eq. (4.4) that is $\forall i \in \mathfrak{R}, \nexists x_i'$ such that.

$$f_i(x_i) \equiv x_i' \tag{4.4}$$

A structure for representing a PO attribute can be a set of tuples/triplets, each with the following information = {timestamp, location, value}

Figure 4.4 A functional mapping between the physical object and LO.

4.7 Internet of things analytics

With the number of connected devices growing, the data generated are hard to be analyed. The data collected are complex, less structured, and generated in larger volumes. New environments need newer IT skills and the constant need to adapt and improve operations without IoT analytics will be a challenge. Leveraging IoT analytics, a manufacturer gets to incorporate a greater set of features into the product. For example, a vacuum cleaner can now be controlled remotely, and battery life can be optimized depending on usage patterns and certain issues can be updated using the software. Thus, IoT analytics helps greater customer satisfaction and improves product design in the long run. A well-designed IoT analytics collects the truly relevant data points and stores them for analysis. With a proper tool applied, there is less use of IT resources and data scientists so that the various data sources, analysis of data, and sharing of insights are less expensive. The modified analytics is ideal for IoT data since the devices generate a lot of data in less time. It is similar to big data, the only difference being the origins are the diverse source. Because of these heterogeneous data sources, data integration is extremely complicated and, hence, IoT analytics plays a role. Data analytics can act as a cornerstone for tasks such as automation, cloud solutions, mobile apps, and a variety of hardware prototypes (Galanopoulos, Valls, Leith, & Iosifidis, 2020).

With a product analytics implementation that is available off-the-shelf, tracking IoT actions is easier. When assessing vendors, companies should look at devices or data sources used to collect data. It should also ensure to track the potential vendor that has open APIs and can collect actions or events to process the data. An ideal IoT analytics will be a simple user interface and a record of accomplishment of working with other IoT devices. Thus, providing the functionality of self-service and reducing the burden of having a data scientist team.

As an example, we investigate the Amazon web service (AWS) that is managed and makes it easy to run and operationalize advanced analytics on huge volumes of data. Thus, making it far easier and less complex to build an analytical engine. AWS analytics powers each of the complicated steps that are needed to analyze data from IoT devices (Pierleoni, Concetti, Belli, & Palma, 2020). It filters, transforms, and cleans that data with metadata that is appliance device-specific before being stored. Secondary analysis can be carried out by running ad-hoc queries by using a built-in structured query language (SQL) engine or using ML inference.

A custom analysis can be used, bundled into a container to execute AWS analytics. It also allows custom-made applications with Juypter Notebook or tools such as MATLAB to be executed on vendor's schedules.

4.7.1 Case studies 1—internet of things devices for mobile link

Improved usability and user experience are based on the design of user applications (apps), which are dependent on connectivity. Depending on the information, future bandwidth can be used to switch network interfaces such as Wi-Fi of the mobiles. Thus, energy-conserving can be achieved by activating the interface upon increased bandwidth (BW) availability. Hence, the functional requirement for higher accuracy projection and developing BW forecast facilities on mobile devices that meet nonfunctional services (Orsini, Posdorfer, & Lamersdorf, 2020). The key requirements include:

- Supporting different time slots and forecasting, which is the average BW in a time interval.
- Conserving resources through the usage of cloud and edge cloud assets.
- Handling a small amount of data.
- Awareness of privacy.

Fig. 4.5 shows the change of context when the subscriber currently at home switches to an automobile network. To allow for uninterrupted

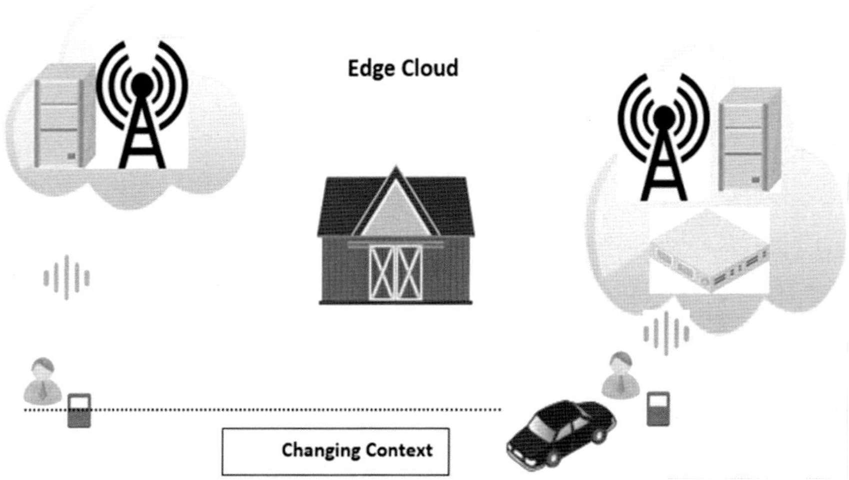

Figure 4.5 Role of edge computing when the change of context occurs.

services, it is necessary to deliver the forecasting on the device itself. Although the forecasting model is carried out on the device, training needs not to be performed on the device. Thus, the forecasting model can be local and/or remote training.

With the ability to offload the training model on a much robust infrastructure, data from the users can also be used to train the model. The common technique is the distributed ML, which uses the recorded data to train forecast models on the respective mobile devices. These models are transferred to a global model that is distributed to all contributing devices. Considering the participating device is unable to process a computation-intensive forecasting model or there is a level of heterogeneity, then "edge-centric" computation is used. This is carried out on a powerful machine after collecting data from the devices and results in model optimization. Further, using the universal approach, the point of the forecast can be based on probability distribution. The minimum and maximum BW can be estimated based on the class interval. Alternatively, classification is another way of forecasting whether there will be a connection or not, with a less desirable misclassification error. There are a variety of approaches to forecasting the context of subscribers and their devices. The selection of algorithms is classification or regression can be used in the domain of context to forecast the BW. Several ML algorithms are commonly used when context data are forecasted. Fig. 4.6 shows the schematic framework of distributed learning in mobile and IoT contexts.

Customized to the domain of mobile devices, Google devised a network called federated learning to create a piece of dispersed information

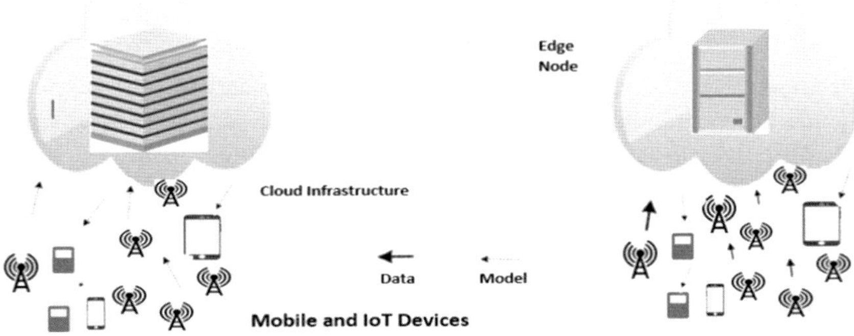

Figure 4.6 Distributed learning—left is the federating learning and right is edge learning.

on the mobile devices by using recorded context data to forecast models. The same models are combined into a global model, that is spread across all sharing devices. The runtime environment for the BW forecasting service is integrated into "Cloud-Aware" which forms the middleware for the device. This middleware provides components such as microservices and discovery in the mobile environment. To save energy, accuracy parameters can also be sent along. Thus, the availability of certain variables like context variables that are used to forecast the future BW of the mobile. This is assuming that the user carries the moving devices with them. For sensors and IoT devices, there might be other metrics that will be relevant for forecasting.

4.7.2 Case study 2—intelligent internet of things -based system studying postmodulation factors

The intensity-modulated radiotherapy (IMRT) model is continuously developed, and the accuracy of the radiotherapy technology has been significantly improved. However, according to the patients and the doctors, the treatment benefits are still limited. IoT technology provides promising approaches targeting precision radiotherapy. Halcyon, which was first launched in China in 2019, is an example of a smart radiotherapy platform. This system offers a variety of advantages including high intelligence, high integration, and high automation, it also upgrades the radiotherapy from conventional to intelligent precise radiotherapy. Halcyon offers the advantage of providing users with informed connection services using artificial intelligence, IoT, and big data.

Cancer radiotherapy methods are becoming more diverse with the continuous development of the equipment and the technology of radiotherapy, which is continuously improving the patients' treatment. This can be a challenge since the software and the hardware of the radiotherapy systems can no longer be a simple planning design. Instead, it should include the entire radiotherapy process. For instance, the radiotherapy system software will not only include a system for treatment planning, but it will also involve the project of radiotherapy plans and information management, the implementation, and the technical services of radiotherapy.

A unique intelligent IoT-based large-scale inverse design approach was developed along with the postmodulation factors, and a pseudo-code flow is used to describe the whole process. The scheme is following two steps: first, obtaining an optimal combination of gantry angles with taking into consideration the requirements of modulation and dose distribution and the constraints. Second, optimizing the power map and smoothing is based on previous knowledge of the specified angles, to obtain a modulation plan.

Radiotherapy system components include radiotherapy hardware such as a radiotherapy simulator, accelerator, and multileaf collimator. The radiotherapy software system is a combination of different systems, where the exchange of data between these systems occurs via a high rate network connection, and the complete arrangement obtains the image data from an external image acquisition machine-like MRI and CT scan (Lan, Li, Li, Yue, & Zhang, 2020). The main parts of the radiotherapy software include:

4.7.2.1 Radiotherapy treatment preparing system

The treatment plan is formulated based on the image data acquired from exterior image procurement and the preset physical constraints of the accelerator. The treatment proposal consists of the factors of the gantry, multileaf collimator, the treatment bed, the distribution of the radiation dose, and the treatment grading. The acquired image data along with the treatment plan would be sent in the form of DICOM files.

4.7.2.2 Radiotherapy database administration system

The database management system comprises three main parts; which are the patient information record, the patient data case, and the database program; which manages and generates the patient information database records and is also used to create DICOM files. After the treatment plan is completed by the treatment planning system, the plan content will be sent to the database along with the related images in a type of DICOM file. Using the DICOM protocol, the database management system will store the received information in the patient information database. Once the IMRT finishes the image acquisition, both X-ray and CT images will be sent to the database in the DICOM format. The data of the treatment plan, related images, and the basic information of the patient are normally deposited in the patient info database. The patients' basic information along with the bed and gantry parameters, which are achieved from the treatment plan file, will be combined with the relevant tables located in the database of the patient information using the database management program. The storage locations of the patients' related data files in the data folder will be recorded to the database of the patient information. The data files of the treatment plan and the DICOM images will be extracted and stored separately by the database management system, and the storage locations will be saved as database records in tables in the database of patient information. Also, the images obtained from X-ray and CT scans will be stored in their data files.

4.7.2.3 Radiotherapy control system

The network connects the control system to the database, so it can acquire the data of the treatment plan and the patient information. The data of the treatment plan are downloaded by the control system from the database that has the patient information, based on the selected information. After the control system gets the data of the treatment plan with the patient's information from the database, according to the choice that the doctor made, the control system then transmits the setup directions for the IMRT system. Once the IMRT response message is received, the control system sends the treatment's basic information to the IMRT system, which will be used as an index to search in the database for related data of the treatment plan, images, and storage locations of the data files and then download this information to the computer using the network. After the position correction of the treatment bed is completed by the IMRT system using registration, the multileaf collimator and the accelerator will be controlled for the patients' treatments using the control system, based on the data of the treatment plan which were acquired before. Eventually, and after the completion of all the treatment operations, the results of the treatments will be collected in the database of the patient report using the network.

Radiotherapy simulation and verification system: The simulation and validation system form a part of the interface with the control system and the planning system. It delivers the reset and verification information to the database management subsystem. The simulation and verification system can recognize the cone-beam functions of the CT reconstruction, the plan verification, and the three-dimensional simulation.

Inverse planning software process: Different factors need to be considered in the early stage of the dose modulation process, which are: the field angle selection, the dose intensity map smoothness and optimization, the subfields number, and the reduction in the tongue-groove and the machine hops. These factors can affect the simplicity and the stability of the dose modulation process, in addition to the possibility of dose leakage during the process. The inverse optimization process consists of two main steps: first, obtaining optimal gantry angles with considering the requirements of modulation and dose distribution, along with the constraints. Second, to optimize and smooth the intensity map based on the specified angles to obtain the complete modulation scheme.

Optimal angle combo using a finite beam angle: Using IoT technology, the initial frame angle and the requirements of the dose distribution and postmodulation are given by the big data (Sohail et al., 2019). The optimization objectives, iteration steps and constraints are used to determine the firing angle through an ant colony algorithm: which is a bionic algorithm used to find the ideal path that is optimal. This algorithm is influenced by the ants' behavior in nature where they can find the optimal path in their foraging process from their nests to the food sources. Therefore, a group of people tends to show intelligent behavior if they collaborate, which makes it possible to solve a complex problem.

A definition of ŋ is proposed in this chapter in the following Eq. (4.5): considering the grading complexity based on the leaf motion direction.

$$ŋ = \frac{\sum \Delta p_x^+}{\int p_x dx} \tag{4.5}$$

where p_x represents the fluence map which is projected in the X direction (see the sketch below). Δp_x^+ represents the fluctuation characteristics (the positive value after difference), this parameter can be reduced by having a large area average irradiation for a single peak, as shown in Fig. 4.7.

To prevent the effect of the tongue groove, the difference in the variance between the projection function is calculated in the XY and YX directions. This value is defined as per Eq. (4.6).

$$\theta = |\sigma_{xy} - \sigma_{yx}|\left(\sum \sum X\right) \tag{4.6}$$

where the fluence map pheromone is defined using the below equation, considering the number of monitor units and the dose distribution objective function value:

$$\Phi = \alpha.ŋ + \beta.\theta + \gamma \tag{4.7}$$

With objective value + λ. $T\ N\ M\ U$, where θ is the tongue-groove parameter, $T\ N\ M\ U$ is the irradiation duration, objective value is the dose conformity, ŋ is the smoothness parameter and it is given as Eq. (4.8)

$$\alpha + \beta + \lambda = 1 \tag{4.8}$$

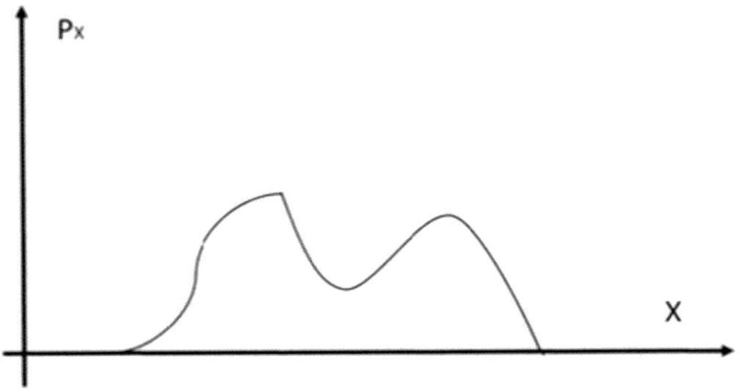

Figure 4.7 A sketch of the fluence map projection.

During the experiments, the tongue-groove parameter value was put at 0.2, the irradiation duration was given as 0.1, the smoothness was set to be 0.2, and the dose conformity was put at 0.5. The results show a lower value of a total number of monitor units (TNMU) because of the consideration of tongue-groove and smoothness parameters, and the higher value, which was set for the dose conformity. After the optimization process of the angle, the objective value is noticed to be varying mainly in the target area, and the target value is decreasing by approximately 10%. In addition, the fluence maps appear to have a smoother distribution caused by the effect of the convex groove and the smoothing. Once the angle is selected, the fluence map is smoothed by PDE with the reduction of the monitor unit's number.

When the optimal gantry angle is determined by the fluence maps comparison, it becomes easy to modulate the fluence map because of the isotropic smooth out; however, after the map is smoothed, it ignores organ at risk (OAR) protection. On the contrary, the fluence maps which are produced by the smoothing and subfield decomposition processes do not show radiation from sizable areas, they only show relatively frequent radiation blocks, which basically show the OAR protection.

The dose-volume histogram (DVH) was compared with the dose distribution on the target area. Also, the DVH was compared with the dose distribution for the nontarget area. The objective value indicated that the target value remains the same for the target area and it decreases by 16% for the nontarget area, while the TNMU remains the same.

The postmodulation key factors were taken into consideration in the early stages of the optimization process, which allows an intelligent

IoT-based huge scale reverse planning approach that was effectively optimized in various cases; such as the choice of the gantry angle, the optimization of the dose distribution, and the smoothing of the fluence map. The large-scale healthcare planning method which is proposed in this article (taking into account the postmodulation key factors) offers an integrated solution that can be used in different inverse planning systems, compared with the other optimization strategies such as the usual divide and conquer and the stepwise strategies.

4.7.3 Case studies 3—internet of things -based vertical plant wall for indoor climate control

Parameters such as volatile organic compounds, humidity, air temperature and poisonous gas concentrations are related to our health and productivity. Maintaining and purifying air quality is a concern. A vertical plant wall system (VPS) is vertical greenery or a living wall that provides indoor climate by purifying the air and an enhancement using lighting and irrigation to the plant wall (Ottelé & Perini, 2017). With IoT and cloud technologies, smarter homes and smarter cities have supported the revolution in such technologies. A formed set of environment sensors is installed to collect the wall sensory data and is transmitted to the Azure cloud for storage and visualization. Using a web-based human–machine interaction (HMI), there is a system of remote controlling the actuators and watering of the VPS to maintain a pure indoor climate. To detect environmental anomalies, the climate data are continuously collected and processed in the cloud server. When anomalies have detected an alarm, the system goes off and a message is sent to the administrator for responsive measures. Early detection of the anomaly from the time series data is critical. There are two categories, namely point anomaly, and contextual anomaly. The point has an outlier that is mainly different from the rest of the environmental data while contextual is a series of data points that are unsuitable for making logical inferences. The cause of this may be external interferences to indoor climate; hence, early detection of such anomalies facilitating responsive operations needs to be taken, Actions such as exploitation of ventilation, lighting, and irrigation actuators in VPS to alleviate the indoor climate need to be taken.

Fig. 4.8 shows the collection of training and test data. The raw data collected are CO_2 from the VPS by continuously sensing CO_2 and streaming it to the cloud and stored every 30 s (as shown in D_{row}). The IoT hub collects the data and sends them to the storage, which is later analyzed or, depending on the infrastructure, analyzed in real-time. The data cleaning is undertaken and a percentage division into D_{train} and D_{test} is performed. Upon this splitting,

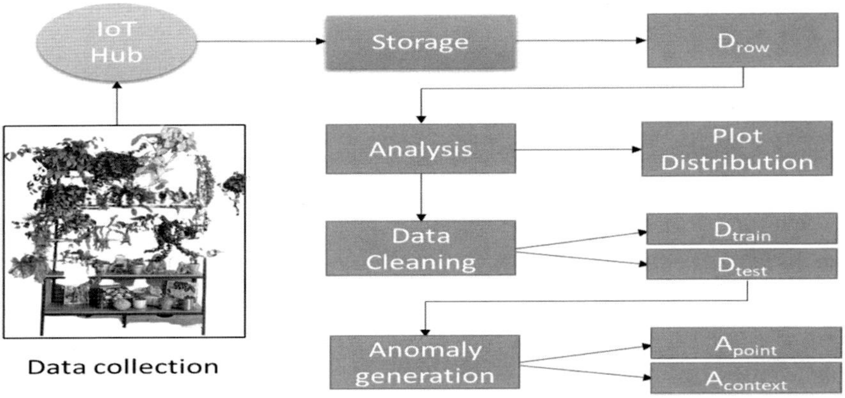

Figure 4.8 Illustration of data collection and preparation process.

the models are used to predict the point based on contextually based anomalies. Further, VPSs has an embedded collection of sensors and actuators to control indoor climate. Using ML, anomaly detection is carried out to improve the predictive maintenance of the indoors. For an effective solution, the development of ML and IoT technologies is particularly important. The benchmarking of prediction and pattern recognition-based approaches by applying models to the CO_2 and temperature dataset collected from real indoor environments is critical to establish the facts. Thus, making it possible to integrate the ML model to the VPS and continuously stream it to the Azure cloud for further analysis of the data. We know that the performance of ML is based on training data that have many parameters which affect its collection. The parameters involved are sampling rate, quality of sensors, and length of the dataset.

VPSs embedded with sensors and actuators are promising in maintaining the health of indoor climate control, using ML algorithms for anomaly detection, for enhancement of automation, and enriching the indoor climate. Prediction-based methods are commonly used in indoor climate anomaly detection. For applications in industrial systems, Neural-Network-based autoencoders may be deployed.

IoT is reflective of connected assets between digital and physical worlds through the internet backbone. It plays a critical role in providing information in a range of fields that comprises structure and environment. In the VPS context, IoT has greatly enhanced the widespread usage of applications. Tracking of indoor climate in real-time is done using the IoT technology using sensors deployed in the plant wall. This affects the growing environments of plants while lighting, ventilation, and irrigation systems can be remotely controlled. The environmental data can be stored centrally in a cloud database and can be used for futuristic analysis. Thus, IoT is an enabler that transforms automation in the direction of digitalization of VPS and the industry.

The goal of IoT uses the building environmental data from the deployed sensors and the intelligence created and extracted from sensory data. The ML helps to map and process the hidden information from the stored data. Cloud computing and NN rely on heavily powerful computing resources that can be utilized by an array of applications in the building environment to optimize operations.

The data analytics for CO_2 concentration fluctuate during the day. The distribution of day-time and night-time are normal distributions to get normalized data and clean the irregular data points. Empirically, point anomalies do not appear in the data collected. One way of correction would be three deviations of the mean values for both daytime and night. This covers 99% of the data points, which filters the $\mu \pm 3\sigma$ from the raw dataset. After cleaning the data, they are divided into say 80% of training and 20% of text dataset. The training data (D_{train}) are used for anomaly detection while D_{test} is used for performance evaluation. The A_{point} and $A_{context}$ are the point and contextual anomalies situations that have to be detected. The true positive rate (TPR) and false-positive rate (FPR) values characterise the anomalies detected. To differentiate between the anomalies, say the point anomaly, a uniform distribution is applied to the D_{test} to select the data point. Once the value is selected, it is replaced by an irregular value that is within the three-sigma according to the time-stamped data points for day and night. For contextual, several data points with variable length are randomly chosen from D_{test}. These consecutive data points are substituted with values from 12 to 18 h far from the original to guarantee contextual variance.

Thus, the study of integrating ML-based techniques for various applications such as telecommunications, radiology, and VPS for data that are collected from various sources through an IoT hub is discussed (Liu, Pang, Karlsson, & Gong, 2020). The need to improve the performance of the algorithms and data investigation is application-dependent and based on sensor quality.

4.8 Discussion

In this chapter, we present in-depth case studies that focus on the challenges and contributions toward IoT-enabled applications that take advantage of off-the-shelf devices. We present the technical components and the idea behind the reduction in terms of the cost of creating a product to end-of-life with the least cost by having a DT.

From our first case study, we see that the limited bandwidth of the device needs to be forecasted, to alleviate the effect of network bottlenecks by either adapting data transfer or delaying to save energy. User mobility is a challenge and needs to optimize the BW forecast by choice of relevant features and evaluation of different cases that make use of it. The future work is tailoring to the changing context and limited computation or even memory of mobile and IoT devices. In our second case study, we look at the medical application with intelligent IoT that effectively optimizes dose distribution using angle selection and fluence map. By using divide and conquer strategies, a large-scale healthcare planning technique considering factors such as postmodulation gives an integrated solution. The third application of IoT is exploring the anomaly detection of the VPS to increase and improve predictive maintenance of indoor climate. In our discussion, we show the benchmarking of prediction-based and pattern-based detection techniques by applying the models to the CO_2 and temperature collected from the practical indoor environment. Using the ML techniques, we see that the data processing from information collected by the sensors is reflective of the assets in the IoT scenario.

As far as DT is concerned, it is well-accepted in industrial and academic scenarios. It has been known that the application domain is dictated by prototypes and/or solutions that are proprietary. Also, some of them address specifically the applications and lack a generic approach. A clear definition to open standards in the application domain and lacking in generality needs to be widely used. There are some standard organizations such as ISO/ETSI and European projects which are extending their platforms. A coordinated effort is needed for different stakeholders to determine a shared definition of a DT and its capabilities.

4.9 Conclusion

We have presented the technical components and the idea of the cost of creating a product to end-of-life with the least cost by using a DT. The knowledge or data flow used in the case of a DT could be bidirectional with a measure and actions flowing from actual to virtual, while the forecasts change the states of flowing in the opposite path. By dealing with such communication paths there is a need for optimization and cost-cutting in new product designs. We look at the data management in the context of IoT and its analytics for enabling the products for interoperability, upgradation, and flexibility. The three case studies from mobile links, that is, in the field of telecommunication to medical application in postmodulation, are a few examples that show the usage of DT to reduce cost and maintain a virtual model for futuristic growth. Entanglement of capabilities, scaling up of DT to millions of objects, and the likelihood of self-management are some of the technical issues that need to be addressed. The ability to contextualize the DT, augmentation of data from sources other than POs such as sensors, and predict the behavior of large systems, and many more practical scenarios need to be documented and considered for a viable DT. The long-term growth of the technology of DT is considered as probable—projected baseline, plausible-happenings and possible—that might happen.

References

Akram, A., Jiadong, R., Rizwan, T., Irshad, M., Noman, S. M., Arshad, J., & Badar, S. U. (2021). A pilot study on survivability of networking based on the mobile communication agents. *International Journal of Network Security*, *23*(2), 220–228.

Asif, R. M., Arshad, J., Shakir, M., Noman, S. M., & Rehman, A. U. (2020). Energy efficiency augmentation in massive MIMO systems through linear precoding schemes and power consumption modeling. *Wireless Communications and Mobile Computing*, *2020*(20), 1–13.

Galanopoulos, A., Valls, V., Leith, D.J., & Iosifidis, G. (2020). Dynamic scheduling for IoT analytics at the edge. In *2020 IEEE 21st international symposium on a world of wireless, mobile and multimedia networks (WoWMoM)* (pp. 157–166). IEEE.

Hsieh, H. C., Chen, J. L., & Benslimane, A. (2018). 5G virtualized multiaccess edge computing platform for IoT applications. *Journal of Network and Computer Applications*, *115*, 94–102.

Huo, C., Hameed, J., Haq, I. U., Noman, S. M., & Chohan, S. R. (2020). The impact of artificial and nonartificial intelligence on production and operation of new products-an emerging market analysis of technological advancements a managerial perspective. *Revista Argentina de Clínica Psicológica*, *29*(5), 69–82.

Irshad, M., Liu, W., Arshad, J., Sohail, M. N., Murthy, A., Khokhar, M., & Uba, M. M. (2019). A novel localization technique using luminous flux. *Applied Sciences*, *9*(23), 5027.

Irshad, M., Liu, W., Wang, L., & Khalil, M. U. R. (2019). Cogent machine learning algorithm for indoor and underwater localization using visible light spectrum. *Wireless Personal Communications*, *116*(2), 993–1008.

Irshad, M., Liu, W., Wang, L., Shah, S.B.H., Sohail, M.N., & Uba, M.M. (2018). Li-local: Green communication modulations for indoor localization. In *Proceedings of the 2nd international conference on future networks and distributed systems* (pp. 1–6).

Jacoby, M., & Usländer, T. (2020). Digital twin and internet of things—Current standards landscape. *Applied Sciences*, *10*(18), 6519.

Lan, Y., Li, F., Li, Z., Yue, B., & Zhang, Y. (2020). Intelligent IoT-based large-scale inverse planning system considering postmodulation factors. *Complex & Intelligent Systems*, 1–15.

Lei, T., Cai, Z., & Hua, L. (2021). 5G-oriented IoT coverage enhancement and physical education resource management. *Microprocessors and Microsystems, 80*, 103346.

Liu, Y., Pang, Z., Karlsson, M., & Gong, S. (2020). Anomaly detection based on machine learning in IoT-based vertical plant wall for indoor climate control. *Building and Environment, 183*, 107212.

Minerva, R., Lee, G. M., & Crespi, N. (2020). Digital twin in the IoT context: A survey on technical features, scenarios, and architectural models. *Proceedings of the IEEE, 108* (10), 1785–1824.

Mohsin, A., & Yellampalli, S.S. (2017). IoT based cold chain logistics monitoring. In *2017 IEEE international conference on power, control, signals and instrumentation engineering (ICPCSI)* (pp. 1971–1974). IEEE.

Muzammal, S. M., Murugesan, R. K., & Jhanjhi, N. Z. (2020). A comprehensive review on secure routing in internet of things: Mitigation methods and trust-based approaches. *IEEE Internet of Things Journal, 8*(6), 4186–4210.

Orsini, G., Posdorfer, W., & Lamersdorf, W. (2020). Saving bandwidth and energy of mobile and IoT devices with link predictions. *Journal of Ambient Intelligence and Humanized Computing, 1*(1), 10.

Ottelé, M., & Perini, K. (2017). Comparative experimental approach to investigate the thermal behaviour of vertical greened façades of buildings. *Ecological Engineering, 108*, 152–161.

Patnaik, S., & Popentiu-Vladicescu, F. (2019). Recent developments in intelligent computing, communication and devices. *Advances in Intelligent Systems and Computing*, 752.

Pierleoni, P., Concetti, R., Belli, A., & Palma, L. (2020). Amazon, Google and Microsoft solutions for IoT: Architectures and a performance comparison. *IEEE Access, 8*, 5455–5470.

Pradhan, B., Vijayakumar, V., Pratihar, S., Kumar, D., Reddy, K. H. K., & Roy, D. S. (2021). A genetic algorithm based energy efficient group paging approach for IoT over 5G. *Journal of Systems Architecture, 113*, 101878.

Rizwan, T., Cai, Y., Ahsan, M., Sohail, N., Nasr, E. A., & Mahmoud, H. A. (2020). Neural network approach for 2-Dimension person pose estimation with encoded mask and keypoint detection. *IEEE Access, 8*, 107760–107771.

Sadiq, M. W., Hameed, J., Abdullah, M. I., & Noman, S. M. (2020). Service innovations in social media & blogging websites: Enhancing customer's psychological engagement towards online environment friendly products. *Revista Argentina de Clínica Psicológica, 29*(5), 677–696.

Sarkar, D., Patel, H., & Dave, B. (2020). Development of integrated cloud-based internet of things (IoT) platform for asset management of elevated metro rail projects. *International Journal of Construction Management*, 1–10.

Shah, S. M. A., Ge, H., Haider, S. A., Irshad, M., Noman, S. M., Meo, J. A., ... Younas, T. (2021). A quantum spatial graph convolutional network for text classification. *Computer Systems Science and Engineering, 36*(2), 369–382.

Sohail, M. N., Jiadong, R., Uba, M. M., Irshad, M., Iqbal, W., Arshad, J., & John, A. V. (2019). A hybrid forecast cost benefit classification of diabetes mellitus prevalence based on epidemiological study on real-life patient's data. *Scientific reports, 9*(1), 1–10.

Song, M. S., Kang, S. G., Lee, K. T., & Kim, J. (2019). Wireless, skin-mountable EMG sensor for human–machine interface application. *Micromachines, 10*(12), 879.

Thibaud, M., Chi, H., Zhou, W., & Piramuthu, S. (2018). Internet of things (IoT) in high-risk environment, health and safety (EHS) industries: A comprehensive review. *Decision Support Systems, 108*, 79–95.

Vandebroek, S.V. (2016). 1.2 Three pillars enabling the internet of everything: Smart everyday objects, information-centric networks, and automated real-time insights. In *2016 IEEE international solid-state circuits conference (ISSCC)* (pp. 14–20). IEEE.

Verma, S., Kawamoto, Y., Fadlullah, Z. M., Nishiyama, H., & Kato, N. (2017). A survey on network methodologies for real-time analytics of massive IoT data and open research issues. *IEEE Communications Surveys & Tutorials*, *19*(3), 1457–1477.

Yi, H. C., & Park, J. W. (2015). Design and implementation of an end-of-life vehicle recycling center based on IoT (Internet of Things) in Korea. *Procedia CIRP*, *29*, 728–733.

Yu, J., Meng, X., Yan, B., Xu, B., Fan, Q., & Xie, Y. (2020). Global navigation satellite system—Based positioning technology for structural health monitoring: A review. *Structural Control and Health Monitoring*, *27*(1), e2467.

CHAPTER 5

Subspace tracking for time-varying direction-of-arrival estimation with sensor arrays

Bin Liao[1], Zhiguo Zhang[2] and Shing Chow Chan[3]
[1]College of Elecronics and Information Engineering, Shenzhen University, Shenzhen, P.R. China
[2]School of Biomedical Engineering, Health Science Center, Shenzhen University, Shenzhen, P.R. China
[3]Department of Electrical and Eletronic Engineering, The University of Hong Kong, Hong Kong, P.R. China

5.1 Introduction

5.1.1 Subspace tracking

Subspace-based methods play a key role in many engineering applications such as sensor array signal processing (Buzzi & D'Andrea, 2019; Chan, Tan, & Lin, 2020; Vaswani, Bouwmans, Javed, & Narayanamurthy, 2018). Usually, the signal/noise subspace is computed by performing the singular value decomposition (SVD) of the data matrix or the eigenvalue decomposition (EVD) of the covariance matrix of the measurements. However, this implicitly requires that the subspace is time-invariant. Moreover, updating the subspace through EVD or SVD online is computationally expensive. Therefore subspace tracking methods with the ability to handle nonstationary scenarios and with low computational complexity have received considerable attention during the last decades (DeGroat, 1992; Rabideau, 1996; Champagne & Liu, 1998; Chan, Wu, & Tsui, 2012).

For subspace tracking, one of the most classical algorithms is the projection approximate subspace tracking (PAST) method proposed by Yang (1995a,b). Based on the assumption that the subspace changes slowly, the PAST algorithm makes use of a "projection approximation" to compute the signal subspace with the recursive least squares (RLS) algorithm. In certain cases, such as array signal processing (Huang & Liao, 2019; Liao, Zhang, & Chan, 2010a,b; Schmidt, 1986), an orthonormal basis of the subspace may be wanted. Hence, an additional step of reorthonormalization needs to be performed. However, this would result in much increase in computational complexity. To this end, an extension of the PAST

IoT and Spacecraft Informatics
DOI: https://doi.org/10.1016/B978-0-12-821051-2.00011-8

algorithm, that is, orthonormal PAST (OPAST), has been proposed in (Abed-Meraim, Chkeif, & Hua, 2000) to straightforwardly achieve an orthonormal subspace with reduced complexity. To improve the performance of PAST and OPAST algorithms, it is proposed in (Liao, Zhang, & Chan, 2012) to use a better subspace approximation to achieve more accurate tracking results. Two modified algorithms, named modified PAST (MPAST) and modified OPAST (MOPAST), are then developed.

It is worth noting that the RLS-based PAST algorithms mentioned above are only suitable for scenarios where the subspace is slowly time-varying. For instance, in subspace-based target detection and localization with sensor arrays, the target should be stationary or moving slowly. When the subspace changes rapidly, the performance of PAST and its modified versions will be far from satisfactory. To deal with this problem, a Kalman filter-based subspace tracking method, called Kalman filter with a variable number of measurements (KFVMs), is presented in Chan, Wen, and Ho (2006), Chan, Zhang, and Zhou (2006), and Liao et al. (2012). This algorithm adopts the Kalman filter, instead of the RLS algorithm, to track the time-varying subspace. The Kalman filtering (KF) algorithm is an optimal recursive state estimator in the minimum mean-square error sense, and it has a better tracking ability than RLS for fast varying scenarios since it uses a state-space model to describe the subspace dynamics.

Generally, the study of subspace tracking has focused on the situation in which the ambient noise is additive white Gaussian noise (AWGN). In practice, the noise may be non-Gaussian, due to the impulsive nature of the man-made electromagnetic interference and a great deal of natural noise. In this case, the measurement data would be corrupted by outliers. It is known that the performance of standard subspace tracking algorithms is sensitive to outliers. Their performance may be considerably deteriorated by even a small number of outliers (Chan Wen, et al., 2006; Chan, Zhang et al., 2006; Liao et al., 2010a; Linh-Trung, Nguyen, Thameri, Minh-Chinh, & Abed-Meraim, 2018; Luan et al., 2017; Thanh, Dung, Trung, & Abed-Meraim, 2021). Towards that end, we proposed a robust estimate of the correlation matrix in impulsive noise and incorporate it into the PAST algorithm to obtain a robust subspace tracking method. In this approach, the impulse-corrupted data vectors are detected using the robust estimator and are prevented from corrupting the subspace estimate. For fast time-varying subspace tracking in impulsive noise, M-estimation in robust statistics (Huber, 1981) is applied to the KFVM algorithm (Liao et al., 2010a). By adaptively detecting the corrupted measurements, the adverse effects of impulsive noise are suppressed by assigning a small or even zero weight to a potentially impulse-contaminated measurement in the KFVM algorithm.

5.1.2 Direction-of-arrival estimation

Direction-of-arrival (DOA) estimation, also known as direction finding, which determines the spatial spectra of the impinging signals, is a functional problem in sensor array signal processing and many engineering applications, such as radar, radio astronomy, sonar, navigation, remote sensing, wireless communications, biomedical engineering and speech processing. Since the early 1900s, a great number of advances have been made in the field of DOA estimation. In particular, the introduction of subspace-based estimation techniques, such as MUSIC, ESPRIT, and their variants, marked the beginning of a new era in the sensor array signal processing literature (Krim & Viberg, 1996; Roy & Kailath, 1989; Hua, Gershman, & Cheng, 2003; Tuncer & Friedlander, 2009).

In most of the subspace-based DOA estimation algorithms, the subspaces are computed from the EVD of the array covariance matrix, which in practice is estimated from a set of array measurements (snapshots). Note that, this implicitly requires that the key parameters, such as the target location, during the period of data collection remain unchanged. However, in practice, the targets are likely to be moving, and therefore the corresponding DOAs (and subspaces) are time-varying. One possible way to deal with this issue is shortening the data collection duration. However, this will lead to insufficient snapshots for achieving a reliable covariance matrix. Moreover, computing the EVD directly online usually involves high arithmetic complexity. Hence, numerous algorithms have been proposed for tracking DOAs (Sastry, Kamen, & Simaan, 1991; Gu, Chan, Zhu, & Swamy, 2013; Wang, Xin, Wang, Zheng, & Sano, 2015; Zhuang et al., 2020) and subspace (Sanchez-Araujo & Marcos, 1999; Yang, 1995a,b).

5.2 Subspace tracking algorithms

5.2.1 Signal model

We consider subspace tracking in an antenna array with M sensors impinged by N narrow-band incoherent signals, the array output

$\boldsymbol{x}(t) \in C^M$ observed at time instant t can be expressed as $\boldsymbol{x}(t) = [x_1(t), x_2(t), \cdots, x_N(t)]^T$. More precisely, it can be written as

$$\boldsymbol{x}(t) = \sum_{k=1}^{r} \boldsymbol{a}(\theta_k) s_k(t) + \boldsymbol{n}(t) = \boldsymbol{A}\boldsymbol{s}(t) + \boldsymbol{n}(t) \tag{5.1}$$

where $\boldsymbol{a}(\theta_k)$ is the steering vector corresponding to the kth DOA for the given array geometry, $(\cdot)^T$ is the transpose operator. For a uniform linear array (ULA), we have $\boldsymbol{a}(\theta_k) = [1, e^{j\omega_k}, \cdots, e^{j(N-1)\omega_k}]^T$ and $\omega_k = 2\pi\lambda^{-1}d\,\sin\theta_k$ with λ, d and θ_k denoting the carrier wavelength, intersensor spacing, and kth DOA, respectively. $\boldsymbol{A}$ is the steering matrix defined as

$$\boldsymbol{A} = [\boldsymbol{a}(\theta_1), \boldsymbol{a}(\theta_2), \cdots, \boldsymbol{a}(\theta_N)] \tag{5.2}$$

and $\boldsymbol{s}(t)$ is the vector of signal waveforms and $\boldsymbol{n}(i)$ is the noise vector which is commonly considered to be an additive white Gaussian noise (AWGN) vector with zero mean and covariance matrix $\sigma^2\boldsymbol{I}$, where $\boldsymbol{I}$ is an identity matrix. Conventionally, the signal subspace $\boldsymbol{U}_s$ is obtained as the N principal eigenvectors of the covariance matrix of array output. Note that $\boldsymbol{U}_s$ spans the same signal subspace as the steering matrix $\boldsymbol{A}$.

5.2.2 Projection approximate subspace tracking

According to the principle of PAST algorithm (Yang, 1995b), the signal subspace can be obtained by minimizing the following objective function:

$$\begin{aligned} J(\boldsymbol{W}(t)) &= \sum_{i=1}^{t} \eta^{t-i} \left\| \boldsymbol{x}(i) - \boldsymbol{W}(t)\boldsymbol{W}^H(t)\boldsymbol{x}(i) \right\|^2 \\ &= \mathrm{trace}(\boldsymbol{\Omega}(t)) - 2\mathrm{trace}\left(\boldsymbol{W}^H(t)\boldsymbol{\Omega}(t)\boldsymbol{W}(t)\right) + \mathrm{trace}\left(\boldsymbol{W}^H(t)\boldsymbol{\Omega}(t)\boldsymbol{W}(t)\boldsymbol{W}^H(t)\boldsymbol{W}(t)\right) \end{aligned} \tag{5.3}$$

where

$$\boldsymbol{\Omega}(t) = \textstyle\sum_{i=1}^{t} \eta^{t-i}\boldsymbol{x}(i)\boldsymbol{x}^H(i) = \eta\boldsymbol{\Omega}(t-1) + \boldsymbol{x}(t)\boldsymbol{x}^H(t) \tag{5.4}$$

and $0 < \eta \leq 1$ is the forgetting factor. The columns $\boldsymbol{W}$ spans the same space as the signal subspace $\boldsymbol{U}_S$, that is,

$$\mathrm{span}\{\boldsymbol{W}\} = \mathrm{span}\{\boldsymbol{U}_S\}. \tag{5.5}$$

As can be seen from (5.3), $J(\boldsymbol{W}(t))$ is a fourth-order function of $\boldsymbol{W}(t)$ and is rather difficult and expensive to minimize directly. Therefore a projection approximation was introduced in Yang (1995a,b) to simplify this problem to the

RLS algorithm with considerately reduced complexity. More precisely, (5.3) is rewritten as

$$J(\boldsymbol{W}(t)) = \sum_{i=1}^{t} \eta^{t-i} \left\| \boldsymbol{x}(i) - \boldsymbol{W}(t)\boldsymbol{y}(i) \right\|^2 \tag{5.6}$$

where

$$\boldsymbol{y}(i) = \boldsymbol{W}^H(t)\boldsymbol{x}(i) \tag{5.7}$$

For slowly varying subspace, $\boldsymbol{y}(i)$ can be approximated by

$$\boldsymbol{y}(i) \approx \boldsymbol{W}^H(i-1)\boldsymbol{x}(i) \tag{5.8}$$

Let $J'(\boldsymbol{W}(t))$ denote the corresponding approximation $J(\boldsymbol{W}(t))$ with $\boldsymbol{y}(i)$ approximated by (5.8). Next, we can approximate the signal subspace by minimizing $J'(\boldsymbol{W}(t))$ with the RLS algorithm, since $J'(\boldsymbol{W}(t))$ is now linear with respect to $\boldsymbol{W}(t)$. This leads to the PAST algorithm. The main steps of PAST algorithm are summarized in Algorithm 5.1, where the operator $Tri\{\cdot\}$ indicates that only the upper (or lower) triangular part of the matrix argument is calculated and its Hermitian transposed version is copied to the lower (or upper) triangular part.

5.2.3 Modified projection approximate subspace tracking

In (5.8), the approximation is performed by replacing $\boldsymbol{W}(t)$ with $\boldsymbol{W}(t-1)$. This means that the obtained subspace estimate in the last PAST iteration is employed for this approximation. Therefore the approximation error would be smaller if a better estimate of $\boldsymbol{W}(t)$ is

---

ALGORITHM 5.1 The projection approximation subspace tracking algorithm.

Initialize $\boldsymbol{P}(0)$ and $\boldsymbol{W}(0)$
For $t = 1, 2, \ldots do$
  $\boldsymbol{y}(t) = \boldsymbol{W}^H(t-1)\boldsymbol{x}(t)$
  $\boldsymbol{h}(t) = \boldsymbol{P}(t-1)\boldsymbol{y}_{(0)}(t)$
  $\boldsymbol{g}(t) = \frac{\boldsymbol{h}(t)}{\eta + \boldsymbol{y}^H(t)\boldsymbol{h}(t)}$
  $\boldsymbol{P}(t) = \frac{1}{\eta}\mathrm{Tri}\{\boldsymbol{P}(t-1) - \boldsymbol{g}(t)\boldsymbol{h}^H(t)\}$
  $\boldsymbol{e}(t) = \boldsymbol{x}(t) - \boldsymbol{W}(t-1)\boldsymbol{y}(t)$
  $\boldsymbol{W}(t) = \boldsymbol{W}(t-1) + \boldsymbol{e}(t)\boldsymbol{g}^H(t)$
end t

---

available. As a matter of fact, the current PAST output provides the updated subspace estimate as

$$\boldsymbol{W}(t) = \boldsymbol{W}(t-1) + \boldsymbol{e}(t)\boldsymbol{g}^H(t) \tag{5.9}$$

In other words, a better estimate of $\boldsymbol{W}(t)$ can be obtained from the current PAST output. Hence, it makes sense to repeat the PAST iteration with $\boldsymbol{W}(t-1)$ replaced by the current PAST iteration, and so on. This results in the so-called MPAST algorithm (Liao et al., 2012). The MPAST algorithm for signal subspace tracking is summarized in Algorithm 5.2, where K is the number of inner iterations used to achieve better estimation of $\boldsymbol{W}(t)$ and the subscript (k) denotes the kth iteration. Particularly, when $K = 1$, the MPAST algorithm reduces to the conventional PAST algorithm.

It is worth noting that the tracking error of the PAST may be large at the initial stage due to a limited number of measurements. To avoid error propagation in the iterations of the MPAST algorithm, the number of iterations K can be made adaptive by monitoring the following error:

$$\boldsymbol{e}_{(k)}(t) = \boldsymbol{x}(t) - \boldsymbol{W}_{(k-1)}(t-1)\boldsymbol{y}_{(k-1)}(t) \tag{5.10}$$

ALGORITHM 5.2 The modified projection approximation subspace tracking algorithm.

Initialize $\boldsymbol{P}_{(0)}(0)$ and $\boldsymbol{W}_{(0)}(0)$
For $t = 1, 2, \ldots do$
  $\boldsymbol{y}_{(0)}(t) = \boldsymbol{W}_{(0)}^H(t-1)\boldsymbol{x}(t)$
  $\boldsymbol{h}_{(0)}(t) = \boldsymbol{P}_{(0)}(t-1)\boldsymbol{y}_{(0)}(t)$
  For $k = 1, 2, \ldots, K$ do
    $\boldsymbol{g}_{(k)}(t) = \frac{\boldsymbol{h}_{(k-1)}(t)}{\eta + \boldsymbol{y}_{(k-1)}^H(t)\boldsymbol{h}_{(k-1)}(t)}$
    $\boldsymbol{P}_{(k)}(t) = \frac{1}{\eta}\mathrm{Tri}\left\{\boldsymbol{P}_{(k-1)}(t-1) - \boldsymbol{g}_{(k)}(t)\boldsymbol{h}_{(k-1)}^H(t)\right\}$
    $\boldsymbol{e}_{(k)}(t) = \boldsymbol{x}(t) - \boldsymbol{W}_{(k-1)}(t-1)\boldsymbol{y}_{(k-1)}(t)$
    if $\|\boldsymbol{e}_{(k)}(t)\| > \|\boldsymbol{e}_{(k-1)}(t)\|$ break; end
    $\boldsymbol{W}_{(k)}(t) = \boldsymbol{W}_{(k-1)}(t-1) + \boldsymbol{e}_{(k)}(t)\boldsymbol{g}_{(k)}^H(t)$
    $\boldsymbol{P}_{(k)}(t-1) = \boldsymbol{P}_{(k)}(t)$
    $\boldsymbol{W}_{(k)}(t-1) = \boldsymbol{W}_{(k)}(t)$
    $\boldsymbol{y}_{(k)}(t) = \boldsymbol{W}_{(k)}^H(t-1)\boldsymbol{x}(t)$
    $\boldsymbol{h}_{(k)}(t) = \boldsymbol{P}_{(k)}(t-1)\boldsymbol{y}_{(k)}(t)$
  end k
  $\boldsymbol{P}_{(0)}(t) = \boldsymbol{P}_{(K)}(t)$
  $\boldsymbol{W}_{(0)}(t) = \boldsymbol{W}_{(K)}(t)$
end t

Once the norm of $\boldsymbol{e}(t)$ at the kth iteration is larger than that of the previous iteration, that is, $||\boldsymbol{e}_{(k)}(t)|| > ||\boldsymbol{e}_{(k-1)}(t)||$, the iteration will be terminated, and $\boldsymbol{W}_{(k-1)}(t)$ will be used as the estimated subspaces at time instant t.

We now discuss the computational complexity of this algorithm. At each time instant, the complexity of PAST algorithm is $O(N^2) + 3MN$. Hence, the complexity of MPAST algorithm is K times of that of PAST, that is, $K[O(N^2) + 3MN]$. As K increases, it is expected that the estimation accuracy will be improved in exchange for increased computational complexity. Fortunately, it is experimentally found that a small K, say, $K = 2$, can achieve a satisfactory performance and complexity tradeoff. In other words, we only need to perform one repetition.

5.2.4 Modified orthonormal projection approximate subspace tracking

Analogous to the PAST algorithm, the obtained subspace by MPAST is not guaranteed to be orthonormal. In cases where orthonormal subspace is required, an additional step of reorthonormalization of $\boldsymbol{W}(t)$ in Algorithm 5.2 should be performed. However, the arithmetic complexity is increased to $K[O(N^2 + MN^2) + 3MN]$. In order to reduce the arithmetic complexity due to reorthonormalization, an orthonormal version of the MPAST algorithm, called the MOPAST algorithm, is derived by extending the OPAST algorithm (Abed-Meraim et al., 2000). The extended algorithm, as summarized in Algorithm 5.3, can directly produce an orthonormal subspace, with a complexity of $K[O(N^2) + 4MN]$. Obviously, the complexity is lower than that of the MPAST algorithm with reorthonormalization.

5.2.5 Kalman filtering

In this subsection, we briefly introduce the Kalman filter, which will be exploited for subspace tracking in the following sections. Under the framework of KF, the following two equations will be introduced to construct a linear state-space model for subspace tracking

$$\boldsymbol{W}^T(t) = \boldsymbol{F}(t)\boldsymbol{W}^T(t-1) + \boldsymbol{\Xi}(t) \tag{5.11a}$$

$$\boldsymbol{x}^T(t) = \boldsymbol{H}(t)\boldsymbol{W}^T(t) + \boldsymbol{\Psi}(t) \tag{5.11b}$$

where $\boldsymbol{W}(t)$ is the subspace to be tracked, $\boldsymbol{x}(t)$ is the observation, $\boldsymbol{F}(t)$ is the state transition matrix and it is chosen as an identity matrix to impose

ALGORITHM 5.3 The modified orthonormal projection approximation subspace tracking algorithm.

Initialize $\boldsymbol{P}_{(0)}(0)$ and $\boldsymbol{W}_{(0)}(0)$
For $t = 1, 2, \ldots do$
  $\boldsymbol{y}_{(0)}(t) = \boldsymbol{W}_{(0)}^{H}(t-1)\boldsymbol{x}(t)$
  $\boldsymbol{h}_{(0)}(t) = \boldsymbol{P}_{(0)}(t-1)\boldsymbol{y}_{(0)}(t)$
  For $k = 1, 2, \ldots, K$ *do*
    $\boldsymbol{g}_{(k)}(t) = \frac{\boldsymbol{h}_{(k-1)}(t)}{\eta + \boldsymbol{y}_{(k-1)}^{H}(t)\boldsymbol{h}_{(k-1)}(t)}$
    $\boldsymbol{P}_{(k)}(t) = \frac{1}{\eta}\mathrm{Tri}\left\{\boldsymbol{P}_{(k-1)}(t-1) - \boldsymbol{g}_{(k)}(t)\boldsymbol{h}_{(k-1)}^{H}(t)\right\}$
    $\boldsymbol{e}_{(k)}(t) = \boldsymbol{x}(t) - \boldsymbol{W}_{(k-1)}(t-1)\boldsymbol{y}_{(k-1)}(t)$
    if $||\boldsymbol{e}_{(k)}(t)|| > ||\boldsymbol{e}_{(k-1)}(t)||$ break; end
    $\tau_{(k)}(t) = ||\boldsymbol{g}_{(k)}(t)||^{-2}\left(\left(1 + ||\boldsymbol{e}_{(k)}(t)||^{2}||\boldsymbol{g}_{(k)}(t)||^{2}\right)^{-2} - 1\right)$
    $\tilde{\boldsymbol{e}}_{(k)}(t) = \tau_{(k)}(t)\boldsymbol{W}_{(k-1)}(t-1)\boldsymbol{g}_{(k)}(t) + \left(1 + \tau_{(k)}(t)||\boldsymbol{g}_{(k)}(t)||^{2}\right)\boldsymbol{e}_{(k)}(t)$
    $\boldsymbol{W}_{(k)}(t) = \boldsymbol{W}_{(k-1)}(t-1) + \tilde{\boldsymbol{e}}_{(k)}(t)\boldsymbol{g}_{(k)}^{H}(t)$
    $\boldsymbol{P}_{(k)}(t-1) = \boldsymbol{P}_{(k)}(t)$
    $\boldsymbol{W}_{(k)}(t-1) = \boldsymbol{W}_{(k)}(t)$
    $\boldsymbol{y}_{(k)}(t) = \boldsymbol{W}_{(k)}^{H}(t-1)\boldsymbol{x}(t)$
    $\boldsymbol{h}_{(k)}(t) = \boldsymbol{P}_{(k)}(t-1)\boldsymbol{y}_{(k)}(t)$
  end k
  $\boldsymbol{P}_{(0)}(t) = \boldsymbol{P}_{(K)}(t)$
  $\boldsymbol{W}_{(0)}(t) = \boldsymbol{W}_{(K)}(t)$
end t

smoothness in the state estimates, $\boldsymbol{\Xi}(t)$ and $\boldsymbol{\Psi}(t)$ are, respectively, the innovation matrix and residual error.

For the sake of simplicity, we first adopt the following state-space model for derivation:

$$\boldsymbol{z}(t) = \boldsymbol{F}(t)\boldsymbol{z}(t-1) + \boldsymbol{w}(t) \tag{5.12a}$$

$$\boldsymbol{u}(t) = \boldsymbol{H}(t)\boldsymbol{z}(t) + \boldsymbol{\delta}(t) \tag{5.12b}$$

where $\boldsymbol{z}(t)$ and $\boldsymbol{u}(t)$ are, respectively, the state vector and observation vector. $\boldsymbol{F}(t)$ and $\boldsymbol{H}(t)$ denote the state transition matrix and observation matrix, respectively; $\boldsymbol{w}(t)$ and $\delta(t)$ are zero-mean Gaussian noise with covariance matrix $\boldsymbol{Q}_{w}(t)$ and $\boldsymbol{R}_{\delta}(t)$, respectively. It is known that the optimal mean square error (MSE) estimator can be obtained by the standard KF recursions as

$$\hat{\boldsymbol{z}}\left(t/t-1\right) = \boldsymbol{F}(t)\hat{\boldsymbol{z}}\left(t-1/t-1\right) \tag{5.13a}$$

$$\boldsymbol{P}(t/t-1) = \boldsymbol{F}(t)\boldsymbol{P}(t-1/t-1)\boldsymbol{F}^T(t) + \mathbf{Q}_w(t) \tag{5.13b}$$

$$\boldsymbol{e}(t) = \boldsymbol{u}(t) - \boldsymbol{H}(t)\hat{\boldsymbol{z}}(t/t-1) \tag{5.13c}$$

$$\boldsymbol{K}(t) = \boldsymbol{P}(t/t-1)\boldsymbol{H}^T(t)\left[\boldsymbol{H}(t)\boldsymbol{P}(t/t-1)\boldsymbol{H}^T(t)+\boldsymbol{R}_\delta(t)\right]^{-1} \tag{5.13d}$$

$$\hat{\boldsymbol{z}}(t/t) = \hat{\boldsymbol{z}}(t/t-1) + \boldsymbol{K}(t)\boldsymbol{e}(t) \tag{5.13e}$$

$$\boldsymbol{P}(t/t) = [\boldsymbol{I} - \boldsymbol{K}(t)\boldsymbol{H}(t)]\boldsymbol{P}(t/t-1) \tag{5.13f}$$

where $\boldsymbol{e}(t)$ denotes the prediction error of the observation vector, $\hat{\boldsymbol{z}}(t/\tau)$, $(\tau = t-1, t)$ represents the estimate of $\boldsymbol{z}(t)$ given the measurements up to time instant τ, that is, $\{\boldsymbol{u}(i), i \leq \tau\}$ and $\boldsymbol{P}(t/\tau)$ is the corresponding covariance matrix of $\hat{\boldsymbol{z}}(t/\tau)$.

5.2.6 Kalman filter with variable number of measurements based subspace tracking

From the KF algorithm above, it is known that the state vector is updated by using only a single measurement. When the system is fast varying, the estimation bias could be small if a single measurement is used. However, the estimation variance will rise correspondingly. On the other hand, when the system is time-invariant or slowly time-varying, more measurements in the past should be used to reduce the estimation variance. In other words, one can achieve a bias-variance trade-off by adaptively choosing the number of measurements for state updates. This concept leads to the so-called KFVM algorithm (Chan Wen, et al., 2006; Chan, Zhang et al., 2006; Liao et al., 2012).

According to the state-space model in (5.12), we assume that the measurements for tracking the state estimate at time instant t are $\boldsymbol{u}(t-L(t)+1), \ldots, \boldsymbol{u}(t-1), \boldsymbol{u}(t)$, where $L(t)$ is the number of measurements used to update the state estimate. As a result, the linear state-space model can be extended as

$$\begin{bmatrix} \boldsymbol{I} \\ \boldsymbol{H}(t-L(t)+1) \\ \vdots \\ \boldsymbol{H}(t-1) \\ \boldsymbol{H}(t) \end{bmatrix} \boldsymbol{z}(t) = \begin{bmatrix} \boldsymbol{F}(t)\hat{\boldsymbol{z}}(t-1/t-1) \\ \boldsymbol{u}(t-L(t)+1) \\ \vdots \\ \boldsymbol{u}(t-1) \\ \boldsymbol{u}(t) \end{bmatrix} + \boldsymbol{\Delta}(t) \tag{5.14}$$

where

$$\boldsymbol{\Delta}(t) = \begin{bmatrix} \boldsymbol{F}(t)\left[\boldsymbol{z}(t-1) - \hat{\boldsymbol{z}}\left(t-1/t-1\right)\right] + \boldsymbol{w}(t) \\ -\boldsymbol{\delta}(t-L(t)+1) \\ \vdots \\ -\boldsymbol{\delta}(t) \end{bmatrix} \tag{5.15}$$

and

$$\begin{aligned} E\left[\boldsymbol{\Delta}(t)\boldsymbol{\Delta}^T(t)\right] &= \begin{bmatrix} \boldsymbol{P}\left(t/t-1\right) & 0 \\ 0 & \mathrm{diag}\left\{\boldsymbol{R}_\delta(t-L(t)+1), \ldots, \boldsymbol{R}_\delta(t-1), \boldsymbol{R}_\delta(t)\right\} \end{bmatrix} \\ &= \boldsymbol{S}(t)\boldsymbol{S}^T(t) \end{aligned} \tag{5.16}$$

where $E[\cdot]$ denotes mathematical expectation and $\boldsymbol{S}(t)$ can be computed from the Cholesky decomposition of $E\left[\boldsymbol{\Delta}(t)\boldsymbol{\Delta}^T(t)\right]$. Multiplying both sides of (5.14) by $\boldsymbol{S}^{-1}(t)$ yields a linear regression as

$$\boldsymbol{X}(t) = \overline{\boldsymbol{H}}(t)\boldsymbol{z}(t) + \boldsymbol{n}(t) \tag{5.17}$$

where $\boldsymbol{n}(t) = -\boldsymbol{S}^{-1}(t)\boldsymbol{\Delta}(t)$, $\boldsymbol{X}(t)$, and $\overline{\boldsymbol{H}}(t)$ are defined as

$$\boldsymbol{X}(t) = \boldsymbol{S}^{-1}(t)\begin{bmatrix} \boldsymbol{F}(t)\hat{\boldsymbol{z}}\left(t-1/t-1\right) \\ \boldsymbol{u}(t-L(t)+1) \\ \vdots \\ \boldsymbol{u}(t-1) \\ \boldsymbol{u}(t) \end{bmatrix}, \quad \overline{\boldsymbol{H}}(t) = \boldsymbol{S}^{-1}(t)\begin{bmatrix} \boldsymbol{I} \\ \boldsymbol{H}(t-L(t)+1) \\ \vdots \\ \boldsymbol{H}(t-1) \\ \boldsymbol{H}(t) \end{bmatrix} \tag{5.18}$$

From (5.17), it is known that we can estimate $\boldsymbol{z}(t)$ by least square (LS) regression as

$$\hat{\boldsymbol{z}}(t) = \left(\overline{\boldsymbol{H}}^T(t)\overline{\boldsymbol{H}}(t)\right)^{-1}\overline{\boldsymbol{H}}^T(t)\boldsymbol{X}(t) \tag{5.19}$$

and the covariance of estimating $\boldsymbol{z}(t)$ is given by

$$E\left[(\boldsymbol{z}(t) - \hat{\boldsymbol{z}}(t))(\boldsymbol{z}(t) - \hat{\boldsymbol{z}}(t))^T\right] = \overline{\boldsymbol{P}}\left(t/t\right) = \left(\overline{\boldsymbol{H}}^T(t)\overline{\boldsymbol{H}}(t)\right)^{-1} \tag{5.20}$$

Since choosing L adaptively can achieve a better bias-variance tradeoff at each time instant, how to determine this quantity is thus of great importance. Following (Chan et al., 2006), let us define

$$\hat{\boldsymbol{e}}(t) = \hat{\boldsymbol{z}}(t-1) - \tilde{\boldsymbol{z}}(t-1) \tag{5.21a}$$

$$\tilde{\boldsymbol{z}}(t) = \eta\tilde{\boldsymbol{z}}(t-1) + (1-\eta)\hat{\boldsymbol{z}}(t-1) \tag{5.21b}$$

where $\hat{\boldsymbol{z}}(t)$ is the state estimate and $\hat{\boldsymbol{e}}(t)$ is its approximated time derivative. η is the forgetting factor $(0 < \eta \leq 1)$ for calculating the smoothed tap weight $\tilde{\boldsymbol{z}}(t)$. When the algorithm is about to converge to the signal subspace in a static environment, the l_1 or l_2 norms of $\hat{\boldsymbol{e}}(t)$ will decrease and converge gradually from its initial value to a very small value. Therefore they serve as a measure of the variation of the signal subspace.

To determine the number of measurements $L(t)$, the variable forgetting factor control scheme can be applied. More specifically, the absolute value of the approximate derivative of $||\hat{\boldsymbol{e}}(t)||$ is first computed as

$$G_e(t) = \left|\left|\left|\hat{\boldsymbol{e}}(t)\right|\right| - \left|\left|\hat{\boldsymbol{e}}(t-1)\right|\right|\right| \tag{5.22}$$

which is smoothed to obtain $\overline{G}_e(t)$ by averaging it over a time window of length T_s. The initial value of $\overline{G}_e(t)$, denoted by $\overline{G}_{e0}$, is obtained by averaging the first T_s data. By normalizing $\overline{G}_e(t)$ with $\overline{G}_{e0}$, we get $\overline{G}_N(t)$, which is a more stable measure of the subspace variation. Therefore the number of measurements $L(t)$ can be updated as

$$L(t) = L_L + \left[1 - g\left(\overline{G}_N(t)\right)\right](L_U - L_L) \tag{5.23}$$

where L_L and L_U are respectively the lower and upper bounds of $L(t)$, and $g(x)$ is a clipping function given by

$$g(x) = \begin{cases} 1; & x \geq 1 \\ x, & 0 < x < 1 \\ 0, & x \leq 0 \end{cases} \tag{5.24}$$

which keeps the range of $\overline{G}_N(t)$ within $[0, 1]$. It is seen that more measurements will be used if the subspace variation measure $\overline{G}_N(t)$ is small and vice versa. To further stabilize the adaptive number of measurements, time-recursive forward smoothing similar to the forgetting-factor-based method can be employed. More precisely $L(t)$ can be recursively estimated as

$$L(t) = \lambda_L L_L + (1-\lambda_L)\left\{L_L + \left[1 - g\left(\overline{G}_N(t)\right)\right](L_U - L_L)\right\} \tag{5.25}$$

where $0 < \lambda_L \leq 1$ is a forgetting factor.

ALGORITHM 5.4 The Kalman filter with a variable number of measurements-based subspace tracking algorithm.

Initialize $\boldsymbol{P}(0)$, $\boldsymbol{W}(0)$, and $L(0)$

For $t = 1, 2, ..., T_s$, *do*

1. Calculate $\hat{\boldsymbol{e}}(t)$ as (5.21a).
2. Calculate $G_e(t)$ as (5.22).
3. Estimate $\boldsymbol{W}(t)$ using the standard Kalman filter.

end t

At time $t = T_s$, obtain $\overline{G}_{e0}$ by averaging first T_s estimates $G_e(t)$.

For $t = T_s + 1, T_s + 2, ...$*do*

1. Calculate $\hat{\boldsymbol{e}}(t)$ as (5.21a).
2. Calculate $G_e(t)$ as (5.22) and obtain $\overline{G}_N(t)$ by normalizing $\overline{G}_e(t)$ with $\overline{G}_{e0}$.
3. Update $L(t)$ as (5.25).
4. Estimate $\boldsymbol{W}(t)$ using the KFVM with $L(t)$ measurements as (5.14)–(5.19).

end t

Consequently, we can employ the state-space model in (5.12) with adaptively chosen $L(t)$ measurements, that is, $\boldsymbol{x}(t - L(t) + 1), ..., \boldsymbol{x}(t - 1), \boldsymbol{x}(t)$, to update the subspace $\boldsymbol{W}(t)$. To be more specific, by associating $\boldsymbol{W}(t)$ with the subspace instead of $\boldsymbol{z}(t)$ in (5.12), then the following state-space model for subspace tracking can be expressed as (5.11). Moreover, the observation matrix $\boldsymbol{H}(t)$ can be approximated as $\boldsymbol{H}(t) = \boldsymbol{x}^T(t)\hat{\boldsymbol{W}}^*(t-1)$. Alternatively, we propose to employ a better estimate of $\boldsymbol{H}(t)$ as follows

$$\boldsymbol{H}(t) = \boldsymbol{x}^T(t)\hat{\boldsymbol{W}}^*\left(t/t-1\right) \tag{5.26}$$

where the superscript $*$ denotes the complex conjugate operation.

The main procedures of the KFVM-based subspace tracking algorithm are summarized in Algorithm 5.4. In each update, the complexity of the KFVM algorithm is $O(L^3 + MN^2) + (L+1)MN + 2LN^2 + L^2N$ if orthonormalization is performed. When $L \gg M$ and $L \gg N$, the complexity is around $O(L^3)$, which will be higher than the PAST-based algorithms. If L is small, the complexity is comparable to them.

5.3 Robust subspace tracking

In this section, we introduce two robust algorithms for subspace tracking in the presence of impulsive noise.

5.3.1 Robust projection approximate subspace tracking

It is known from the PAST algorithm given in Section 5.2.2 that, if the measurement $\boldsymbol{x}(t)$ is corrupted by impulsive noise, then $\boldsymbol{y}(t)$, $\boldsymbol{h}(t)$, $\boldsymbol{g}(t)$, $\boldsymbol{P}(t)$, $\boldsymbol{e}(t)$ and $\boldsymbol{W}(t)$ will be affected in turn. The corrupted matrices, $\boldsymbol{P}(t)$ and $\boldsymbol{W}(t)$, will be used to further update these matrices. This would cause hostile effects on the subspace estimate and requires many iterations to recover. To deemphasis the effect of impulsive noise, the following robust distortion measure is defined for the PAST algorithm (Liao et al., 2010a):

$$J_P(\boldsymbol{W}(t)) = \sum_{i=1}^{t} \eta^{t-i} \rho_r\left(\left\|\boldsymbol{e}(i)\right\| - \mu_e\right) \cdot \left\|\boldsymbol{e}(i)\right\|^2 \tag{5.27}$$

where μ_e is the robust location or mean estimator of $\boldsymbol{e}(i)$, which is defined as

$$\boldsymbol{e}(i) = \boldsymbol{x}(i) - \boldsymbol{W}(t)\boldsymbol{y}(i) \tag{5.28}$$

and $\rho_r(\cdot)$ is chosen as the derivative of an M-estimate function (Huber, 1981). For the modified Huber M-estimate, we have

$$\rho_r(v) = \begin{cases} 1 & |v| \leq \gamma \\ 0 & \text{otherwise} \end{cases} \tag{5.29}$$

where γ is a threshold to be estimated continuously. Specifically, at time instant t, γ is estimated from the "impulsive-free" error variance $\hat{\sigma}(t)$ as

$$\gamma(t) = 1.96 \cdot \hat{\sigma}(t) \tag{5.30}$$

with

$$\hat{\sigma}^2(t) = \lambda_\sigma \hat{\sigma}^2(t-1) + 1.483\left(1 + \frac{5}{N_\omega - 1}\right)(1 - \lambda_\sigma)\text{med}\left(A^2\left(\Delta e_\mu(t)\right)\right) \tag{5.31}$$

where λ_σ is a forgetting factor, $A(x(t)) = \{x(t), \cdots, x(t - N_\omega + 1)\}$, N_ω is the length of the estimation window and is chosen between 5 and 11 to reduce the number of operations required by the median filter, which is symbolized as med$(\cdot)$. $\Delta e_\mu(t)$ is given by

$$\Delta e_\mu(t) = \left|\left\|e(t)\right\| - \hat{\mu}(t)\right| \tag{5.32}$$

where $\hat{\mu}(t) = \lambda_\mu \hat{\mu}(t) + (1 - \lambda_\mu)\left\|e(t)\right\|$ and λ_μ is a forgetting factor.

By taking advantage of the robust distortion measure, it is known that if the measurement $\boldsymbol{x}(t)$ is corrupted by impulsive noise, $\left\|\boldsymbol{e}(t)\right\|$ will become abnormally large. In this case, $\rho_r(\left\|\boldsymbol{e}(t)\right\| - \mu_e)$ will be zero, and therefore impulse-corrupted measurement is prevented from entering into the minimization. This is the fundamental principle of the robust PAST algorithm. In Algorithm 5.5, the main steps of the robust PAST algorithm are summarized.

ALGORITHM 5.5 The robust projection approximate subspace tracking algorithm.

Initialize $\boldsymbol{P}(0)$, $\boldsymbol{W}(0)$, $\hat{\sigma}^2(0)$ and $\hat{\mu}(0)$

For $t = 1, 2, \ldots do$

$\boldsymbol{y}(t) = \boldsymbol{W}^H(t-1)\boldsymbol{x}(t)$

$\boldsymbol{h}(t) = \boldsymbol{P}(t-1)\boldsymbol{y}_{(0)}(t)$

$\boldsymbol{g}(t) = \frac{\boldsymbol{h}(t)}{\eta + \boldsymbol{y}^H(t)\boldsymbol{h}(t)}$

$\boldsymbol{e}(t) = \boldsymbol{x}(t) - \boldsymbol{W}(t-1)\boldsymbol{y}(t)$

$\hat{\sigma}^2(t) = \lambda_\sigma \hat{\sigma}^2(t-1) + 1.483\left(1 + \frac{5}{N_\omega - 1}\right)(1 - \lambda_\sigma) med\left(A^2\left(\Delta e_\mu(t)\right)\right)$

$\hat{\mu}(t) = \lambda_\mu \hat{\mu}(t-1) + \left(1 - \lambda_\mu\right) med\left(A\left(\left|\left|\boldsymbol{e}(t)\right|\right|\right)\right)$

$\gamma(t) = 1.96 \cdot \hat{\sigma}(t)$

$\boldsymbol{P}(t) = \left(1 - \rho_r(|\Delta e_\mu(t)|)\right)\boldsymbol{P}(t-1) + \rho_r\left(|\Delta e_\mu(t)|\right)\frac{1}{\eta} Tri\left\{\boldsymbol{P}(t-1) - \boldsymbol{g}(t)\boldsymbol{h}^H(t)\right\}$

$\boldsymbol{W}(t) = \boldsymbol{W}(t-1) + \rho_r(|\Delta e_\mu(t)|)\boldsymbol{e}(t)\boldsymbol{g}^H(t)$

end t

From the algorithm, it is obvious that if $\Delta e_\mu(t)$ is large, we have $\rho_r\left(\left|\Delta e_\mu(t)\right|\right) = 0$, and hence, $\boldsymbol{P}(t) = \boldsymbol{P}(t-1)$ and $\boldsymbol{W}(t) = \boldsymbol{W}(t-1)$. In other words, the subspace is kept unchanged if the current measurement is corrupted. Otherwise, $\Delta e_\mu(t)$ will be small, we have $\rho_r\left(\left|\Delta e_\mu(t)\right|\right) = 1$. The subspace $\boldsymbol{W}(t)$ will be updated as the PAST algorithm.

5.3.2 Robust Kalman filter with variable number of measuremen

In the KFVM algorithm, the state $\boldsymbol{z}(t)$ is estimated from the past $L(t)$ measurements. Thus any impulse-contaminated measurement $\boldsymbol{x}(k)$ with $t - L(t) + 1 \le k \le t$ will impair the estimate of $\boldsymbol{z}(t)$ and the determination of $L(t)$. In M-estimation, an M-estimate cost function is employed instead of the LS solution in (5.19) to combat the adverse effect of the outliers. Since we only consider outliers in the measurements, the M-estimate cost function is only applied to the measurement equations to yield the following cost function:

$$J(\boldsymbol{z}(t)) = \frac{1}{2}\left|\left|\boldsymbol{e}_o(t)\right|\right|^2 + \sum\nolimits_{k=t-L(t)+1}^{t} \rho\left(\left|\left|e(k)\right|\right|\right) \tag{5.33}$$

where

$$\boldsymbol{e}_o(t) = \boldsymbol{S}_o^{-1}(t)(\boldsymbol{F}(t)\hat{\boldsymbol{z}}(t-1) - \boldsymbol{z}(t)) \tag{5.34}$$

$$\underline{e}(k) = \boldsymbol{S}_k^{-1}(t)(\boldsymbol{x}(k) - \boldsymbol{H}(k)\boldsymbol{z}(t)) \tag{5.35}$$

with $\boldsymbol{S}_o(t)$ and $\boldsymbol{S}_k(t)$ being the Cholesky decomposition of $\boldsymbol{P}(t/t-1)$ and $\boldsymbol{R}_\delta(k)$ in (16), respectively. $\rho(\cdot)$ is an M-estimation function chosen as

$$\rho(v) = \begin{cases} v^2/2 & |v| < \xi \\ \xi^2/2 & \text{otherwise} \end{cases} \tag{5.36}$$

where ξ is a threshold parameter used to control the suppression of outliers and it needs to be estimated recursively. Similar to the robust PAST algorithm in Section 5.3.1, at time instant t, we can estimate ξ from the "impulsive-free" error variance $\overline{\sigma}(t)$ as $\xi(t) = 1.96\overline{\sigma}(t)$, where

$$\overline{\sigma}^2(t) = \lambda_\sigma \overline{\sigma}^2(t-1) + 1.483\left(1 + \frac{5}{N_\omega - 1}\right)(1-\lambda_\sigma)\mathrm{med}(A_\sigma(t)) \tag{5.37}$$

$$\text{with } A_\sigma(t) = \{||\boldsymbol{e}(t-N_\sigma+1)||^2, \cdots, ||\boldsymbol{e}(t)||^2\}.$$

A necessary condition for the optimal solution of (5.33) is $\nabla_z J(z(t)) = \boldsymbol{0}$, that is,

$$(\boldsymbol{S}_o^{-1}(t))^T \boldsymbol{e}_o(t) + \sum_{k=t-L(t)+1}^{t} q(||\boldsymbol{e}(k)||_2)(\boldsymbol{S}_k^{-1}(t)\boldsymbol{H}(k))^T \boldsymbol{e}(k) = \boldsymbol{0} \tag{5.38}$$

where $q(v)$ is defined as

$$q(v) = \frac{\rho'(v)}{v} = \begin{cases} 1 & 0 \le |v| < \xi \\ 0 & \xi \le |v| \end{cases} \tag{5.39}$$

and $\rho'(e)$ is the derivative of $\rho(e)$. From (5.38), it can be derived that

$$\hat{\boldsymbol{z}}(t) = (\overline{\boldsymbol{H}}^T(t)\boldsymbol{\Omega}(t)\overline{\boldsymbol{H}}(t))^{-1}\overline{\boldsymbol{H}}^T(t)\boldsymbol{\Omega}(t)\boldsymbol{X}(t) \tag{5.40}$$

where $\boldsymbol{X}(t)$ is given by (5.18) and $\boldsymbol{\Omega}(t)$ is a weight matrix:

$$\boldsymbol{\Omega}(t) = \mathrm{diag}\left\{\boldsymbol{I}, q\left(||\boldsymbol{e}(t-L(t)+1)||_2\right)\boldsymbol{I}, \cdots, q(||\boldsymbol{e}(t)||_2)\boldsymbol{I}\right\} \tag{5.41}$$

Note that (5.40) is a system of nonlinear equations, and its solution usually requires an iterative method such as the iteratively reweighted LS (IRLS) method. In the IRLS method, the previous iteration is used to compute the weight matrix in (5.41). By doing so, (5.29) becomes a weighted LS solution. This process is repeated until convergence. Fortunately, the a priori state estimate $\boldsymbol{F}(t)\hat{\boldsymbol{z}}(t-1)$ is usually a good estimate of $\boldsymbol{z}(t)$ and hence it can be used to compute the weight at (5.41). Hence, a satisfactory result can be obtained usually after one iteration. Basically, the function (5.39) assigns a weight of one to ordinary samples and a zero weight to those samples with very large prediction error. For the similar reason mentioned above, it would be simpler to use $\boldsymbol{\delta}(t)$ instead of $\boldsymbol{e}(t)$ in computing the weight matrix in (5.41).

We now consider the estimation of $\mathbf{Q}_w(t)$ and $\boldsymbol{R}_\delta(t)$. As shown in (Chan et al., 2006), the noise covariance matrices can be updated as

$$\mathbf{Q}_w(t) = \lambda_w \mathbf{Q}_w(t-1) + (1-\lambda_w)\tilde{\mathbf{Q}}_w(t) \tag{5.42}$$

$$\boldsymbol{R}_\delta(t) = \lambda_\delta \boldsymbol{R}_\delta(t-1) + (1-\lambda_\delta)\tilde{\boldsymbol{R}}_\delta(t) \tag{5.43}$$

where λ_w and $\lambda_\delta \in (0,\ 1]$ are forgetting factors, $\tilde{\mathbf{Q}}_w(t)$ is estimated from the errors with a window length of N_w, that is, $\hat{\boldsymbol{w}}(k) = \hat{\boldsymbol{z}}(k/k) - \boldsymbol{F}(k)\hat{\boldsymbol{z}}(k-1/k-1)$, $t - N_w + 1 \le k \le t$, and $\tilde{\boldsymbol{R}}_\delta(t)$ is estimated from the errors with a window length of N_δ, that is, $\hat{\boldsymbol{\delta}}(k) = \boldsymbol{x}(k) - \boldsymbol{H}(k)\hat{\boldsymbol{z}}(k/k)$, $t - N_\delta + 1 \le k \le t$. Typical values of N_w and N_δ are between 1 and 5. Whenever $q(||e(k)||_2)$ is found to be zero, $\hat{\boldsymbol{\delta}}(k)$ is likely to be corrupted by impulsive noise and these corrupted measurements will be excluded from updating the covariance matrices $\mathbf{Q}_w(t)$ and $\boldsymbol{R}_\delta(t)$ in (5.42 and 5.43). By applying KF to each row of $\boldsymbol{W}(t)$ yields the robust KFVM subspace tracking algorithm as shown in Algorithm 5.6. It

ALGORITHM 5.6 The robust Kalman filter with variable measurements -based subspace tracking algorithm.

Stage 1. Initialize $\boldsymbol{W}(0)$, $\boldsymbol{R}_\delta(0)$, $\mathbf{Q}_w(0)$ and $L(0)$

Stage 2. *For* $t = 1, 2, \ldots, T_s$ *do*

1. Calculate $\hat{\boldsymbol{e}}(t)$ as (5.21a).
2. Calculate $G_e(t)$ as (5.22) and $\overline{G}_{e0}$ by averaging first t estimates $G_e(t)$.
3. Calculate $L(t)$ as (5.25) and update as min $min\{L(t), t\}$.
4. Examine the estimating error $\hat{\sigma}(t)$ and update the weight matrix $\boldsymbol{\Omega}(t)$ as (5.41).
5. Estimate $\boldsymbol{W}(t)$ using the robust KFVM as (5.40).
6. Update the noise covariance matrix $\boldsymbol{R}_\delta(t)$ and $\mathbf{Q}_w(t)$ as (5.42 and 5.43).

end t

At time $t = T_s$, obtain $\overline{G}_{e0}$ by averaging first T_s estimates $G_e(t)$.

Stage 3. For $t = T_s + 1, T_s + 2, \ldots$ *do*

1. Calculate $\hat{e}(t)$ as (5.21a).
2. Calculate $G_e(t)$ as (5.22) and obtain $\overline{G}_N(t)$ by normalizing $\overline{G}_e(t)$ with $\overline{G}_{e0}$.
3. Update $L(t)$ as (5.25).
4. Examine the estimating error $\hat{\sigma}(t)$ and update the weight matrix $\boldsymbol{\Omega}(t)$ as.
5. Estimate $\boldsymbol{W}(t)$ using the robust KFVM as (5.40).
6. Update the noise covariance matrix $\boldsymbol{R}_\delta(t)$ and $\mathbf{Q}_w(t)$ as (5.42 and 5.43).

end t

should be noted that, at the beginning of the algorithm, the selected $L(t)$ may be larger than the number of available measurements. Hence, $L(t)$ should be chosen as $min\{L(t), t\}$.

5.4 Subspace-based direction-of-arrival tracking

At time instant t, the signal subspace matrix $\boldsymbol{W}(t)$ can be applied to determine the DOAs, by using subspace-based high-resolution methods. For the implementation of the MUSIC algorithm, the subspace should be orthonormal. Let $\boldsymbol{U}_S(t)$ be the orthonormalized signal subspace (equal to $\boldsymbol{W}(t)$ if it is orthonormal), and $\boldsymbol{U}_N(t)$ be the noise subspace, then we have

$$\boldsymbol{U}_S(t)\boldsymbol{U}_S^H(t) + \boldsymbol{U}_N(t)\boldsymbol{U}_N^H(t) = \boldsymbol{I} \tag{5.44}$$

Therefore the MUSIC spectrum is given by

$$P(\theta, t) = \frac{1}{\boldsymbol{a}^H(\theta)\boldsymbol{U}_N(t)\boldsymbol{U}_N^H(t)\boldsymbol{a}(\theta)} = \frac{1}{\boldsymbol{a}^H(\theta)\left(\boldsymbol{I} - \boldsymbol{U}_S(t)\boldsymbol{U}_S^H\right)(t)\boldsymbol{a}(\theta)} \tag{5.45}$$

When the array geometry has a shift-invariant property, more computationally efficient DOA estimation approaches can be applied. For instance, for ULAs, the ESPRIT algorithm can be used. First, we construct two submatrices $\boldsymbol{U}_{S1}(t)$ and $\boldsymbol{U}_{S2}(t)$ composed of the first and last $M-1$ columns of $\boldsymbol{W}(t)$, then we carry out the EVD of the following matrix:

$$\left(\boldsymbol{U}_{S1}^H(t)\boldsymbol{U}_{S1}(t)\right)^{-1}\boldsymbol{U}_{S1}^H(t)\boldsymbol{U}_{S2}(t) \tag{5.46}$$

to obtain N eigenvalues $\nu_1(t), \cdots, \nu_N(t)$. Next, we can obtain the DOA estimate as

$$\hat{\theta}_n(t) = asin\left(\lambda\frac{\text{angle}\left(\nu_{n(t)}\right)}{2\pi d}\right), \quad n = 1, \ldots, N \tag{5.47}$$

where λ and d denote the carrier wavelength and intersensor spacing, respectively.

Note that when the sensors suffer from imperfections, such as unknown mutual coupling, subspace-based array calibration approaches can also be applied to estimate the DOAs based on the subspace tracking algorithms (Liao et al., 2012).

5.5 Simulation results

In this section, we show some numerical examples to demonstrate the performance of the aforementioned subspace tracking algorithms for DOA estimation and tracking. A ULA with $M = 10$ sensors with interelement spacing being half wavelength is considered. We assume that $N = 2$ uncorrelated narrow-band signals impinge on the array from the far-field. For DOA estimation and tracking, the ESPRIT algorithm is applied based on the tracked subspace.

5.5.1 Subspace and direction-of-arrival tracking in Gaussian noise

In this subsection, we consider subspace and DOA tracking in Gaussian noise. First, θ_1 is assumed to be invariant while θ_2 is changing linearly and slowly according to the following model

$$\theta_2(t) = 30 - 1.5 \times 10^{-2}t, \quad 0 \leq t \leq 800. \tag{5.48}$$

The SNR is 20 dB. The forgetting factors of the PAST-based algorithms are set to $\eta = 0.98$, and hence the window length is approximately equal to 50. $\boldsymbol{P}(0)$ and $\boldsymbol{W}(0)$ of the algorithm are initialized to identity matrices. For MPAST, we set $K = 2$. Note that the ESPRIT algorithm does not need orthonormal subspaces, and it is experimentally found that MPAST and MOPAST lead to the same DOA estimation performance. Thus the results of MOPAST are omitted. In the KFVM algorithm, the state transition matrix $\boldsymbol{F}(t)$ is chosen as an identity matrix to impose smoothness in the state estimates. In the KFVM algorithm, $T_s = 100$, $L_L = 1$, $L_U = 50$ and $\lambda_L = 0.9$. If we fix the number of measurements to be unchanged, the KFVM algorithm reduces to the so-called Kalman Filter with multiple measurements (KFMM) algorithm. We set $L = 50$ for KFMM.

Fig. 5.1 shows the DOA tracking results of PAST and MPAST using a single experimental run. It can be clearly seen that the MPAST algorithm achieves a better performance than the PAST algorithm when the subspace is slowly time-varying. Moreover, it is noticed that when the DOA is time-invariant, the tracking results for this angle have no much difference.

Fig. 5.2 compares the DOA tracking results of MPAST and KFMM and KFVM. It is observed that all these three algorithms provide similar performance in this case.

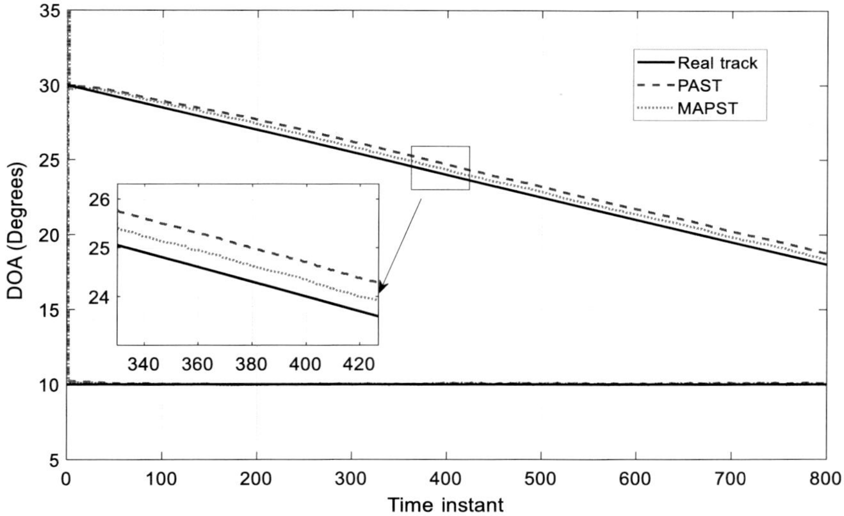

Figure 5.1 Comparison of direction-of-arrival tracking results of projection approximation subspace tracking and modified projection approximation subspace tracking in a slowly varying environment.

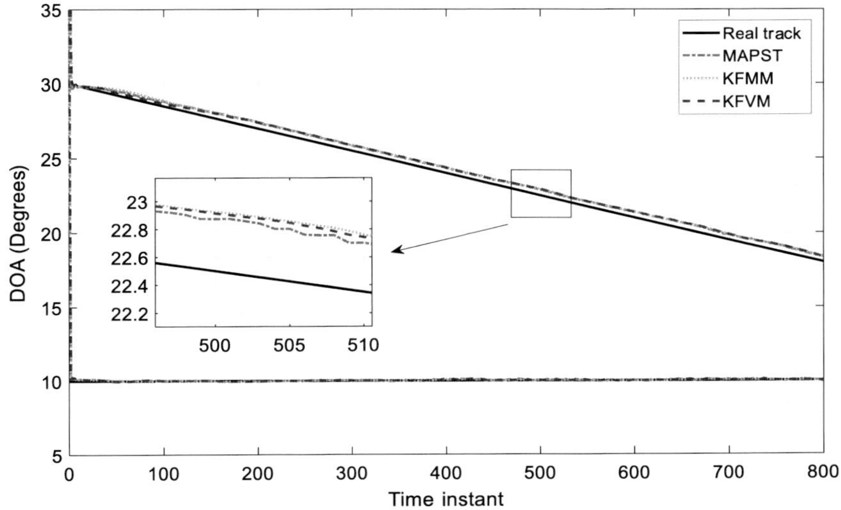

Figure 5.2 Comparison of direction-of-arrival tracking results of modified projection approximation subspace tracking, Kalman filter with multiple measurements, and Kalman filter with a variable number of measurement in a slowly varying environment.

We next compare the tracking performances of different algorithms in fast varying DOA environments. For illustration, θ_2 is assumed to undergo a sharp change within the time interval [600, 620] according to the following model

$$\theta_2(t) = \begin{cases} 30 - 1.5 \times 10^{-2}t, & 0 \le t < 600 \\ 21 - 1.5 \times 10^{-1}(t - 600), & 600 \le t \le 620 \\ 18 - 7.9 \times 10^{-3}(t - 620), & 620 < t \le 800 \end{cases} \tag{5.49}$$

Figs. 5.3 and 5.4 depict the DOA tracking results of different algorithms. In specific, we notice from Fig. 5.3 that, the MPAST algorithm can also offer improved performance in a rapid-changing DOA environment. However, the performance is limited when the DOA changes fast. From Fig. 5.4, it is observed that the KFMM performs similarly to the MPAST algorithm, since the number of measurements is fixed, even when the DOA is fast varying. On the contrary, it can be seen that the KFVM algorithm achieves a good performance in both fast-varying and slow-varying DOA environments. This lies in the fact that the KFVM algorithm can choose the number of measurements adaptively. If the

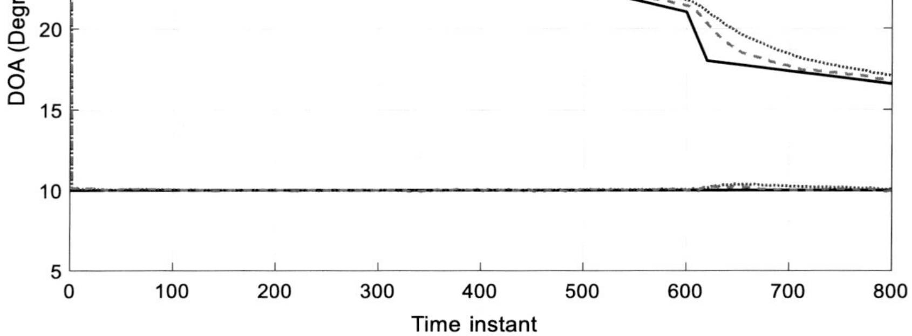

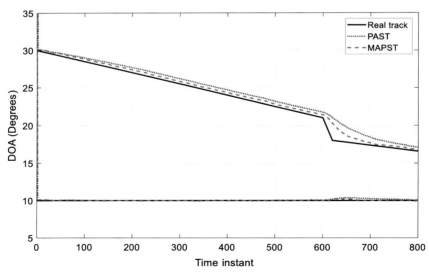

Figure 5.3 Comparison of direction-of-arrival tracking results of projection approximation subspace tracking and modified projection approximation subspace tracking in a fast varying environment.

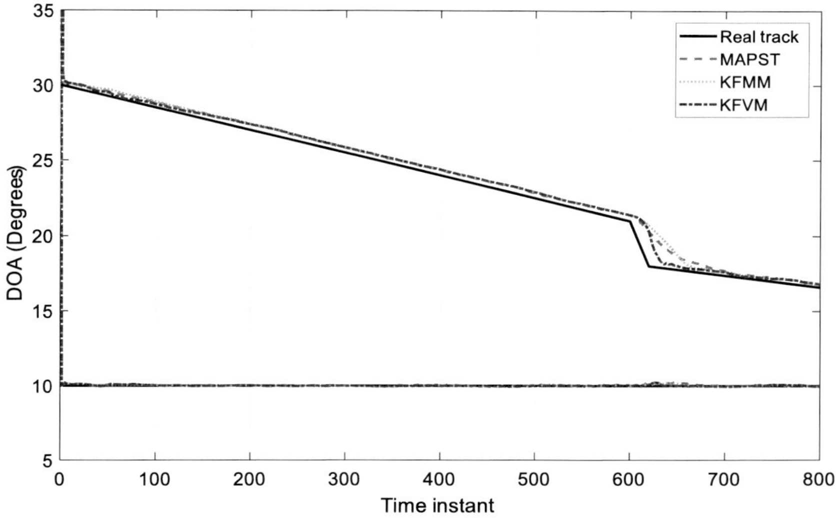

Figure 5.4 Comparison of direction-of-arrival tracking results of modified projection approximation subspace tracking, Kalman filter with multiple measurements, and Kalman filter with a variable number of measurement in a fast varying environment.

DOA is fast varying, less measurements will be used to update the subspace, since more previous measurements will result in a larger estimation bias.

5.5.2 Subspace and direction-of-arrival tracking in impulsive noise

In this subsection, the first DOA is assumed to be time-invariant at $\theta_1 = 10$ degrees, while the second one is time-varying as specified by

$$\theta_2(t) = \begin{cases} 40 - 1 \times 10^{-2}t, & 0 \leq t < 400 \\ 36 - 2.5 \cdot 10^{-1}(t - 400), & 400 \leq t \leq 440 \\ 26 - 7.5 \cdot 10^{-3}(t - 440), & 440 < t \leq 1000 \end{cases} \tag{5.50}$$

where θ_2 changes slowly at the time interval [0, 400] and [440, 1000], and changes rapidly at the time interval [400, 440].

The ambient noise is assumed to be contaminated Gaussian (CG) (Chan et al., 2006). For simplicity, we assume the nominal noise is zero-mean white Gaussian distributed with a power of 0 dB, while the impulsive noise has the same distribution with a power of 30 dB occurring at the time interval [301, 303], [424, 425] and [726, 727]. The locations are fixed so that their effects can be visualized more clearly. Fig. 5.5 shows the noise powers at each time instant.

The PAST and robust PAST algorithms are tested for comparison and the forgetting factors of these two algorithms are set to be 0.98. The parameters of the KFVM algorithm are $\lambda_z = \lambda_w = \lambda_\delta = \lambda_\sigma = 0.99$, $T_s = 100$, $N_w = N_\delta = 5$, and $N_\sigma = 10$

The DOA tracking results of nonrobust KFVM-based subspace tracking algorithm and the PAST are shown in Fig. 5.6. We notice that both methods are significantly affected by impulsive noise, and it takes a long time to get rid of these adverse effects.

Conversely, due to the robustness of the robust PAST and robust KFVM algorithms, the impulsive noise can be suppressed and subspace is accurately tracked, as shown in Fig. 5.7. Furthermore, the robust KFVM algorithm can offer a better tracking performance than the robust PAST algorithm, since a dynamic model of the fast varying subspace is employed.

Fig. 5.8 shows the number of measurements selected by KFVM and robust KFVM algorithms. It shows that the selection in the LS-based KFVM algorithm is significantly affected by the impulsive noise, while the robust KFVM algorithm is able to suppress the

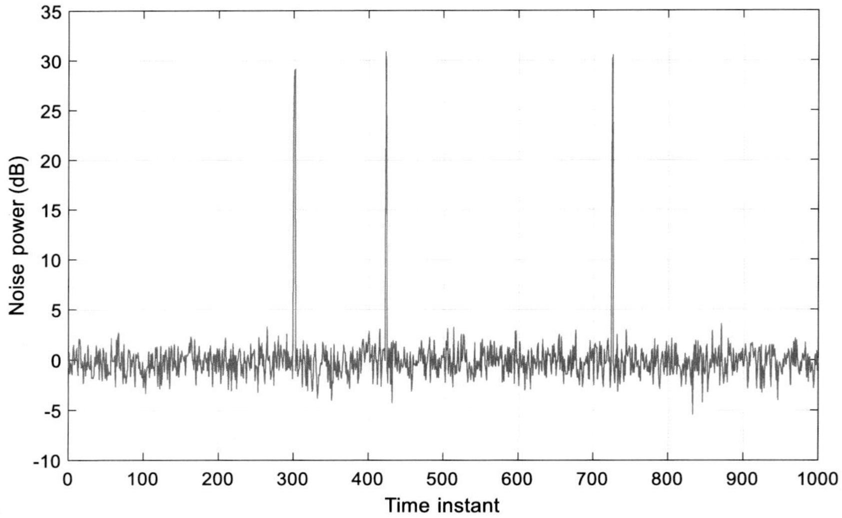

Figure 5.5 Illustration of the noise power.

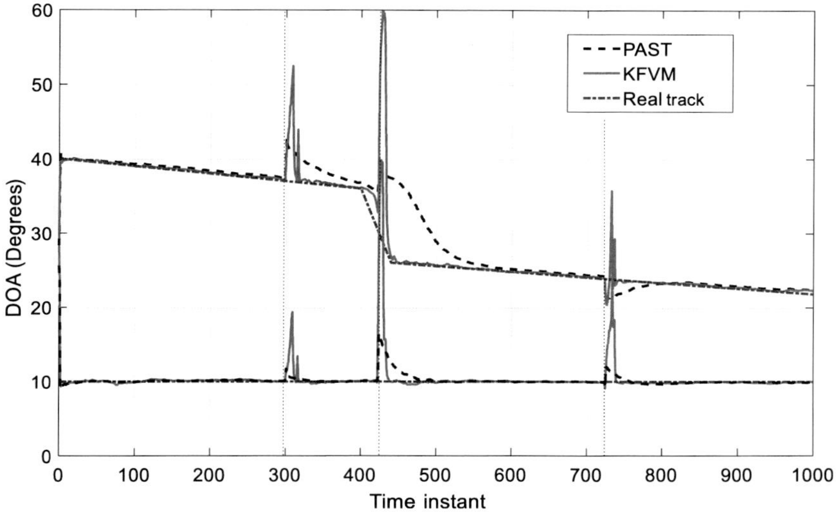

Figure 5.6 Comparison of direction-of-arrival tracking results in impulsive noise using Kalman filter with a variable number of measurement and projection approximation subspace tracking.

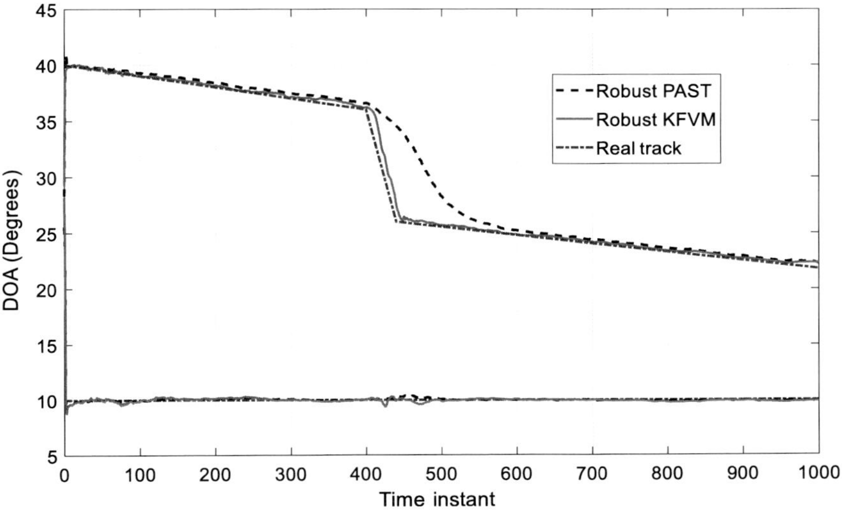

Figure 5.7 Comparison of direction-of-arrival tracking results in impulsive noise using robust projection approximation subspace tracking and robust Kalman filter with variable number of measurement.

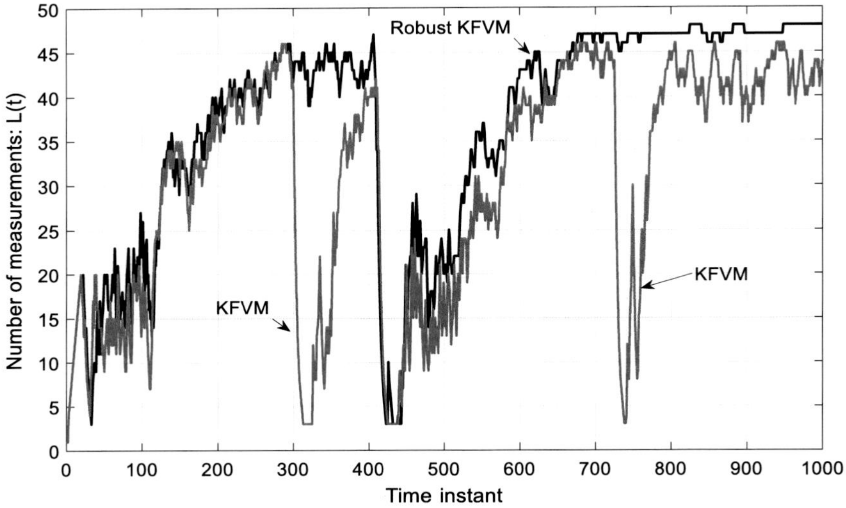

Figure 5.8 Comparison of numbers of measurements $L(t)$ used in Kalman filter with variable number of measurement and robust Kalman filter with variable number of measurement in impulsive noise environment.

impulsive noise effectively and select the measurements more appropriately.

Next, impulsive noise is added to the measurements at three randomly selected time instants and each impulsive noise lasts for two or three times instants. 500 Monte-Carlo simulations are run, and the SNR is 10 dB. The root mean squared error (RMSE) at each time instant is calculated as

$$\text{RMSE} = \sqrt{\frac{\sum_{i=1}^{M_c} \sum_{n=1}^{N} (\theta_n - \hat{\theta}_{i,n})^2}{M_c N}} \tag{5.51}$$

where M_c is the total number of Monte-Carlo experiments, N is the number of signals, θ_n is the nth DOA and $\hat{\theta}_{i,n}$ denotes the n-th estimated DOA in the ith Monte-Carlo experiment.

The RMSEs of the DOA at each time instant estimated by the robust KFVM and robust PAST are illustrated in Fig. 5.9. The robust FMVM method is seen to achieve a better performance than robust PAST when the subspace is fast time-varying. Also, we can see that the impulsive noise can be satisfactorily suppressed by the robust algorithms.

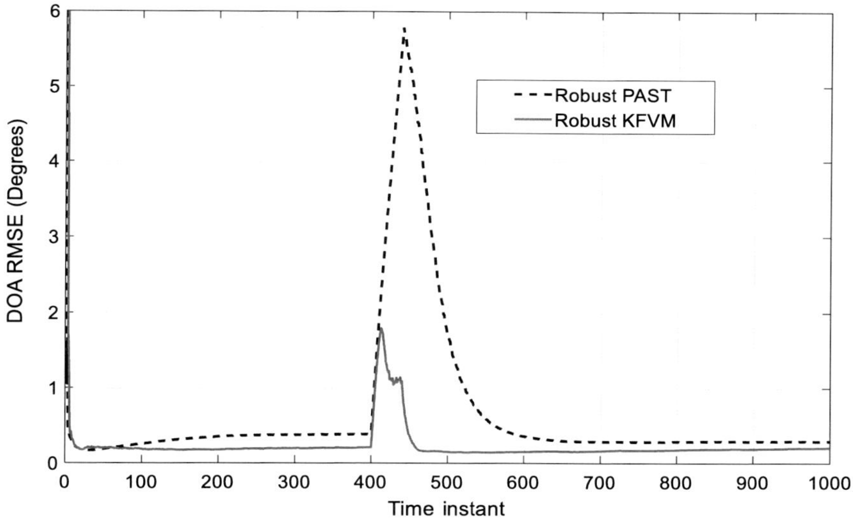

Figure 5.9 Comparison of direction-of-arrival root mean squared errors in impulsive noise using robust Kalman filter with a variable number of measurement and robust projection approximation subspace tracking.

5.6 Conclusions

In this chapter, we introduced several algorithms for subspace tracking and examined the application of these algorithms to subspace-based DOA estimation with sensor arrays. For slowly varying subspace, the MPAST and MOPAST algorithms adopt a better approximation by repeating the update procedure. Thus better tracking performance can be achieved. However, since the forgetting factor in the PAST-based algorithms is not adaptively chosen, they cannot provide satisfactory performance when the subspace is fast varying. To tackle this problem, a KFVM algorithm is applied. This algorithm achieves a bias-variance trade-off by adaptively choosing the number of measurements for state updates. In addition, to deal with the problem of subspace tracking in impulsive noise, which brings outliers to the measurements and leads to serious performance degradation, we developed robust PAST and KFVM algorithms by taking advantage of robust statistics. To examine the performance of these algorithms, various examples of DOA estimation have been carried out.

References

Abed-Meraim, K., Chkeif, A., & Hua, Y. (2000). Fast orthonormal PAST algorithm. *IEEE Signal Processing Letters*, 7(3), 60–62.

Buzzi, S., & D'Andrea, C. (2019). Subspace tracking and least squares approaches to channel estimation in millimeter wave multiuser MIMO. *IEEE Transactions on Communications*, *67*(10), 6766–6780.
Champagne, B., & Liu, Q. G. (1998). Plane rotation-based EVD updating schemes for efficient subspace tracking. *IEEE Transactions on Signal Processing*, *46*(7), 1886–1900.
Chan, S. C., Wen, Y., & Ho, K. L. (2006). A robust past algorithm for subspace tracking in impulsive noise. *IEEE Transactions on Signal Processing*, *54*(1), 105–116.
Chan S.C., Zhang Z.G., & Zhou Y. (2006). A new adaptive Kalman filter-based subspace tracking algorithm and its application to DOA estimation. In *Proceedings of the IEEE international symposium on circuits and systems* (pp. 129–132). Island of Kos, Greece.
Chan, S. C., Wu, H. C., & Tsui, K. M. (2012). Robust recursive Eigendecomposition and subspace-based algorithms with application to fault detection in wireless sensor networks. *IEEE Transactions on Instrumentation and Measurement*, *61*(6), 1703–1718.
Chan, S. C., Tan, H., & Lin, J. (2020). A new variable forgetting factor and variable regularized square root extended instrumental variable PAST algorithm with applications. *IEEE Transactions on Aerospace and Electronic Systems*, *56*(3), 1886–1902.
DeGroat, R. D. (1992). Noniterative subspace tracking. *IEEE Transactions on Signal Processing*, *40*(3), 571–577.
Gu, J., Chan, S. C., Zhu, W., & Swamy, M. N. S. (2013). Joint DOA estimation and source signal tracking with Kalman filtering and regularized QRD RLS algorithm. *IEEE Transactions on Circuits and Systems II: Express Briefs*, *60*(1), 46–50.
Hua, Y., Gershman, A., & Cheng, Q. (2003). *High-resolution and robust signal processing*. New York: Marcel Dekker.
Huang, X., & Liao, B. (2019). One-bit MUSIC. *IEEE Signal Processing Letters*, *26*(7), 961–965.
Huber, P. J. (1981). *Robust statistics*. New York: Wiley.
Krim, H., & Viberg, M. (1996). Two decades of array signal processing research: The parametric approach. *IEEE Signal Processing Magazine*, *13*(4), 67–94.
Liao, B., Zhang, Z. G., & Chan, S. C. (2010a). A new robust Kalman filter-based subspace tracking algorithm in an impulsive noise environment. *IEEE Transactions on Circuits and Systems II: Express Briefs*, *57*(9), 740–744.
Liao B., Zhang Z. G., & Chan S. C. (2010b). A subspace-based method for DOA estimation of uniform linear array in the presence of mutual coupling. In *Proceedings of the IEEE international symposium on circuits and systems* (pp. 1879–1882). Paris.
Liao, B., Zhang, Z. G., & Chan, S. C. (2012). DOA estimation and tracking of ULAs with mutual coupling. *IEEE Transactions on Aerospace and Electronic Systems*, *48*(1), 891–905.
Linh-Trung, N., Nguyen, V. D., Thameri, M., Minh-Chinh, T., & Abed-Meraim, K. (2018). Low-complexity adaptive algorithms for robust subspace tracking. *IEEE Journal of Selected Topics in Signal Processing*, *12*(6), 1197–1212.
Luan, S., Qiu, T., Yu, L., Zhang, J., Song, A., & Zhu, Y. (2017). BNC-based projection approximation subspace tracking under impulsive noise. *IET Radar Sonar & Navigation*, *11*(7), 1055–1061.
Rabideau, D. J. (1996). Fast, rank adaptive subspace tracking and applications. *IEEE Transactions on Signal Processing*, *44*(9), 2229–2244.
Roy, R., & Kailath, T. (1989). ESPRIT-estimation of signal parameters via rotational invariance techniques. *IEEE Transactions on Acoustics, Speech, and Signal Processing*, *37*(7), 984–995.
Sanchez-Araujo, J., & Marcos, S. (1999). An efficient PASTd-algorithm implementation for multiple direction of arrival tracking. *IEEE Transactions on Signal Processing*, *47*(8), 2321–2324.

Sastry, C. R., Kamen, E. W., & Simaan, M. (1991). An efficient algorithm for tracking the angles of arrival of moving targets. *IEEE Transactions on Aerospace and Electronic Systems*, *39*, 242–246.

Schmidt, R. O. (1986). Multiple emitter location and signal parameter estimations. *IEEE Transactions on Antennas and Propagation*, *AP-34*, 276–280.

Thanh, L. T., Dung, N. V., Trung, N. L., & Abed-Meraim, K. (2021). Robust subspace tracking with missing data and outliers: Novel algorithm with convergence guarantee. *IEEE Transactions on Signal Processing*, *69*, 2070–2085.

Tuncer, E., & Friedlander, B. (2009). *Classical and modern direction-of-arrival estimation*. Elsevier.

Vaswani, N., Bouwmans, T., Javed, S., & Narayanamurthy, P. (2018). Robust subspace learning: Robust PCA, robust subspace tracking, and robust subspace recovery. *IEEE Signal Processing Magazine*, *35*(4), 32–55.

Wang, G., Xin, J., Wang, J., Zheng, N., & Sano, A. (2015). Subspace-based two-dimensional direction estimation and tracking of multiple targets. *IEEE Transactions on Aerospace and Electronic Systems*, *51*(2), 1386–1402.

Yang, B. (1995a). An extension of the PASTd algorithm to both rank and subspace tracking. *IEEE Signal Processing Letters*, *2*(9), 179–182.

Yang, B. (1995b). Projection approximation subspace tracking. *IEEE Transactions on Signal Processing*, *43*(1), 95–107.

Zhuang J., Tan T., Chen D., & Kang J. (2020). DOA tracking via signal-subspace projector update. In *Proceedings of the IEEE international conference on acoustics, speech and signal processing (ICASSP)* (pp. 4905–4909).

CHAPTER 6

An overview of optimization and resolution methods in satellite scheduling and spacecraft operation: description, modeling, and application

Andrew W.H. Ip[1,2], Fatos Xhafa[3], Jingyi Dong[4] and Ming Gao[4]
[1]Department of Industrial and Systems Engineering, The Hong Kong Polytechnic University, Hong Kong SAR, P.R. China
[2]Department of Mechanical Engineering, University of Saskatchewan, Saskatoon, SK, Canada
[3]Universitat Politècnica de Catalunya, Barcelona, Spain
[4]School of Management Science and Engineering, Key Laboratory of Big Data Management Optimization and Decision of Liaoning Province, Dongbei University of Finance and Economics, Dalian, P.R. China

6.1 Introduction

Mission operations often occur in the process of communication and coordination between satellite station, space station, detector, and ground station. The communication between the spacecraft and the ground operation team is through requesting a specific time window or multiple window antennas at the ground station. In the given time window, the satellite should complete the corresponding tasks. Then, with the popularity of satellites, more and more satellites are used for research and education purposes by institutions, small and medium-sized enterprises, universities, and so on. At the same time, the number of task requests is also increasing each time. Therefore, in the face of requests such as allocating customers, different requests may conflict with satellite communication because they occur in the same window.

6.1.1 Background

As an advanced space-based information platform, the satellite has played an irreplaceable role in the economy, military, and other fields. With the continuous development of satellite technology and the popularization of satellite applications in China, users from all fields have put forward a large

IoT and Spacecraft Informatics
DOI: https://doi.org/10.1016/B978-0-12-821051-2.00002-7

number of task requirements to satellite management and control departments. To meet the large-scale and diversified task requirements, satellite scheduling has become the primary content of satellite management and control.

- *Satellite Task Scheduling*: In the process of the satellite on-orbit operation, the satellite management and control department allocates satellite payload and ground management and control resources and formulates satellites on-orbit operation and ground management and control plans to maximize the revenue of satellite tasks. The satellite task scheduling problem not only involves satellite payload resources but also the allocation of satellite payload resources. It also involves ground management and control resources to ensure its normal operation. It includes not only satellite mission tasks but daily maintenance requirements. It can be seen that satellite task scheduling involves many subproblems and covers a wide range, so a reasonable and effective modeling method is of great significance for the description and solution of satellite task scheduling problems. It is necessary to decide the order of task execution and allocate the resources needed for task execution (Du, Xing, & Chen, 2019).
- *Spacecraft Operation*: There are also various optimization problems in satellite deployment systems. In 2015, the Hong Kong Polytechnic University developed a microsatellite platform and deployment system to promote the implementation of low-cost space tests (POLYU, 2016). Space technology can be used in many research fields, such as aviation, materials, education, and so on (POLYU, 2020). For air transportation, the platform provides additional resources and the platform can supervise and track the air traffic situation, which can reduce the difficulty of accident prevention and investigation to a certain extent.

In this chapter, we summarize the research status of satellite scheduling and spacecraft optimization, analyze different types of satellite scheduling problems, and other problems that may exist in the process of satellite schedule. The mathematical formulas of satellite scheduling are listed and analyzed. The satellite scheduling problem is a kind of time window scheduling problem with multiobjective. At the same time, the computational complexity of this kind of problem is high, and it has a high degree of restriction, and the goal conflict caused by the accessibility and visibility of the time window of all kinds of task planning. In this paper, we list the local search and population-based methods and propose a two-stage heuristic algorithm to determine the observation sequence and generate a feasible scheduling scheme. Then, we develop and test the time-based greedy algorithm, weight-based greedy algorithm, and the first two algorithms, the improved DE algorithm is used to get the feasible solution and further optimize it. Finally, we discuss the optimization problems that may exist in the operation of spacecraft. After analyzing the complexity of the problems, we discuss and propose relevant solutions.

The remainder of the chapter is organized as follows. We present and discuss in Section 6.2 the basic concepts about ground stations, spacecraft/satellites, and the satellite scheduling problems. Optimization problems from spacecraft operations are presented in Section 6.3. In Section 6.4, we discuss several resolution methods and a useful simulation toolkit together with a benchmark of instances presented in Section 6.6. We collect some algorithms, strategies, frameworks, and models about satellite scheduling problems in recent years. We end the paper with some conclusions and future work in Section 6.7.

6.1.2 Literature review and classification of scheduling problems

With the rapid development of satellite technology, the number of satellite mission planning and operation has increased greatly, and an intelligent scheduling system is needed to automatically and optimally deal with this kind of mission planning, that is, to assign tasks to space missions through ground station services. The core of satellite task planning is the satellite scheduling problem, whose solution is to optimize the allocation of user requests, to carry out effective communication between ground and spacecraft system operation teams. This kind of problem belongs to NP-hard problem; it is difficult to get the optimal solution and the computational complexity is high, so we need to use some algorithms to find the optimal solution, and use tools to evaluate the algorithm to solve the optimization problems in satellite scheduling, spacecraft design, operation, and satellite deployment system.

In this section, we present the common scheduling problem formulation and several variants of satellite schedule. The application scope, algorithm idea, algorithm architecture, tools of different variants are analyzed and evaluated, and their application situation and practical significance are pointed out. In the general literature of scheduling and planning, they can be classified into the following development stages:

Scheduling and planning are two interrelated processes: planning refers to the process of finding the "sequence of actions that transfer the world from some initial state to a desired state," while scheduling refers to the process of "assigning a set of tasks to a set of resources subject to a set of constraints," according to Bartak, Salido, and Rossi (2010, p. 5). Most real-world problems can be regarded as highly constrained, integrated planning and scheduling problems. Limited resources, such as but not limited to people, machines, materials. and assets need to be allocated to optimize particular objective(s) and with the consideration of satisfying several constraints, such as time, budget, and space. Over the last decades, planning and scheduling have received ever-increasing attention. Policy makers, entrepreneurs, and researchers have paid much attention and substantial effort in exploiting various techniques for solving real-life planning and scheduling problems.

Planning problems can be found in various areas, ranging from controlling of autonomous submarines (McGann, Py, Rajan, Ryan, & Henthorn, 2008), to manufacturing (Ruml, Do, & Fromherz, 2005), manpower planning (De Bruecker, Van den Bergh, Belien, & Demeulemeester, 2015) and home healthcare services (Nickel, Schroder, & Steeg, 2012). Classical scheduling and planning problems address the finding of a sequence of actions that "transfer the world from some initial states to a desired state" (Bartak et al., 2010, p. 5). The majority of the researches assume (1) finite state space, (2) deterministic effect of action, (3) static environment, (4) instantaneous actions, and (5) sequential actions. Variations and extensions of planning problems may deal with parallel actions with uncertain effects.

6.1.3 The scheduling problems

Scheduling problems address the allocation of resources to activities, that is, finding schedules, for optimizing some performance objectives (Bartak et al., 2010). Resources could be machinery and manpower. Activities could be manufacturing operations and duties, while performance objectives could be minimization of resource utilization and minimization of staffing costs. Scheduling problems have been extended from traditional machine scheduling to university timetabling, broadcast scheduling (Lin & Wang, 2013), and healthcare professional scheduling (Brunner & Edenharter, 2011; Maenhout & Vanhoucke, 2013). Similar to planning problems, there may be some constraints needed to be satisfied by scheduling problems. Examples of constraints include precedence relationship between jobs and start time of processing.

Scheduling problems have been studied since the 1950s (Leung, 2004). Scheduling problems studied at the time were simple and mainly dealt with the problems related to operations and manufacturing processes in factories. In the late 1960s, the focus of scheduling problems shifted toward the efficient utilization of scarce and expensive computational resources. Production systems are one of the most prominent and important research fields in scheduling for both practitioners and academia (Abedinniaa, Glocka, Grossea, & Schneider, 2017). In a production system, jobs need to be processed, with unknown sequence to be determined, on machine(s) to optimize objective(s) without violating constraint (s) (Graves, 1981). Effective scheduling plays an important role in reducing idle time, increasing production efficiency, and reducing cost, even in modern production systems, for instance, semiconductor manufacturing plants (Monch, Fowler, Dauzere-Peres, Mason, & Rose, 2011).

Solving machine scheduling problems in production is challenging in most cases, as modifying any assumption often leads to a new problem that requires new solution approaches. Thus, machine scheduling problem in production is one of the most frequently studied optimization problems in management and engineering (Abedinniaa et al., 2017). The earlier researches focused on machine scheduling problems in production, which can be characterized according to machine environment, constraints, and objectives. Machine scheduling problem in production is widely applied in modern manufacturing systems, for instance, flexible manufacturing systems, robotic (Levner, Kats, Alcaide Lopez de Pablo, & Cheng, 2010) and semiconductor manufacturing (Monch et al., 2011), and wafer fabrication (Sarin, Varadarajan, & Wang, 2011). There is a wide variety of solution approaches used to solve specific machine scheduling problems in production, ranging from exact algorithms (Kis & Pesch, 2005) and approximation algorithms (Gupta, Koulamas, & Kyparisis, 2006) in the past, to heuristic and metaheuristics as well as artificial intelligence solution approach (Ouelhadj & Petrovic, 2009). The solution being NP-hard in nature, research in scheduling problems is paying increasing attention to applying approximation and optimization algorithms to the problems. Unfortunately, the majority of the existing studies focus on proposing a specific scheduling approach in addressing a specific scheduling problem. This makes the scheduling approach far distinct from the planning approach, where the former deals with specific scheduling problems while the latter deals with generic planning problems.

On the other hand, manpower scheduling has found wide applications. According to Ernst, Jiang, Krishnamoorthy, and Sier (2004, p. 3), staff scheduling and rostering is the process of "constructing work timetables for its staff so that an organization can satisfy the demand for its goods or services." This process involves firstly the determination of the number of staff required to meet the service demand, and subsequently the allocation of shifts to meet the required staffing level. The staff scheduling and rostering problem is also a highly constrained and complex problem and, thus, very difficult to solve. Difficulties increase exponentially with the number of constraints and objectives taken into consideration. Approaches dealing with the problem have evolved with time, evolving from spreadsheets to mathematical models and algorithms. In general, industry-specific and application area-specific mathematical models and algorithms are developed to deal with the uniqueness of different industries, and cannot be easily transferred to another industry or application area. Classic applications of staff scheduling can be found in airline crew scheduling (Day & Ryan, 1997), call centers (Segal, 1974), and nurse scheduling (Maenhout & Vanhoucke, 2013). More recent applications of staff scheduling and rostering include physicians scheduling (Brunner & Edenharter, 2011) and home healthcare routing and scheduling (Fathollahi-Fard, Hajiaghaei-Keshteli, & Tavakkoli-Moghaddam, 2018; Shi, Boudouh, & Grunder, 2017). People are the most valuable resource in an organization (Jorn, Jeroen, Philippe, Erik, & Liesje, 2013; Zulch, Rottinger, & Vollstedt, 2004) and the most flexible factor for most of the industries, such as airlines (Petrovic & Berghe, 2007), call centers (Aksin, Armony, & Mehrotra, 2007) and healthcare (Burke & Petrovic, 2004; Ernst et al., 2004), human resources are an important factor contributing to the long-term competitiveness of organizations (Zulch et al., 2004). In particular, manpower planning is a major concern for organizations facing a dilemma in balancing resources, quality, productivity, and time. Manpower planning is a fundamental and important part of an organization's strategic and human resources planning processes (Ernst et al., 2004; Jorn et al., 2013; Zulch et al., 2004). Implementing an effective manpower plan has proven to be beneficial in reducing labor costs while providing high quality or services within a specific period (Ernst et al., 2004; Koeleman, Bhulai, & van Meersbergen, 2012; Nickel et al., 2012).

6.1.4 Integrating scheduling in the big data environment

Driven by the exponential growth of data generated in daily lives, such as social networks and healthcare platforms, the processing, and analysis of data in cloud environments have become important and challenging problems (Soualhia, Khomh, & Tahar, 2017). There are some multitasking frameworks widely used by large organizations for distributed storage and distributed processing of big data in cloud platforms (Jian, Shicong, Xiaoqiao, & Li, 2013). Examples include Hadoop (Kurazumi, Tsumura, Saito, & Matsuo, 2012),

Storm (Peng, Hosseini, Hong, Farivar, & Campbell, 2015), and Mesos (Hindman et al., 2011). According to Soualhia et al. (2017, p. 170), these frameworks have a "pluggable architecture that permits the use of schedulers optimized for particular workloads and applications," which highly affect the computation time and resources utilization. However, due to the dynamic nature of the cloud environment, efficient task scheduling is very challenging. Different algorithms have been proposed for frameworks to improve task submission, task scheduling, and task recovery.

Hadoop is a well-known processing platform that implements the MapReduce programming model (Lee, Lee, Choi, Chung, & Moon, 2012). Spark is an in-memory computing framework that can be run on Hadoop. Storm is a distributed computation framework for real-time applications. Spark (Zaharia, Chowdhury, Franklin, Shenker, & Stoica, 2010) and Storm (Peng et al., 2015) can implement the MapReduce programming model. These platforms can be deployed in a cluster and managed by cluster managers, such as Mesos (Hindman et al., 2011) and yet another resources negotiator (YARN) (Liu, Yang, Sun, Jenkins, & Ross, 2015). Fig. 6.1 shows the relationships between MapReduce, Hadoop, Spark, Storm, and Mesos.

In general, the classification of classical planning and scheduling problems mentioned earlier refers generally to the process and decisions of deciding what to do; they refer generally to the process and decisions of when and how to do it. In other words, planning problems deal with strategic goals and a series of actions to transform states, while scheduling problems deal with operational constraints such as temporal and resource. In the traditional hierarchical approach, planning and scheduling are two separate modules that take place sequentially: planners formulate plans and schedulers generate schedules while ensuring practicability and optimality of plans formulated. The traditional distinction between them may result in strategic and operational conflicts, or suboptimality (Hassani, Barkany, Jabri, & Abbassi, 2018). Among the existing studies, there are three major approaches to deal with the integrated planning and scheduling problem. The first stream of studies distinguishes the separable parts and the nonseparable parts. The separable parts are treated as planning and scheduling problems separately and the nonseparable parts are solved collectively (Halsey, Long, & Foz, 2004). The second stream of studies tries to decompose the integrated problem into distinctive planning and scheduling problems as much as possible (Halsey et al., 2004). The last stream of studies deals with the integrated problem.

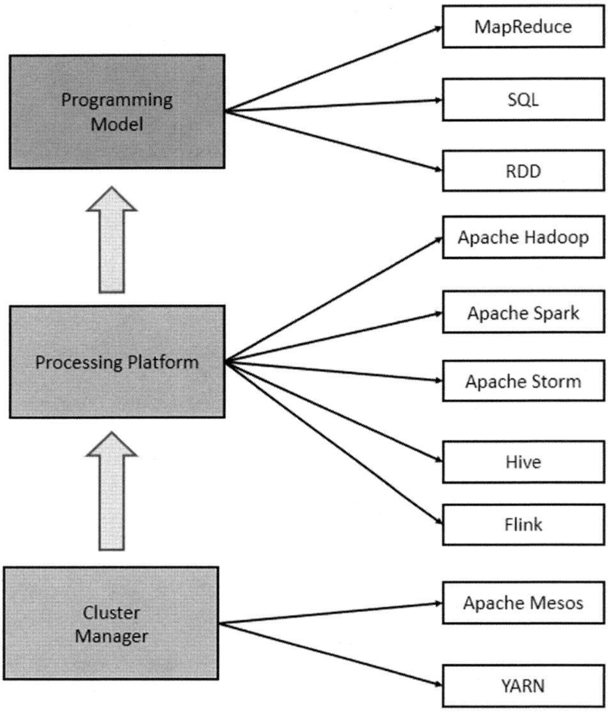

Figure 6.1 Relationships among MapReduce, Hadoop, Spark, Storm, and Mesos.

6.2 Satellite scheduling problems

According to our discussion of different classifications of scheduling methods and development, satellite scheduling problems can be divided into the following specific problems, as shown in Table 6.1.

6.2.1 Satellite range scheduling

It aims to schedule satellite requests to GS according to their time windows so that the profit from the scheduled requests is maximized. Again, with the increasing number of requests for satellite services, the problem becomes more intractable and Luo, Wang, Li, and Li (2017) developed a conflict-resolution technique (Xhafa & Ip, 2021).

6.2.2 Satellite downlink scheduling

Satellite downlink scheduling aims at optimizing the communication traffic from the satellite to the GSs, which is crucial to achieving the efficiency of

Table 6.1 Classification.

Classification	Question	Problem description
Characteristic	Satellite range Scheduling	It aims to schedule satellite requests to GS according to their time windows so that the profit from the scheduled requests is maximized (Xhafa & Ip, 2021).
	Satellite downlink scheduling	Satellite downlink scheduling aims at optimizing the communication traffic from the satellite to the GSs, which is crucial to achieving the efficiency of the system for acquiring high-quality images of the Earth surface (Xhafa & Ip, 2021).
	Satellite broadcast scheduling	Satellite broadcast scheduling is formulated for maximizing broadcasting to various GSs under certain constraints and to maximize the number of time intervals used for broadcasting (Xhafa & Ip, 2021).
	Satellite scheduling data download	The goal in the process is to minimize the total amount of data downloaded and transmitted to the Earth.
Scale	Satellite scheduling at a large scale	Current scheduling of operations for spacecraft to ground stations communications has shown limitations, and still needs human coordination and manual labor-intensive activity. Manual computations of satellite mission planning, even when partially employed, are slow, prone to errors, and costly. The problem becomes more challenging with the increasing number of submitted requests (Xhafa & Ip, 2021).
	Satellite scheduling at a small scale	The number of ground stations is smaller compared to the number

(*Continued*)

Table 6.1 (Continued)

Classification	Question	Problem description
		of mission planning requests; therefore, the aim is to achieve maximum usage of ground stations to support as many as possible requests (Xhafa & Ip, 2021).
	Multisatellite scheduling	In the scheduling procedure, it involves multiple satellites and multiple orbits. In the view of the same resource, there are several targets and one target can be observed by more than one resource contemporaneously (Chen, Reinelt, Dai, & Wang, 2018) (Chen, 2018).
Multisatellite & ground station	Multisatellite, multistation TT & C scheduling	Multisatellite and multistation TT & C scheduling is of great significance to improve the utilization efficiency of TT & C resources, better meet the needs of Satellite TT & C, and give full play to the comprehensive efficiency of the TT & C system.
	Ground station scheduling	Ground station scheduling problem arises in spacecraft operations and aim to allocate ground stations to spacecraft to make possible the communication between operations teams and spacecraft systems. The problem belongs to the family of satellite scheduled for the specific case of mapping communications to ground stations. Ground stations are terrestrial terminals designed for extra-planetary communications with spacecraft (Xhafa et al., 2012).

(Continued)

Table 6.1 (Continued)

Classification	Question	Problem description
Special Satellite	Low-earth-orbit satellite scheduling	Earth observation satellite scheduling deals with the case of low orbit satellites to collect information about the Earth (Xhafa et al., 2012).
Others	Computational complexity of satellites scheduling	Satellite scheduling problem is actually NP-hard. That is, the best solution cannot be found. Satellite scheduling problem is highly constrained, so it is not easy to solve them, and it is not easy to find a feasible solution (that is, to allocate all task requests).
	Satellite deployment systems	Satellite deployment system (simulation system) plays an important role in evaluating the performance of optimal scheduling solutions.

the system for acquiring high-quality images of the Earth's surface (Xhafa & Ip, 2021). As an application problem, downlink scheduling is characterized by many constraints, which make it difficult to optimize the schedule and even produce feasible solutions. Effective scheduling of image acquisition and downlink plays an important role in satellite mission planning. These operations are usually interconnected and resolved by scheduling heuristics (Karapetyan, Daniel, Mitrovic, Snezana, & Malladi, 2015).

6.2.3 Satellite broadcast scheduling

Satellite Broadcast Scheduling is formulated for maximizing for broadcasting to various GSs under certain constraints and to maximize the number of time intervals used for broadcasting (Xhafa & Ip, 2021). The characteristics of a satellite communication system are the number of satellites flying around the Earth at high or low altitudes, several fixed ground terminals on the earth, and several communication requests, one for each satellite, specifying how long the satellite must broadcast to perform communication tasks. Some potential advantages of low altitude satellites include lower power requirements and fewer portable antennas, smaller propagation delay, higher image resolution, and lower launch cost than satellites launched at high geostationary altitude. Because the low-altitude satellites cannot always be seen from the ground terminal, to meet the communication request, the handover between satellites must be carried out. This handover constitutes the satellite broadcast scheduling problem of low altitude satellite communication systems. Therefore, the optimization of broadcast time from a group of satellites to a group of ground terminals has become an important design problem in satellite communication systems (Salman, Ahmad, and Omran, 2015). As we all know, the satellite broadcast scheduling optimization problem is an NP-complete problem. The goal of SBS is to find an effective broadcast scheduling mode and maximize the number of broadcast slots under certain restrictions. According to the specific limitations of the field, the term "satellite scheduling" has been applied to different aspects of satellite services, such as remote sensing and communication services. Salman et al. (2015) proposed a differential evolution algorithm, which is shown to be effective and scalable (Xhafa & Ip, 2021).

6.2.4 Satellite scheduling data download

In many applications of satellites, how to effectively download and transfer the data of multiple satellites to the ground station network leads to scheduling problems. The goal in the process is to minimize the total amount of data downloaded and transmitted to the Earth. Castaing (2014) proposed a heuristic scheduling algorithm based on a greedy algorithm, and provided the model to obtain a reasonable approximation. This type of satellite scheduling is suitable for low-cost small satellites to collect data from space (Xhafa & Ip, 2021).

6.2.5 Satellite scheduling at large scale

Large-scale satellite scheduling problems usually occur in large space organizations, such as the European Space Agency (ESA), NASA, etc. - and mainly support other tasks according to customer service requirements. The number of planning requests for missions continues to rise (Barbulescu, Howe, Watson, & Whitley, 2002). ESA uses the network of ESTRACK- the ESA tracking stations (Damiani, Dreihahn, Nizette, & Calzolari, 2007). This is a global ground station system, which can realize the connection between the operation control center and the satellite in orbit. However, at present, there are limitations in the operation plan of spacecraft and ground communication. They are not completely automated and still need to add human factors, such as human coordination and labor-intensive activities. The labor cost is high and the process is efficient.

6.2.6 Satellite scheduling at small scale

Small scale satellite mission planning usually exists in the research projects of European, American, Chinese and other institutions, as well as small and medium-sized enterprises and universities. Small-scale satellite mission planning can refer to the need to automate the allocation of ground station services to spacecraft missions (Xhafa & Ip, 2021). Now, with the rapid development of satellite networks, the application of satellite communication in many fields is also increasing. The number of ground stations is smaller than the number of mission planning requests. So, our goal is to make the most of the ground station to accept as many requests as possible. The ground station usage is the time during which the ground station is communicating with some spacecraft. Alternatively, maximizing the ground station usage is equivalent to minimizing its idle time (Xhafa & Ip, 2021).

The goal of satellite scheduling is to assign tasks to ground stations, to ensure the spacecraft complete the task.

6.2.7 Multisatellite scheduling

Earth-observing-satellite (EOS) plays a representational role in repaid reaction to observation requests on Earth's surface. A resource can only carry out the observation when the target is visible to it. In the scheduling procedure, it involves multiple satellites and multiple orbits. In the view of the same resource, there are several targets and one target can be observed by more than one resource contemporaneously. Chen et al. (2018) address that the allocation of visible time windows is dramatically overlapped in the whole scheduling period, which makes the combinational characteristic of the problem more prominent and credible. The operational constraints of diverse resources may be different as well, rendering uniform modeling of the problem difficult. Therefore, the multiple EOSs scheduling problem is NP-Complete.

6.2.8 Multisatellite, multistation TT & C scheduling

With the rapid development of space technology, more and more spacecraft is launched into orbit, and the task of ground measurement and control network becomes even more. Multisatellite and multistation TT & C scheduling is of great significance to improve the utilization efficiency of

TT & C resources, better meet the needs of Satellite TT & C, and give full play to the comprehensive efficiency of the TT & C system.

Considering the conflicts among visible arcs and the requirement of ground station setup transition time, a combinational optimization model for the multisatellite, multistation TT & C scheduling problem is designed by using whether to perform TT & C in the visible arcs as design variables. The objective function is a combination of the satellite priority, the ground station priority, the efficiency of ground stations, and the length of the scheduled visible arcs.

6.2.9 Ground station scheduling

We describe here the problem characteristics, identify the input problem instance and its optimization objectives types. Then, the output instance computed from resolution methods is formally given (see Xhafa, Herrero, Barolli, & Takizawa, 2013a, 2013b; Xhafa, Sun, Barolli, Biberaj, & Barolli, 2012).

Basic Concepts and Terminology: Here are some basic concepts and terminology about ground stations (GS), satellites, and spacecraft (SC). Ground stations are terrestrial terminals designed for extra-planetary communications with spacecraft (SC). Any satellite communication line includes sending and receiving ground stations, uplink, and downlink lines, and communication satellite repeaters. The ground station is an important part of the satellite communication system. The basic function of the ground station is to transmit signals to satellites and receive signals transmitted by other ground stations through satellites. The ground stations, for various purposes, are slightly different, but the infrastructure is the same (Xhafa et al., 2012; Xhafa et al., 2013b) .

SCs are extra-planetary crafts, such as satellites, probes, space stations, orbiters, etc. GSs communicate with a spacecraft by transmitting and receiving radio waves in high-frequency bands and usually contain more than one satellite dish. In mission planning, a dish is usually assigned to a space mission; however, through the scheduling from the control center, dishes can handle and switch among mission spacecraft (Xhafa et al., 2012; Xhafa et al., 2013b).

Problem Instance in Input. The problem instance in input or the problem input data is the information that defines a concrete instance of the problem. The input data are given by the values of a series of parameters shown in Table 6.2 (Xhafa et al., 2012; Xhafa et al., 2013b).

Optimization Objectives: In a general setting, scheduling problems are multiobjective optimization problems, that is, several optimization objectives can

Table 6.2 Problem data in the input (Xhafa et al., 2012; Xhafa et al., 2013b).

Data	Meaning
***SC*[*i*]**	List of SCs that participate in the planning
***GS*[*g*]**	List of GSs that participate in the planning
Ndays	Total number of days for the schedule
***T AOS-V IS*[*i*][*g*]**	Visibility times of GSs to SCs
***T LOS-V IS*[*i*][*g*]**	Information on timed when a GS loses signal from an SC
***T Req*[*i*]**	List of required communication time for SCs to complete a task

be established. This is also the case for satellite schedule for which we can formulate several optimization objectives. Among these objectives, the most relevant ones are (1) maximizing matching of visibility windows of SCs to communicate with GSs; (2) minimizing the time window clashes of different SCs to one GS; (3) maximizing the communication time of SC with Gs, and, (4) maximizing the usage of GSs. Solving an optimization problem under various objectives is more challenging than single optimization objective problems (Xhafa & Ip, 2021).

Problem Solution in Output: For the solution of the objective problem, all the solving methods need to get the value of the variable when the objective is optimal in the output. In Table 6.3, we list the variables to be calculated. However, due to the high computational complexity of the problem we studied, the decomposition method can only get the approximate optimal solution, and the accurate decomposition method can only get the optimal solution under the limited condition of small-scale cases.

Multi-fitness Types and Their Combination: Based on the above discussion, four fitness functions can be defined for this kind of problem, which are: (1) maximize the time window of communication between satellites and ground stations; (2) minimize the conflict between different satellites and the same ground station; (3) maximize the communication time between satellites and ground stations; (4) maximize the utilization of ground stations. In the case of multiple fitness problems, we should consider how to calculate the total fitness function according to the specific fitness function. At the same time, the evaluation order of fitness function also needs to be considered.

In optimization theory, there are usually two models, the hierarchical optimization model, and the simultaneous optimization model. At the same time, the optimization model considers that all the objectives are

Table 6.3 Variables values computed in output (Xhafa & Ip, 2021).

Variable	Meaning
$\boldsymbol{T_{Start}[i][g]}$	Starting time of the communication between SC[*i*] and GS[g]
$\boldsymbol{T_{Dur}}[i][g]$	Duration time of the communication between SC[i] and GS[g]
SC_GS[*i*]	The list of GS assigned to SC[i].
$\boldsymbol{Fit_{LessClash}}$	The fitness function to minimize the collision of two or more SCs to the same GS for a given period (range 0−100).
$\boldsymbol{Fit_{TimeWin}}$	Fitness function to maximize time window access for every pair spacecraft to ground stations GS − SC (range 0−100).
$\boldsymbol{Fit_{Req}}$	Fitness function measuring the satisfaction of the communication time requirement (range 0−100).
$\boldsymbol{Fit_{GSU}}$	Fitness function to maximize the usage of all GS for a planning (range 0−100)

equally related, so all the fitness functions are optimized at the same time (Xhafa & Ip, 2021). The optimization of *n* different fitness functions can be reduced to the optimization of a single fitness function. But there are often fitness functions in different ranges, or there are parameters that cannot be summed together. To solve this problem, the Pareto-front model is usually used for simultaneous optimization. For the ground station scheduling problem, we define four fitness functions and use the method of simultaneous optimization to sum them into one fitness function. Next, we describe and define four fitness functions and their combinations.

Access Window Fitness Function. In satellite schedule, the spacecraft is not always available, only during the time window. During this period, the GS can establish a communication link with the SC to ensure the communication between the GSS as much as possible.

To formally define the corresponding fitness function, let $W(g,i)$ denote the set of access windows for a GS g and SC i. Furthermore, let $T_{\mathrm{Start}}(s)$ and $T_{\mathrm{End}}(s)$ be respectively the starting and ending time of each access window. Then, we express $W(g,i)$ in Eq. (6.1) (Xhafa & Ip, 2021).

$$AWg, i = \cup_{s=1}^{S}[T_{\mathrm{AOS}(g,i)}(s), T_{\mathrm{LOS}(g,i)}(s)] \qquad (6.1)$$

By using the expression of $W(g,i)$ in Eq. (6.1), we calculate the total access window fitness of the scheduling (or mission planning) solution, denoted by Fit_{AW}, as given in Eq. (6.3), where n corresponds to an event number, N is the total number of events of an entire schedule, g is a GS and i an SC (see Fig. 6.2A).

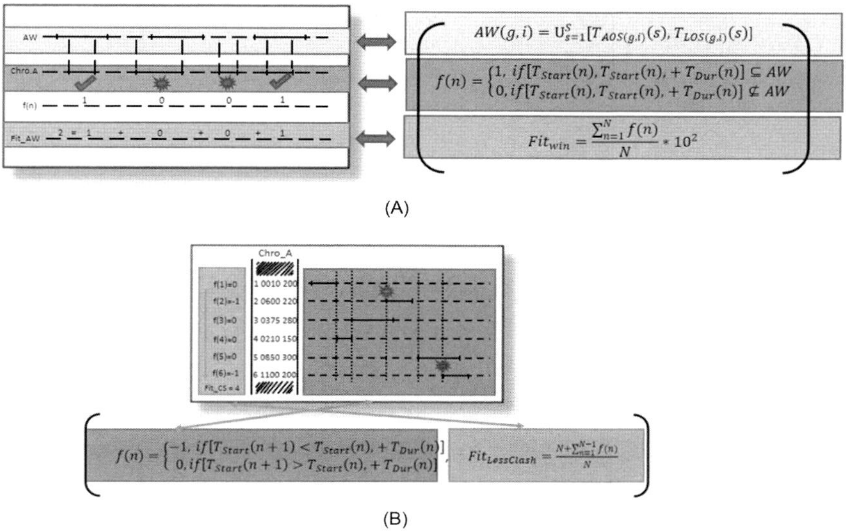

Figure 6.2 Access window fitness and communication clash fitness function. (A) Access window fitness function. (B) Communication clashes fitness (Xhafa & Ip, 2021).

It should be noted that we normalize the fitness value of the access window in the range of 0–100, which is conducive to the subsequent summary of different fitness functions (Xhafa & Ip, 2021).

$$f_{AW}(n) = \begin{cases} 1, & if[T_{\text{Start}}(n), T_{\text{Start}}(n) + T_{\text{Dur}}(n) \subseteq AW(n_g, n_i) \\ & 0, \text{otherwise} \end{cases} \tag{6.2}$$

$$Fit_{AW} = \frac{\sum_{n=1}^{N} f_{AW}(n)}{N} \cdot 100, \tag{6.3}$$

Communication Clash Fitness Function. Each event or task in a scheduling activity has a time to start communication and a time to end communication. When the request to assign a task to a ground station arrives, a conflict may occur, that is, the start time of one communication task of the same ground station is before the end time of another communication task. Such a scheduling scheme with conflicts is not desirable. Therefore, the purpose of scheduling is to minimize the number of communication conflicts between different satellites to a specific ground station. In the final scheduling solution, conflicts should be eliminated and the number of conflicts should be minimized, so as to maximize the number of communications between each satellite and the ground station (Xhafa & Ip, 2021).

To formally express the number of clashes, first, we sort SCs according to their starting communication time. A clash is produced when:

$$T_{\text{Start}}(n+1) < T_{\text{Start}}(n) + T_{\text{Dur}}(n), 1 \leq n \leq N-1 \tag{6.4}$$

where n represents an event number and N is the total number of events in the schedule. As discussed earlier, when a clash happens, to bring the scheduling to a feasible solution, the communication fitness has to be reduced by one (see Eq. (6.5)), and one of the entries that provoked the clash is removed from the solution (a graphical representation is shown in Fig. 6.1B). We compute, therefore, the total fitness of communication clashes, as given in Eq. (6.6) (Xhafa & Ip, 2021).

$$f_{SC}(n) = \begin{cases} -1, \text{if } T_{\text{Start}}(n+1) < T_{\text{Start}}(n) + T_{\text{Dur}}(n) \\ 0, \text{otherwise} \end{cases} \tag{6.5}$$

$$\text{Fit}_{CS} = \frac{N + \sum_{n=1}^{N} f_{SC}(n)}{N} \cdot 100 \tag{6.6}$$

Communication Time Requirement Fitness Function: When the task is assigned to the ground station, it needs to meet the short time requirements of telemetry, tracking, and command. Especially when the satellite downloads large image data, the data download task is very important, it needs to spend more time to link and communicate with the ground station. Therefore, we can define a fitness function, here called FIT_{TR} (Eq. 6.7) (Xhafa & Ip, 2021), aiming to maximize the communication time of SCs with GS under the requirement that each SC(i) will communicate at least $T_{\text{req}}(i)$ time with a GS. This fitness function is formally expressed by summing up all the communication link durations of each SC, and dividing them in the required period to check if the allocated times according to the schedule satisfy communication time requirements (for a graphical representation, see Fig. 6.3A) (Xhafa & Ip, 2021).

Ground Station Usage Fitness Function: The number of ground stations is far less than the number of mission planning requests. Therefore, the goal we want to achieve is to maximize the utilization of the ground station, to support mission planning requests as much as possible. Busy time is defined as the use time of the ground station, that is, the time when the ground station communicates with a satellite. The goal can also be described as minimizing the idle time of the ground station when it does

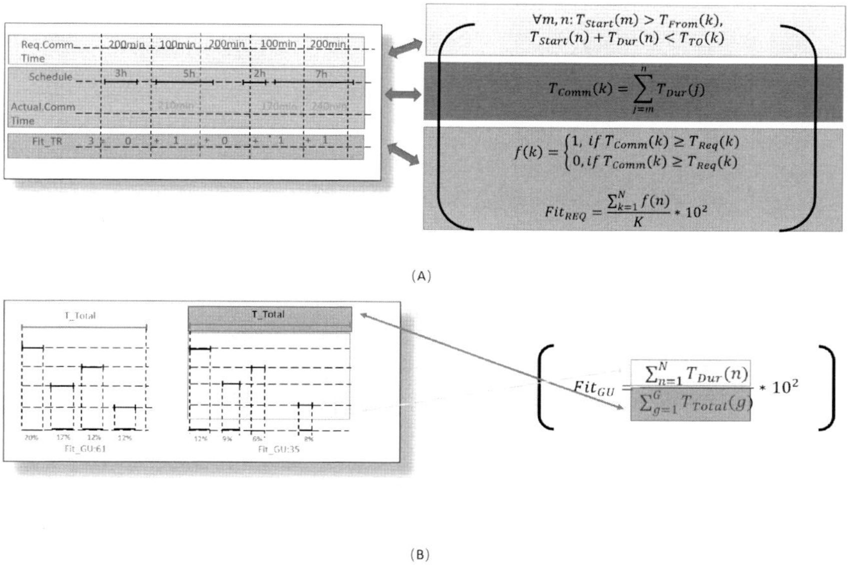

Figure 6.3 Communication time requirements fitness and ground station usage fitness function. (A) Communication time requirements fitness. (B) Ground station usage fitness (Xhafa & Ip, 2021).

not communicate with any satellite (for a graphical representation, see Fig. 6.3B) (Xhafa & Ip, 2021).

$$
\begin{gathered}
T_{Start}(m) > T_{From}(k) \\
T_{Start}(n) + T_{Dur}(n) < T_{TO}(k) \\
T_{Comm}(k) = T_{Dur}(j) \\
f_{TR}(k) = \begin{cases} 1, \textit{if } T_{Comm}(k) \geq T_{REQ}(k), \\ \quad 0 \textit{ otherwise.} \end{cases}
\end{gathered}
\tag{6.7}
$$

$$
\text{Fit}_{TR} = \frac{\sum_{k=1}^{K} f_{TR}(k)}{N} \cdot 100
$$

In this paper, the fitness function for the use of ground stations is defined as the ratio of the peak time of ground stations to the total amount of possible communication time between ground stations. It can be seen that the larger the appropriate function value of the ground station usage value is, the better the scheduling scheme is, and the larger the communication time between the satellite and the ground station is realized.

$$\text{Fit}_{GU} = \frac{\sum_{n=1}^{N} T_{Dur}(n)}{\sum_{g=1}^{NG} T_{Total}(g)} \cdot 100 \tag{6.8}$$

where n is the task or event number, N represents the number of tasks scheduled, NG is the number of ground stations and $TTotal(g)$ is the total available time of a GS (Xhafa & Ip, 2021).

Combination of Fitness Objectives: We have defined four adaptive functions: Fit_{AW}, Fit_{CS}, Fit_{TR}, Fit_{GU}. They include different requirements for efficiency and schedule optimization. The fitness function can be used in the design phase of the scheduler to form more fitness functions (Xhafa & Ip, 2021). The fitness module is the grouping of fitness functions and is divided into two types: serial module and parallel module. Serial fitness (serial-FM) module adopts serial mode based on interdependence. The parallel fitness (parallel-FM) module uses parallel computing.

All available fitness functions in fitness modules are finally combined into a single fitness function by giving weights to different fitness modules (serial or parallel) (Xhafa & Ip, 2021):

$$\text{Fit} = \sum_{i=1}^{n} w_i \cdot \text{Fit}_s(i) + \sum_{j=1}^{m} w_j \cdot \text{Fit}_{p(j)} \tag{6.9}$$

where n and m are the number of fitness modules, resp., w_i, w_j denote the weights assigned to fitness functions, $\text{Fit}_S(i)$ from Serial-FMs, and $\text{Fit}_P\ (j)$ from Parallel-FMs. For the ground station scheduling, and assuming that only the main four fitness functions (Fit_{AW}, Fit_{CS}, Fit_{TR}, Fit_{GU}) will be used, we combine them as shown in Eq. (6.10) (Xhafa & Ip, 2021).

$$\text{Fit}_{\text{TOT}} = \lambda \cdot \text{Fit}_{\text{Win}} + \text{Fit}_{\text{Req}} + \frac{\text{Fit}_{\text{LessClash}}}{10} + \frac{\text{Fit}_{\text{GSU}}}{100} \tag{6.10}$$

for some $\lambda > 0$ parameter. Clearly, the coefficients used in Eq. (6.10) aim to give priority to the certain fitness function, for instance for some $\lambda > 1$ value[1] windows fitness would be given more priority in the schedule than other fitness functions (Xhafa et al., 2013a; Xhafa et al., 2013b).

6.2.10 Low-earth-orbit satellite scheduling

Earth observation satellite scheduling deals with the case of low orbit satellites to collect information about the Earth. Maverick entrepreneur Elon Musk's

[1] We set $\lambda = 1.5$ for the experimental study reported in the references.

interstellar link satellite, launched for the first time, is an example of using the so-called LEO satellite, which represents low Earth orbit.

LEO satellites usually orbit the Earth at a much lower altitude than traditional satellites, about 500–2000 km above the Earth's surface. Traditional satellites usually call their homes 36,000 km objects. The reason for the skimming nature of the LEO satellite is that the delay is lower, which means that the delay of data transmission is much smaller, so it has an advantage in terms of Internet connection speed. The other example is the London-based global communications company One Web: another LEO ambitious organization, which measured the delay of the LEO satellite, with only 32 milliseconds of data to make a round trip, compared with nearly 600 milliseconds. This makes the connection speed of the LEO satellite equal to that of a landplane. In China, the Red Cloud project plans to build an LEO satellite constellation with hundreds of people by 2022 to provide the night sky with the potential of high-speed long-distance Internet connection. Musk said it was estimated that a moderate Internet connection would require about 800 satellites, while Leo Internet connection would require many satellites because their low orbit orbits the Earth at an extremely fast speed in two hours.

Each satellite can only provide a short-term connection at a certain point on the Earth's surface, so many satellites need to be produced to maintain a stable connection. Recently, investment in Leo technology has soared, and governments around the world, including the Chinese government, are keen to invest. Advances in the technology needed to build small satellites and reusable rockets have reduced the cost of this venture, thereby increasing participation. In China, the government attaches great importance to LEO satellites to provide services for Internet connection in remote areas. In the next decade, the LEO satellite may become a common sight in our night sky. This technology has been around since the late 1990s, but now the scale and interest in the field are unprecedented. We can see that various countries all over the world hope to combine this rapidly developing technology field with the benefits of high-speed remote Internet connection for everyone via satellite connection.

6.2.11 Computational complexity of satellites scheduling

Previously we mentioned various types of satellite scheduling formulas are reported in the literature, such as ground station scheduling, satellite range scheduling, AFSCN scheduling, LEO satellite scheduling, etc. (Barbulescu, Watson, Whitley, & Howe, 2004; Pemberton & Galiber 2000; Zufferey, Amstutz, & Giaccari, 2008). Other scheduling models related to the use of satellite technology in air information are also reported (Waluyo, Rahayu, Taniar, & Srinivasan, 2011). The most critical point of the satellite scheduling problem is whether the problem can be solved effectively and the best utilization of the ground station can be achieved, to support the maximum number of task planning requests. However, for other time window scheduling problems, satellite scheduling problem is actually NP-hard. That is, the best solution cannot be found. The satellite scheduling problem is highly constrained, so it is not easy to solve them, and it is not easy to find a feasible solution (that is, to allocate all task requests).

6.2.12 Satellite deployment systems

In the design and development of satellite technology, satellite deployment systems (simulation systems) play an important role in evaluating the performance of optimal scheduling solutions (Xhafa & Ip, 2021). For example, in 2005, the Hong Kong Polytechnic University developed a microsatellite platform and a deployment system to facilitate low-cost space experiments. CubeSat, the first batch of microsatellites launched by China, is also the first microsatellite platform and deployment system successfully developed in Hong Kong. Its weight is only about 2 kg, which greatly reduces the cost of developing and producing microsatellites carrying small payloads and instruments into space (Xhafa & Ip, 2021). For the aviation field, the platform provides a convenient and efficient solution for the investigation of flight incidents. We also propose a strength benchmark generated by a satellite tool kit (STK), which is effective for evaluating and judging the solution of the problem.

6.3 Spacecraft optimization problems

Satellite deployment system (simulation system) plays an important role in evaluating the performance of optimal scheduling solutions. The platform provides a convenient and efficient solution for the investigation of flight incidents. In the satellite research background, many scholars have also done research in this area. Some research works of interest in these contexts are by Weiss and Yung (2009), Qian et al. (2017). Ko and Yung (2006) proposed the functional deployment model (FDM) and linear programming optimization method to solve the design problems of the multifunctional sampling instrument used by ESA.Weiss and Yung (2009) pointed out that it is necessary to design a landing strategy support system. Through three different system architectures, rover, landing station, and installed probe, matching landing point, the landing strategy of the lunar exploration robot is designed based on the review of the target. Qian et al. (2017) proposed and evaluated an algorithm to optimize the space station layout in their research. It can be used in the layout design of the space station.

6.4 Computational complexity resolution methods

Given the rapid development of satellite technology, the number of satellite mission planning and operation has increased greatly, and an intelligent scheduling system is needed to automatically and optimally deal with this kind of mission planning, that is, to assign tasks to space missions through ground station services. The core of satellite task planning is the satellite scheduling problem, whose solution is to optimize the allocation of user requests, to carry out effective communication between ground and spacecraft system operation teams. This kind of problem belongs to NP-hard problem, it is difficult to achieve the optimal solution and the computational complexity is high; the complexity results suggest the use of heuristics and *meta*heuristics to cope with the problem's complexity for efficiently computing high-quality solutions (near-optimal/optimal solutions) for practical applications (Xhafa et al., 2013a; Xhafa et al., 2013b).

Search methods have a complete system, and there are many kinds of classification. In Fig. 6.4, the search method can be divided into random search and enumeration search based on the search type. The main difference between random search and enumeration search lies in the search efficiency and the optimality of the search scheme. The random search method cannot guarantee the optimal solution as does the enumeration method.

In the classification results of Fig. 6.4, the importance of local search method and population-based method are emphasized. The local search method is from simple searches such as mountain climbing and simulated annealing (SA) to more complex search such as Tabu search and variable neighborhood search. The decentralized search and reconnection path as well as hybridization produce a higher-level search method (Xhafa et al., 2013a; Xhafa et al., 2013b). The key lies in neighborhood search, that is, jumping

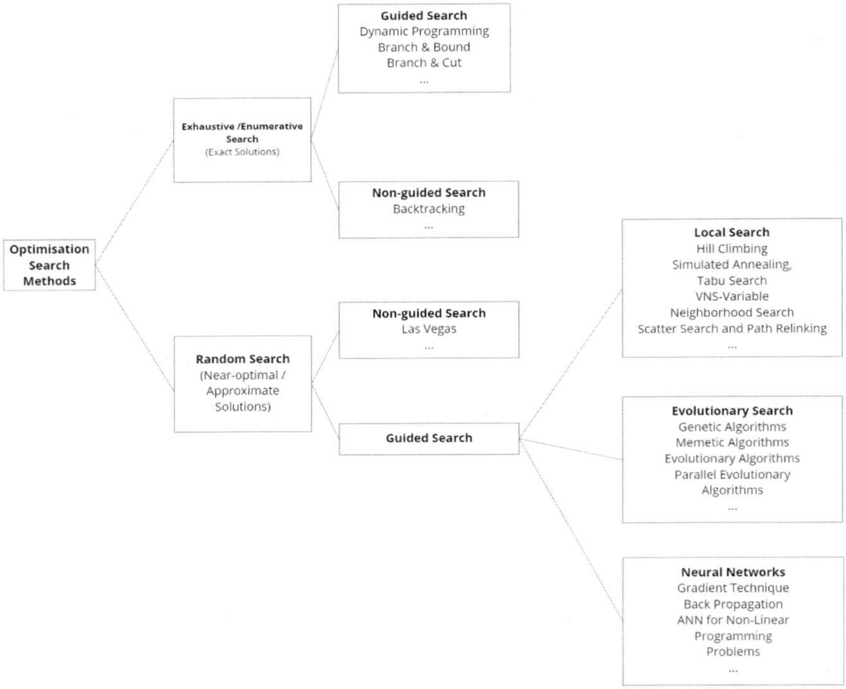

Figure 6.4 Classification of search methods (Xhafa & Ip, 2021).

from one feasible solution to another until any stop criterion is satisfied. But there may be premature convergence to the local optimal solution. Tabu search or variable neighborhood search can overcome the premature convergence and find the global optimal solution (Xhafa & Ip, 2021).

The other branch of search methods comprises Genetic Algorithms (GAs), Memetic Algorithms (MAs), and Evolutionary Algorithms (EAs), in general. Different from local search methods that deal with one solution at an iteration, the EAs use a set of feasible solutions at the same time at any iteration, known as a population of solutions or a generation. For this reason, these search methods are referred to as population-based methods. In contrast to the exploitation strategy of local search, for population-based methods the search strategy is an "exploration" process—here a larger space comprising a set of solutions is explored. The premature convergence is an issue also for EAs, namely, it becomes challenging to know how many generations of populations are needed to converge to the optimal solution(s) (Xhafa & Ip, 2021).

Finally, when solving the objective optimization problem, we should first consider the most suitable search method, such as local search, evolutionary search, or the organic combination of search methods. There is no theoretical basis to show that a certain kind of question is more suitable for a certain search method, but we can get corresponding suggestions from a large number of empirical literature.

6.4.1 Local search methods

The local search method is efficient and simple. Through local search, feasible and high-quality solutions can be found in a short time. This kind of algorithm is suitable for solving large-scale problems, that is, when the number of task planning requests is large, to solve the scheduling problem between satellites and ground stations.

Hill-Climbing algorithm is one of the simplest forms of local search and is a simple greedy search algorithm, which selects an optimal solution from the adjacent solution space of the current solution as the current solution each time until it reaches a locally optimal solution. The Hill Climbing algorithm is very simple to implement, its main disadvantage is that it will fall into the local optimal solution, and not necessarily be able to search the global optimal solution. Assuming that point C is the current solution, the Hill-Climbing algorithm will stop searching for the local optimal solution of point a, because no matter which direction the point a moves to, it cannot get a better solution. SA algorithm is also a greedy algorithm, but its search process introduces random factors, including the cooling mechanism, to avoid premature convergence. When iterating to update the feasible solution, it will accept a worse solution than the current solution with a certain probability, so it may jump out of the local optimal solution and reach the global optimal solution. Tabu search is a more complex but effective local search method. Tabu search algorithm is a kind of metaheuristic random search algorithm. It starts from an initial feasible solution, selects a series of specific search directions (moves) as a trial, and selects the moves that make the value of a specific objective function change the most. To avoid falling into the local optimal solution, a flexible "memory" technology is used in TS search, which records and selects the optimization process that has been carried out, and guides the next search direction, which is the establishment of the Tabu table.

6.4.1.1 Hill climbing

A standard template of Hill Climbing can be found in null Algorithm 6.1.

ALGORITHM 6.1 Hill Climbing (maximizing fitness function f) (Xhafa & Ip, 2021).

Input: Problem instance

Output: Best found solution and its fitness value

Initial solution: Compute initial solution s_0;

s: $= s_0$; $s^*:= s_0; f^*:= f(s_0)$

repeat

Move to Next Neighbor solution: Compute a move m = move(s);

Evaluate and apply move: #δ function computes fitness value variation

if δ(s,m) ≥ 0 then

s': = apply(m,s);

s: $= s'$;

end if

Update current best solution:

if $f(s') > f(s^*)$ then

$f^*:= f(s')$;

$s^*:= s'$

end if

Return s^*, f^*;

until (stopping condition is met)

end

Initial Solution: The initial solution is the starting point of the solution path constructed by the search path process. The initial solution can be generated either by random calculation or in a specific way (Xhafa et al., 2013a; Xhafa et al., 2013b).

Fitness Function Evaluation: In fact, the search process is to calculate the value of fitness function. When a new adjacent solution with a higher fitness value is calculated, the new adjacent solution is the current solution (Xhafa et al., 2013a; Xhafa et al., 2013b).

Neighboring Solution Selection and Move Types: A neighborhood of a feasible solution s, denoted $N(s)$, is the set of solutions that are reachable from s by a local move. Clearly, $N(s)$ depends on the move type and there are as many neighborhoods as move types can be defined. One can see a move as a local small perturbation of solution s (Xhafa et al., 2013a; Xhafa et al., 2013b)

Solution Acceptability Criterion: When the current solution moves to the next solution, we usually use a criterion called the solution acceptability criterion. The criterion consists of three parts: (1) simple ascent, when the fitness value is larger than the current fitness value, it can move to the newly found solution. (2) the steepest ascent, if the fitness value is larger than the current fitness value, can move to the newly found solution among all the evaluated neighboring solutions. (3) randomly moving to the newly calculated neighboring solution (Xhafa et al., 2013a; Xhafa et al., 2013b).

6.4.1.2 Simulated annealing

The earliest idea of SA was put forward by Metropolis in 1953. In 1983, Kirkpatrick et al. successfully introduced the idea of annealing into combinatorial optimization. It is a stochastic optimization algorithm based on Monte Carlo iterative strategy. Its starting point is based on the similarity between the annealing process of solid matter in physics and general combinatorial optimization problems. The SA algorithm starts from a higher initial temperature, with the continuous decline of temperature parameters, combines with the probability jump characteristics to randomly find the global optimal solution of the objective function in the solution space, that is, the local optimal solution can jump out of the probability and finally tend to the global optimal solution. SA algorithm is a general optimization algorithm. Theoretically, it has the global optimization performance of probability. It has been widely used in engineering, such as VLSI, production scheduling, control engineering, machine learning, neural network, signal processing, and so on.

SA generalizes Hill Climbing with the aim to overcome the premature convergence to local optima. The SA algorithm comes from the principle of solid annealing. The solid is heated to a high enough temperature and then cooled slowly. When heated, the particles inside the solid become disordered with the temperature rise, and the internal energy increases. When cooled slowly, the particles become orderly, reaching equilibrium at each temperature, and finally reaching the ground state at room temperature, and the internal energy decreases to the minimum. According to the Metropolis criterion, the probability of a particle approaching equilibrium at temperature T is $e\ (-\Delta E/(KT))$, where E is the internal energy at temperature T, ΔE is its change, and K is the Boltzmann constant. The internal energy E is simulated as the objective function value f, and the temperature T is evolved into the control parameter t, that is, the SA algorithm for solving the combinatorial optimization problem is obtained: starting from the initial solution I and the initial value t of the control parameter, the iteration of "generate a new solution → calculate the difference of the objective function → accept or discard" is repeated for the current solution, and the T value is gradually attenuated, and the current solution at the end of the algorithm is obtained This is a heuristic random search process based on Monte Carlo iterative method. The annealing process is controlled by the cooling schedule, including the initial value t of the control parameter and its attenuation factor ΔT, the number of iterations L for each t value, and the stop condition s. It should be noted, therefore, that tuning the cooling procedure directly affects the convergence of the SA algorithm. A standard template of SA can be found in Algorithm 6.2.

Initial Solution, Fitness Evaluation, and Move Types: As SA is in essence an extended HC, with the difference of being more flexible at accepting a new solution, SA computes similarly an initial solution, uses δ function for fitness evaluation, and generates new moves as explained in Subsect. 4.1.1 for HC (Xhafa & Ip, 2021).

ALGORITHM 6.2 Simulated annealing (Xhafa & Ip, 2021)

Input: Problem instance
Output: Best found solution
t: = 0
Initialize T; #Initial temperature value
s_0: = Initial solution()
v_0: = Evaluate(s_0) Initial solution
while (not stopping condition) do
  while t mod MarkovChainLen = 0 do
    t: = t + 1;
    s_1: = Generate(s_0, T) Compute a new move
    v_1: = Evaluate(s_1) Evaluate fitness corresponding to new move
    if Accept(v_0, v_1,T) then
      s_0: = s_1;
      v_0: = v_1;
    end if
  end while
  T: = Update(T) #Update temperature according to cooling schedule
until (stopping condition is met)
end

Acceptability Criterion: The acceptability criterion uses an accepting threshold value. Let t_k be a succession such that $t_k > t_{k+1}$, $t_k > 0$, and $t_k \to 0$ as k tends to ∞. Intuitively, we can use t_k values to select among two solutions s_i and s_j, for instance, if *fitness value*(s_j) *fitness value*(s_i) $< t_k$, then solution s_j is accepted (Xhafa & Ip, 2021).

6.4.1.3 Tabu search method

The purpose of Tabu search is to realize adaptive search. In order to achieve better performance, we use some specific mechanisms: more flexible choice of the first solution; avoid wandering between the visited solutions; restart the search in other parts of the solution space to prevent falling into local optimum. A standard template of the TS method can be seen in Algorithm 6.3.

Neighborhood Exploration: For a given solution s, neighbor solutions in $N(s)$ are computed as indicated (Xhafa & Ip, 2021).

Short- and Long-Term Memory: To avoid entering the visited solution during the loop, TS implements two kinds of memory, namely short-

---

ALGORITHM 6.3 Tabu search (Xhafa & Ip, 2021).

```
Input: Problem instance
   Output: Best found solution
   Compute an initial solution s;
   Set current solution ŝ: = s;
   Set tabu and aspiration conditions;
   while (not termination − condition met) do
      Choose the best s′ ∈ N*(s); N*(s) ⊆ N(s) s.t.(tabu conditions are not violated) or (aspiration criteria hold)
      s: = s′
      if improved(s′, ŝ) then
      ŝ: = s′
      end if
        Update recency memory and frequency memory;
        if (intensification condition) then
           intensification_proceducre();
        end if
        if (diversification condition ) then
          diversification_procedure();
        end if
   end while
   return ŝ
```

---

term memory and long-term memory. Short-term memory has also become a Tabu list, a data structure that stores the characteristics of recently accessed solutions. TS uses short-term memory to discard the recently accessed solution. Long-term memory collects statistical information about the solution during the search process. This kind of information provides an opportunity for further exploration of solutions that have not been accessed in the future (Xhafa & Ip, 2021).

Tabu Status: A solution is tagged as *tabu* if it has been explored recently aiming to forbid its selection during the next iterations unless some special conditions hold (Xhafa et al., 2013a; Xhafa et al., 2013b).

Aspiration Criteria: These are some logical conditions that serve to the method to waive or not the *tabu* status of a solution (Xhafa et al., 2013a; Xhafa et al., 2013b).

Intensification and Diversification Conditions: When there is evidence that there may be high-quality solutions in a certain area, priority should be given to search in this neighborhood. The purpose of diversification is to search again in the new area of the solution space, and after many iterations, there is evidence that it can be activated without improvement, which can be regarded as a mechanism of escape or exploration (Xhafa et al., 2013a; Xhafa et al., 2013b)

6.4.1.4 Genetic algorithms

EAs constitute another large family of optimization search method. In this family of population-based methods, GAs are the most paradigmatic (Holland, 1975). Parallel versions of them have been greatly studied in the literature (Algorithm 6.4) (Xhafa & Ip, 2021)

Genetic Encoding—The Chromosome: The first step to implementing a genetic algorithm is to encode a series of individual features. The success of the genetic algorithm depends on the effective use of coding and powerful transmission of genetic information and genetic algebra in the search process. The mechanism of the chromosome will also have a direct impact on the efficiency of the genetic algorithm. In Fig. 6.5, chromosome A uses binary code and chromosome B uses decimal code to mark the start time and duration of communication between each satellite code and satellite (Xhafa & Ip, 2021).

Fitness Function: Fitness function again is the one defined in Eq. (6.10) (Xhafa & Ip, 2021).

Selection Operators: For the case of ground station scheduling the selection operators such as Linear Ranking Selection, Best Selection, and

ALGORITHM 6.4 Genetic algorithm(Xhafa & Ip, 2021)

Input: Problem instance

Output: Best found solution

1: t: = 0;

2: Compute P^0 -initial population of size μ;

3: Evaluate fitness of individuals in P^0;

4: while (not termination − condition met) do

5: Select the parental pool T^t of size λ; T^t: = Select (P^t);

6: Crossover pairs of individuals in T^t with probability p_c;P^t_c: = Cross(T^t);

7: Mutate individuals in P^t_c with probability p_m; P^t_m: = Mutate(P^t_c);

8: Evaluate fitness of individuals in P^t_m

9: Let P^{t+1} be the new population of size μ from individuals in P^t and/or P^t_m;

10: P^{t+1}: = Replace(P^t; P^t_m)

11: end while

12: return Best found solution;

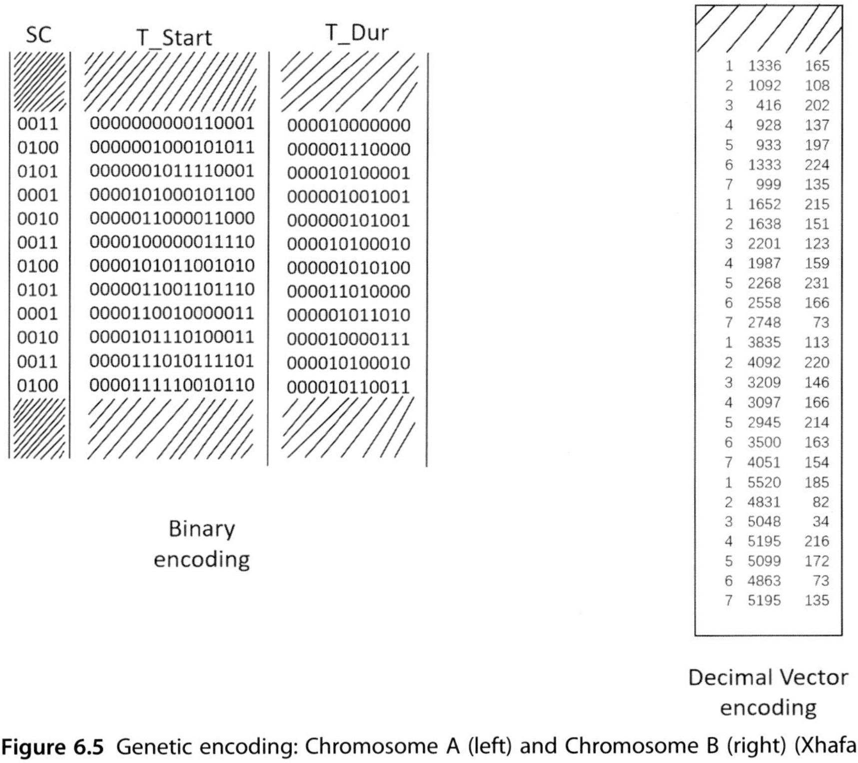

SC
T_Start
T_Dur
0011 0000000000110001 000010000000
0100 0000001000101011 000001110000
0101 0000001011110001 000010100001
0001 0000101000101100 000001001001
0010 0000011000011000 000000101001
0011 0000100000011110 000010100010
0100 0000101011001010 000001010100
0101 0000011001101110 000011010000
0001 0000110010000011 000001011010
0010 0000101110100011 000010000111
0011 0000111010111101 000010100010
0100 0000111110010110 000010110011
Binary encoding
1 1336 165
2 1092 108
3 416 202
4 928 137
5 933 197
6 1333 224
7 999 135
1 1652 215
2 1638 151
3 2201 123
4 1987 159
5 2268 231
6 2558 166
7 2748 73
1 3835 113
2 4092 220
3 3209 146
4 3097 166
5 2945 214
6 3500 163
7 4051 154
1 5520 185
2 4831 82
3 5048 34
4 5195 216
5 5099 172
6 4863 73
7 5195 135
Decimal Vector encoding

Figure 6.5 Genetic encoding: Chromosome A (left) and Chromosome B (right) (Xhafa et al., 2013a).

Tournament Selection were implemented according to the so-called *implicit fitness re-mapping* technique (Xhafa & Ip, 2021).

Crossover Operators: These operators are the most important operators in genetic algorithms, because they are directly related to the efficiency of genetic information transmission. In the genetic algorithm, based on the premise that the newly generated individuals contain the best genetic characteristics of the parents' chromosomes, we input two chromosomes, and the crossover operator generates a new bullet spring, which makes the search develop toward a better individual development direction and get a better problem solving. In addition, the crossover operator should also keep the diversity of the population, to help the algorithm to achieve better convergence.

Mutation Operator: Together with crossover operators, mutation operators aim at the efficient transmission of genetic information along with the generation of the search process. Likewise, mutation operators help in maintaining population diversity and contribute to avoiding premature convergence (Xhafa & Ip, 2021).

6.4.1.5 Two-stage heuristic

With the analysis of the feasibility of resources and the conflict factors of visible time windows, Chen et al. (2018) future developed several approaches based on priority and conflict-avoidance and proposed a method called two-stage heuristic. The first stage of this method is to determine the observing sequence and generate a feasible scheduling scheme. Then they developed and tested a time-based greedy algorithm, a weight-based greedy algorithm, and an improved DE algorithm. The next stage is to do the further optimization, which is based on the obtained feasible solution to prevent the results from being trapped into a local optimum (Chen et al., 2018).

Time-based Greedy Algorithm (Chen et al., 2018): Repeat the process that firstly arranges the tasks having free subintervals and removes them and all the corresponding visible time windows until all the tasks have no free subinterval. When scheduling the remaining tasks, two strategies can be considered: (1) arrange the remaining tasks with the earliest visible time on every resource until we occupy all the feasible time intervals, in case of the earliest feasible time of resources. The above implementation process corresponds to the first strategy of the algorithm. (2) try to arrange all the remaining tasks with the earliest feasible time of every interval until there is no feasible time remaining, in case of the availability of every feasible time interval on the resource. The above implementation process corresponds to the second strategy of the algorithm.

Weight-based Greedy Algorithm (Chen et al., 2018): In the weight-based greedy algorithm, the priority of candidate tasks is decided by the weight, the number of remaining feasible opportunities, and the assignment flexibility. When selecting the observation, we should consider the duration and conflict indicators. The preprocessing and initialization of the weight-based greedy algorithm are the same as that of the time-based greedy algorithm. There are also two strategies in scheduling the remaining tasks: (1) try to arrange all the remaining tasks and choose an observation time window in the visible time window set, in case of the importance of tasks, such as the weight and the flexibility of visible time window. (2) considering the exact contention effect of each candidate's visible time window.

Step 1: Initialization.

Step2: Calculate CS, CP and FI of each time window. Check $\exists FI_{i,j}^{k} \geq D_i$. If yes, then assign mission M_i in a subinterval of $FI_{i,j}^{k}$ and remove all the corresponding $tw_{i,j}^{k}$ *of* M_i and repeat such procedure, until it meets the requirement and turn to Step 3. Else, turn the Step 3.

Step3: Encoding missions and generate initial population.

Step4: Compute the conflict indicator CIW.

Step5: Rearrange the conflict-mission with lower PLW_i and select t_i with higher $\text{TW}S_{i,j}^{k}$.

Step6: Mutation.

Step7: Crossover.

Step8: Calculate the fitness value of each individual base on Constraint Satisfaction.

Step9: Select.

Step10: Generate next population.

Step11: Finished optimization. If yes, then turn to step12. Else, turn back to step 4.

Step 12: Conflict eliminate and feasibility verification.

Step 13: Obtain optimal solution.

Step 14: End.

Improved Differential Evolution Algorithm (Chen et al., 2018): DE algorithm is one of the most competitive EAs to solve real number coding optimization problems. Aiming at the characteristics of large potential solution search space, and considering the diversity and convergence of the population, an improved DE algorithm based on priority and conflict avoidance strategy is proposed. Feasible solutions are coded as individuals, where each gene corresponds to the execution state of the task (selected resource and associated observation time window). In the evolution process, the main parts of obtaining the potential optimal solution in each generation will be described in detail, including initialization, conflict avoidance, mutation operator, crossover operator, fitness function, selection operator, and constraint satisfaction.

Initialization (Chen et al., 2018): Initialize the population into two aspects and each individual represents one feasible scheduling scheme. According to the greedy algorithms, encode the tasks and assign the remaining unfinished tasks in a visible time window of a random resource randomly. On the other hand, encode the tasks which have free subintervals and assign the remaining tasks randomly in a visible time window of *q* random resource.

Constraint Satisfaction (Chen et al., 2018): While encoding individuals, it is important to make sure the individual maintains feasibility subject to all constraints, only except for observation time utilization constraint. At the end of the process generating, we need to take some strategies about conflict and remove and eliminate the tasks which have the highest conflict indicators to others iteratively, until it satisfies all the constraints.

Conflict Avoidance (Chen et al., 2018): In the process of the optimization process, if there appears resource contention conflict between two tasks, it is advised to evaluate the indicators of tasks firstly. Through some special and certain principles, keep the tasks that are qualified and rearrange another observation time for the other tasks.

Mutation Operator (Chen et al., 2018): This operator aims to obtain more possible solutions and expand the diversity of the population. With the passage of generations, the mutation probability changes from a larger value to a lower value. In each generation, multiple genes of an individual may mutate in randomly selected time intervals. For each gene, tasks that conflict with other tasks have a higher mutation probability. Newly generated individuals will also be added to the current population. It is an exploratory strategy, that is, to explore the neighborhood area around the current promising solution, to generate a possibly better solution next to the current solution. Local optimal search capabilities are improved.

Crossover Operator (Chen et al., 2018): This operator means to exchange some genes between two individuals, thereby increasing the diversity of the population. Similar to the mutation operator, the crossover probability changes from large to small with the change of generations. In each generation, multiple genes of an individual will be exchanged in randomly selected time intervals. For each gene, tasks that conflict with other tasks have a higher mutation probability. These two newly generated individuals are also added to the current population. It is an effective mechanism to prevent the solution from converging to the local optimum, thereby improving the global optimum search capability.

Fitness Function (Chen et al., 2018): It is used to evaluate personal profits. In the optimization process, we allocate the most suitable resources and observation time for all tasks. Therefore, the solution may not be feasible under the limitation of observation time. Therefore, we iteratively "remove" tasks that have the highest conflicting impact on other tasks until the individual is feasible. Use the improved feasible solution to calculate the fitness value of each individual. It is the sum of the number/weight of all assigned tasks in the feasible solution. It is actually a forward-looking strategy that reflects the effectiveness of individual missions and the priorities of individual missions.

Selection Operator (Chen et al., 2018): First, calculate the fitness value of each individual, and then maintain the current optimal solution among all candidate individuals. Adopt a tournament selection strategy based on individual fitness.

Repeat the evolution process until the optimal solution between two consecutive generations is not improved. The individual with the largest profit is selected as the scheduling plan, and the feasibility of the optimal solution is verified. The experimental results indicate that the improved DE algorithm demonstrates a significant advantage in all cases, especially when the targets are distributed clustered and the resources are overoccupied (Chen et al., 2018).

To analyze the performance of methods proposed above, Chen et al. (2018) gathered some different classes of classes of test instances according to the resource utilization of a mission conflict, each of which is defined by a set of available resources, requested missions, and the eligibility of a resource for a mission in a given scheduling period (Chen et al., 2018). In summary, both in maximizing the total number of assigned missions and maximizing the total weights of assigned missions, all of the results presented above illustrate the effectiveness of the proposed heuristic strategies (Chen et al., 2018). The improved DE algorithm demonstrates a significant advantage in all cases, especially when the targets are clustered (centralized-distributed) and the resources are overoccupied. The weight-based greedy approach also contributes to a near-optimal solution in an acceptable time. The conflict indicators introduced in weight-based greedy algorithms and improved DE algorithms have a big impact in assigning missions and selecting observation times. They are sophisticated in resolving resource contentions (Chen et al., 2018). It also effectively avoids the results from being trapped into local optimal solutions by taking conflict avoidance strategy and fitness function into account (Chen et al., 2018). Additionally, the conflict indicators analysis of feasible time intervals also can be used in dynamic real-time scheduling problems; the retraction decision of common tasks to incorporate newly-arrived tasks, and the reassignment decision of high conflict mission in the scheduling process. The selection strategies also help to adjust the mission having the highest conflict indicator and to select a suitable resource and a corresponding observing time window when an emergency task comes (Chen et al., 2018).

6.4.1.6 An Improved differential evolution algorithm

In the improved DEA, the constraint satisfaction optimization model of multisatellite joint scheduling planning is established. Based on the standard differential evolution algorithm, the imaging process and working principle of imaging satellites are analyzed (Peng, Bai, & Chen, 2020). The imaging satellite scheduling process is divided into three stages: scheduling preprocessing, task planning, and scheduling optimization (Peng et al., 2020). Among them, the scheduling preprocessing operation is to screen satellite system resources according to user needs, determine the optional resources of each observation task and the time window that can be allocated; in the process of task planning, the imaging satellite scheduling problem is solved by differential evolution algorithm based on a heuristic idea, the individual fitness evaluation function is redefined, and the task conflict resolution method is designed According to the system performance evaluation results, the optimization process mainly optimizes the unfinished tasks due to resource conflicts and the tasks with time delay due to time constraints. The model is developed as follows (Peng et al., 2020):

Problem Description: The scheduling problem of multiple imaging reconnaissance satellites can be described as the scheduling of N observation tasks (activity set M) in M different remote sensing devices (resource set R). For each activity $M_i \in M$, only the subset of resources R (M_i) $\in R$ can meet its execution requirements. The completion of the activity needs to occupy resources $R_j \in R$, and the occupation time is [Begi, Endi]. In addition, activity M_i has a set of disjoint time window set constraints when occupying resource R_j, and it can only be executed in one of the time windows without interruption. If activity M_i and activity M_i' occupy the same resource R_j in the process of execution, and activity M_i is executed before activity M_i', then after the completion of the activity M_i execution, a transition time δ_j must be passed before activity M_i' can be executed. Due to the limitation of resources and time, activities may not be arranged. Therefore, each activity M_i has a weight w_i, which represents the benefit value of the activity arrangement. An optimal scheduling scheme should meet the following conditions: (1) each activity can only be executed in its own time window, otherwise, it is considered that the activity is not scheduled; (2) each activity can only occupy one resource in the resource set that meets its requirements, and the execution process cannot be interrupted; (3) each resource can only meet the needs of one activity at any time; (4) the total weight of scheduling activities is maximum (Peng et al., 2020).

6.4.1.6.1 Symbol definition

See Tables 6.4 and 6.5.

Formal Description of Complex Operational Constraints: See Table 6.6.

The optimization goal is to maximize the total benefit of task execution (Table 6.7).

$$\max \sum_{M_i \in M} \sum_{R_j \in R(M_i)} \sum_{k \in \{1,2,\ldots,N_{i,j}\}} w_i \cdot x_{i,j}^k \tag{6.11}$$

Algorithm Introduction (Peng et al., 2020): This algorithm aims at solving the NP-complete problem of imaging satellite scheduling: to improve the efficiency of solving the problem, we can first classify the problem and then use the corresponding algorithm to solve it. In general, when the scale of the problem is large, it is difficult to get a group of feasible solutions. At this time, it can only be solved by some typical intelligent algorithms. The differential evolution (DE) algorithm is a kind of postheuristic algorithm for optimization problems, which is a greedy genetic algorithm with the idea of optimal preservation based on real number coding (Peng et al., 2020). In the related work, a large number of studies show that DE is a very good method to solve real coding optimization problems. As regards the heuristic method. to improve the superiority of the scheduling solution and the efficiency of the algorithm, this paper adds scheduling preprocessing operation and some heuristic ideas to solve the imaging satellite scheduling problem based on the standard method (Peng et al., 2020).

Table 6.4 Aggregate (Peng et al., 2020).

Aggregate	Meaning
M	Activity collection. $M = \{M_1, M_2. \ldots, M_n\}$, where the ith task is represented by M_i.
R	Resource collection. $R = \{R_1, R_2, \ldots, R_m\}$, where the jth resource is represented by R_j.
$M(\boldsymbol{R_j})$	a collection of all activities that can be scheduled to be executed on resource R_j. $R_j \in R$, $M(R_j) \in M$.
$R(\boldsymbol{M_i})$	A set of resources that meet the MI requirements of observation activities. $M_i \in M$, $R(M_i) \in R$.
$\boldsymbol{TW}_{ij}$	The resource R_j occupied by the activity M_i is the set of allowed time windows. N_{ij} represents the number of visible time windows allowed when the activity M_i occupies the resource R_j. $TW_{ij} = \{tw_{i,j}^1, \quad tw_{i,j}^2, \ldots, \quad tw_{i,j}^{N_{i,j}}\}, tw_{i,j}^k = \left[ws_{i,j}^k, we_{i,j}^k\right], k \in \{1, 2, \ldots, N_{ij}\}$, where $ws_{i,j}^k$ represents the start time of the kth time window allowed when the activity M_i occupies the resource R_j, and $we_{i,j}^k$ represents the end time of the kth time window allowed when the activity M_i occupies the resource R_j.
$\mathrm{RT}\boldsymbol{W_j}$	The set of effective execution intervals of resource R_j indicates the time period in which the resource is available. $\mathrm{RT}W_j = \{\mathrm{rt}w_j^1, \mathrm{rt}w_j^2, \ldots, \mathrm{rt}w_j^{N_j}\}$, $\mathrm{rt}w_j^k = [\mathrm{rws}_j^k, \mathrm{rwe}_j^k], \mathrm{k} \in \{1, 2, \ldots, N_j\}$.

Table 6.5 Parameter (Peng et al., 2020).

Parameter	Meaning
RP_j	Represents the image type of the resource. $RP_j = 1$ for visible imaging, $RP_j = 2$ for infrared imaging, $RP_j = 3$ for multispectral imaging, and $RP_j = 4$ for radar imaging.
Δ_j	Indicates the stable time before the resource R_j imaging
A_j	Represents the maximum execution time of resource R_j in the whole simulation cycle.[[_]]
w_i	The weight of activity M_i. $M_i \in M, w_i > 0$.
D_i	The duration of the activity M_i to complete. $M_i \in M, TD_i > 0$.
TP_i	The type of image required for active M_i imaging. $TP_i = 0$ means that the imaging type is not required, $TP_i = 1$ means visible light imaging, $TP_i = 2$
SC_{Beg}	means infrared imaging, $TP_i = 3$ means multispectral imaging, $TP_i = 4$
SC_{End}	means radar imaging.
Beg_i	Scene start timeScene end time
End_i	The start execution time constraint of activity M_i. Where $M_i \in M$, $Beg_i \geq SC_{Beg}$.The execution end time constraint of activity MI. Where $M_i \in M, End_i \leq SC_{End}$.

Table 6.6 Optimization variables (Peng et al., 2020).

Variable	Meaning
$x^t_{i,k}$	The resource R_j is occupied by the activity M_i during execution, and the resource R_j is occupied by the activity M_i during execution. If the time window used is t, then $x^t_{j,k} = 1$; Otherwise, $x^t_{j,k} = 0$. Where $i \in J$, $K \in M_i$, $t \in \{1, 2, \ldots, N_{i,k}\}$. All undefined $x^t_{j,k}$ are 0.
dst_i	Indicates the start execution time when the activity M_i is scheduled.

Similarly, when the approximate algorithm is used to solve the problem, there are often many unfinished tasks. At this time, the deterministic algorithm can be used again to try to reselect the time window for the unfinished task in the remaining time window, to further achieve the definition of the optimal solution of the problem. In the process of evolution, several important operations involved in the algorithm (scheduling pretreatment, conflict elimination, benefit value calculation, secondary optimization of feasible solution) are as follows (Peng et al., 2020):

Table 6.7 Constraint condition (Peng et al., 2020).

Variable	Meaning
$\sum_{R_j \in R(M_i)} \sum_{k \in \{1,2,\ldots,N_{i,j}\}} x_{i,j}^k \leq 1$	Each task M_i can only be executed once at most, and can only be executed in one-time window, or even if it is executed many times, the benefits can only be calculated once
$\sum_{R_j \in R(M_i)} \sum_{k \in \{1,2,\ldots,N_{i,j}\}} D_i \cdot x_{i,j}^k \leq A_j$	The maximum usage time of resources. The constraint is used to represent the total number of sideway times and start-up working time in a given period of time or orbit period shall not exceed the allowable upper limit
if $\sum_{R_j \in R(M_i)} \sum_{k \in \{1,2,\ldots,N_{i,j}\}} x_{i,j}^k = 1$ then $SC_{Beg} \leq dst_i \leq SC_{End} - D_i Beg_i \leq dst_i \leq End_i - D_i$	Task execution time constraint. Each task needs to be assigned in its planning time interval, which is usually used to represent the week in the emergency task or communication task with periodic coverage or constrained by light.
$\forall M_i \in M, \forall R_j \in R(M_i), k \in \{1, 2, \ldots, N_{i,j}\}, if \quad x_{ij}^k = 1$ then $ws_{ij}^k \leq dst_i \leq we_{ij}^k - D_i$	The observation window assigned to each task needs to meet the availability constraints of imaging resources and the observation duration constraints of tasks. In the scheduling process, if the resource R_j and the corresponding visible time window $tw_{i,j}^k$, j are allocated to the task M_i, then the observation period of the task must fall completely within the allocated visible time window.
$\forall R_j \in R \forall M_i, M_{i'} \in M(R_j), if \quad x_{i,j}^k = 1 \quad and \quad x_{i'j}^k = 1$ then $dst_i \geq dst_{i'} + D_{i'} + \cdot_j \quad or \quad dst_{i'} + D_i + \cdot_j$	The minimum transition time constraint between two consecutive observation tasks on the same resource. The above analysis shows that the satellite is in the middle of the mission.In the process, the yaw angle and rotation angle of the satellite borne sensor change with different start time. Therefore, the minimum conversion time between two adjacent tasks on the same resource should be considered, so that the satellite or the satellite borne sensor can adjust to the correct attitude.

$if \quad x_{i,j}^{k} = 1 then \quad TP_i = 0或RP_j = TP_i$	Imaging resource type selection constraints. When the task M_i selects the resource R_j and the corresponding visible time window $tw_{i,j}^{k}$, it must satisfy that the resource R_j is available in the current situation, and the load image type corresponding to the resource R_j can meet the imaging type requirements of the activity.
$\forall M_i \in M, R_j \in R(M_i), k \in \{1, 2, \ldots, N_{i,j}\}, x_{i,j}^{k}\{0, 1\}$	Optimization variable value constraints.

Scheduling Preprocessing (Peng et al., 2020): Firstly, the satellite system resources and user requirements are analyzed, and the constraints that are only related to the static coverage performance in the process of multisatellite joint ground coverage are given priority, and all the visible time window sets that meet the requirements of satellite to ground observation are calculated. Only when the observation demand has both available resources and available time, it is considered that the observation demand may be completed, and further scheduling is needed to determine whether it is implemented, the satellites that implement the observation demand, and the imaging time period allocated for it (Peng et al., 2020). Secondly, when a task has at least one visible time window and is likely to be executed in the scheduling process, the conflict degree of resource contention in the visible time window of the task is analyzed, and the priority of the task with idle resources is considered, to reduce the search space of the problem and improve the efficiency of the algorithm (Peng et al., 2020). For a hypothetical scheduling scenario, the constraint set of visible time windows between satellite and target can be generated after scheduling preprocessing (Peng et al., 2020), as shown in.

In , $tw_{i,j}^k$ represents the kth visible time window of task M_i on resource R_j, where $k \in \{1,2\ldots, N_{i,j}\}$ represents the total number of visible time windows of task M_i for all resources in the whole simulation cycle. If task M_i has no visible time window on resource R_j or resource R_j does not meet the imaging requirements of task M_i, $N_{i,j} = 0$, $TW_{i,j}^k = \varnothing$。 The time window is the visible time window between satellite and ground. For a specific task execution sequence, the execution time allocated to the task is only a small part of the selected interval which meets the imaging time constraint. On this basis, through the selection of observation resources and execution time window, the execution expression of each task in the scheduling planning result can be obtained: M_i, R_{j_i}, otw_{i,j_i}, where otw_{i,j_i} represents the execution time interval of task M_i on resource R_{j_i} (Peng et al., 2020).

Each locus in a population is encoded as a task. Each individual in the population corresponds to a feasible scheduling solution, including the task attributes, whether the task is scheduled, the scheduled observation resources, and the observation time window. The initialization of individuals is mainly divided into two aspects: (1) the feasible solution is obtained by a greedy algorithm based on task execution time priority or task execution revenue priority, and some tasks in the individuals are coded, while the remaining tasks in the scheduling scheme are randomly selected in their visible time window set for coding; (2) Firstly, the tasks with idle time interval are coded, and then the remaining tasks are randomly selected for coding in all visible time windows (Peng et al., 2020).

Conflict resolution: When multiple tasks in an individual compete for resources to produce resource conflicts, how to eliminate resource conflicts is usually to reselect the time window randomly by using the conflict gene of one individual. However, in the case of large resource conflicts, the reselected time window will often cause more resource conflicts with other tasks, It is also very important to select a new resource and execution time window for the locus of resource conflict. Using the heuristic idea, when the resources and execution time allocated to two imaging tasks in an individual does not meet the constraints, it is necessary to eliminate one task and reallocate the resources and execution time for it. To maximize the benefit value of the optimized sequence, we can select the eliminated loci in the following ways: (1) The conflict benefit value of the conflict loci; (2) The length of the visible time window and the conflict degree of the time window of the conflict loci were compared. For tasks M_1, M_i, M_j and M_k executed on the same resource, the execution time windows of tasks M_i, M_j and M_k conflict with each other. Among them, the conflict degree of task M_i and M_k is 1, while the conflict degree of task M_j is 2, and the visible time window length of the task is the largest, so the time window is reselected for the task M_j. When the time window is reselected, the first time window of the task M_j may cause conflict between task M_i and M_k, so the execution time window reallocated for task M_j is the third visible time window (Peng et al., 2020).

Benefit Value Calculation: In general, the calculation of benefit value is directly related to the number of tasks completed or individual conflict degree, that is, the calculation of fitness function value has nothing to do with individual benefit value, but is only related to the number of tasks completed and conflict degree, that is, the benefit values of two individuals with the same number of conflicts are equal, and the benefit values calculated in the following cases are also equal. Assuming that tasks M_i, M_j and M_k have the same benefit value, it is obvious that, for task execution state 1 shown in Fig. 6.6 (Peng et al., 2020), only task M_j needs to be eliminated in the scheduling scheme to complete two tasks M_i and M_k, while for task execution state 2, shown in Fig. 6.7 (Peng et al., 2020), no matter how conflicts are eliminated, only one task can be completed in the scheduling scheme. For the above situation, we can take away of

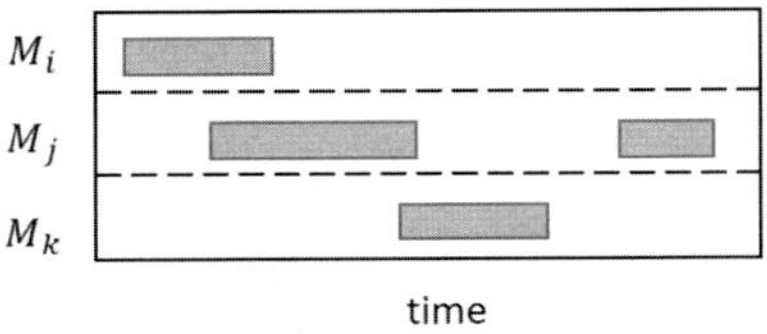

Figure 6.6 Mission execution conflict of case 1 (Peng et al., 2020).

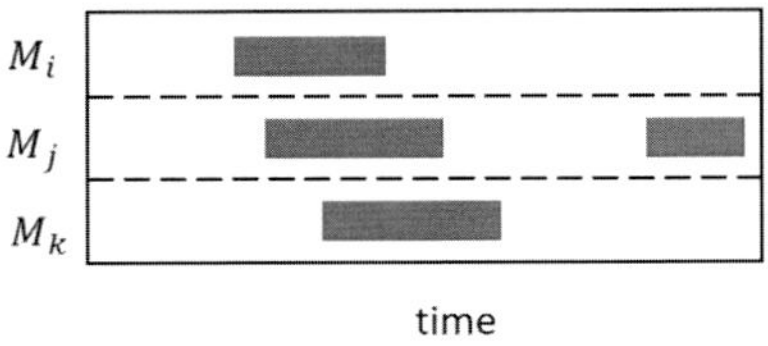

Figure 6.7 Mission execution conflict of case 2 (Peng et al., 2020).

individual task conflict degree to eliminate the conflict. The individual fitness evaluation function is Peng et al. (2020).

$$f = \frac{\sum_{M_i \in M} \sum_{R_j \in R(M_i)} \sum_{k \in \{1,2,\ldots,N_{i,j}\}} w_i \cdot x_{i,j}^k}{1 + \sum_{M_i \in M} \sum_{R_j \in R(M_i)} \sum_{k \in \{1,2,\ldots,N_{i,j}\}} V_i \cdot w_i \cdot x_{i,j}^k} \tag{6.12}$$

where, the benefit values of tasks M_i, M_j and M_k are w_i, w_j and w_k respectively, so the conflict degrees of tasks M_i, M_j and M_k are V_i, $V_j$and V_k respectively, $V_i = w_j + w_k$, $V_j = w_i + w_k$ and $V_k = w_i + w_j$, which are sorted in descending order according to the task conflict degree and the tasks are deleted in turn until there is no task conflict in the individual. At this time, the total benefit of all tasks in the individual assigned time window is calculated. In this way, the tasks with larger benefit value or weight can be retained to a greater extent, and the individuals with larger benefit value can be selected more effectively to generate a better scheduling scheme (Peng et al., 2020).

Quadratic Optimization of Feasible Solution (Peng et al., 2020): When the approximate algorithm is used to solve the satellite constellation scheduling problem, for the scheduling scheme generated by each solution, the deterministic algorithm is used to optimize the scheduling result. According to the scheduling results of a given satellite system, the scheduling scheme is evaluated from three aspects: task completion rate, resource utilization rate, and timeliness. For the current scheduling results, if there is a task that cannot be completed due to resource conflicts or delays in the time allocated to the tasks allocated, the scheduling optimization operation can be optimized in three aspects of the mission, completion rate, resource utilization, and timeliness, to further solve and optimize the scheduling results, resulting in better scheduling results And can meet the needs of users. (1) It optimizes the time of all the tasks with assigned time windows in the scheduling result. Find out the tasks that have been assigned time windows, and move forward the assigned time windows of all tasks according to the preprocessing results, and try to choose a better execution time window for the tasks with time delay. (2) Research the time window for the unfinished tasks due to resource conflict in the scheduling scheme. According to the previous operation, the scheduling results after time optimization are analyzed to find out all the tasks that cannot be scheduled due to resource conflict, traverse all the visible time windows, and try to reselect the execution time window for them (Peng et al., 2020).

By analyzing the imaging process and working principle of imaging satellites Peng et al. (2020) established a constraint satisfaction optimal scheduling model for multiple imaging satellites. On this basis, an improved De solution for joint mission planning of imaging satellites is designed by using a heuristic algorithm. Through the test case, it is found that, in the iterative process of the improved De, the idea of a heuristic algorithm is used to eliminate the conflict of tasks in individuals and calculate the fitness value of individuals, which can quickly generate a better scheduling scheme. The implementation of scheduling optimization operation optimizes each scheduling result, which can further improve the task completion rate, resource utilization rate, and timeliness of the scheduling scheme. At the same time, the realization of scheduling optimization operation can also be realized, to apply to the emergency scheduling, to generate a better scheduling scheme (Peng et al., 2020).

6.4.1.7 Multisatellite task prescheduling algorithm based on conflict imaging probability

This algorithm is a variety of rule-based heuristic algorithms, which is suitable for distributed satellite task planning systems. Its goal is to consider the energy constraints in the prescheduling stage, resolve the conflict between tasks, complete the optimization goal, and reduce the computing time by dispersing the computing load from a single node to multiple nodes (Xu & Wang, 2020). The prescheduling algorithm is easy to implement and has strong generality. It only needs the prior information of satellites and tasks. The algorithm of the subsequent single satellite autonomous scheduling phase can be selected freely according to the actual situation, and there is no need to write a feedback interface (Xu & Wang, 2020).

Multisatellite Imaging Task Scheduling Model Input and Data Preprocessing: The multisatellite scheduling problem can be expressed as five tuples $\langle S, T, C, F, O\rangle$, where the satellite set S can be expressed as $\{S_k\}_{k=1}^{N_s}$, and N_s is the number of satellites; Task set t can be expressed as $\{T_k\}_{k=1}^{N_T}$, $N_T$is a number quantity task; C is the constraint set; F is the optimization objective function; O is the scheduling result. The input parameters and meanings of the multisatellite scheduling problem are shown in Table 6.8 (Xu & Wang, 2020).

According to the characteristics of the image satellite, the following assumptions are made (Xu & Wang, 2020).

1. After the user submits the task, it cannot be withdrawn or changed;
2. Users cannot submit tasks during task planning;
3. A mission may have several satellites to perform it, but only one of them can perform it in the end;

Table 6.8 Main parameters of satellites and missions.

Input	Parameter type	Parameter	Parameter description
Satellite S	Identifier	i	Satellite identifier, $i \in \{1, 2, \ldots, N_s\}$
		in_i	Inclination of satellite i
		aop_i	The angular distance of satellite i
	Track parameters	sa_i	Semi-major axis of satellite i
		$raan_i$	Right ascension of satellite i
		b_i	Battery capacity of satellite i
	Load capacity	m_i	Storage energy of satellite i
		oc_i	Maximum switching times of satellite i
Task T	Identifier	j	Task identifier, $j \in \{1, 2, \ldots, N_T\}$
		ρ_j	Priority of task j
	Attribute	$(long_j, lat_j)$	Longitude and latitude of observation target point of task j
	Time requirement	(a_i, d_i)	Observation time range, the earliest start time and the latest end date of task j
		md_i	Minimum duration of task execution

4. In one operation cycle, the satellite targets a mission. There can only be one observation opportunity at most;
5. Satellite observation is atomic, once the observation task starts, it cannot be interrupted;
6. At a time point, the satellite can only observe one mission target point.

Based on the above assumptions, this paper studies the multisatellite problem. The main parameters of satellite and mission are shown in Table 6.8.

After getting the input, the system preprocesses and gets the result through simulation.

Obtain the time range of the target point visible to the satellite and then calculate the overlap part between the time range and the specified observation time range of the task, which is called the available time window, expressed as (Xu & Wang, 2020).

$$TW = \{tw_{ij} | 1 \leq i \leq N_{SC}, 1 \leq j \leq N_T\} \tag{6.13}$$

Among them, (tw_ $start_{ij}$, tw_end_{ij}) is the period when satellite i can complete task j; mpc_{ij} is the minimum power consumption of satellite i to complete task j; mmc_{ij} is the minimum memory consumption for satellite i to complete task j.

Output and optimization objective of multisatellite imaging task scheduling model. After prescheduling, the allocation matrix O is obtained (Xu & Wang, 2020).

$$O = \left\{o_{ij}|, 1 \leq i \leq N_s + 1, 1 \leq j \leq N_T\right\} \tag{6.14}$$

$$o_{ij} = \begin{cases} 1, \text{Task } \; j \text{ is assigned to satellite } \; i \\ 0, \text{else} \end{cases}$$

In essence, multisatellite observation scheduling is a scheduling machine problem with constraints, which greatly restricts the scheduling process. In this paper, we use the overordering method, that is, under a given degree of relaxation, the number of tasks allocated to the satellite can exceed the satellite's execution capacity; σ_i represents the overallocation rate of the system (Xu & Wang, 2020).

$$\forall k \in [1, \ldots, N_T], \sum_{i=1}^{N_S+1} o_k \leq 1 \tag{6.15}$$

$$\forall k\in[1,\ldots,\quad N_T],\quad \sum_{i=1}^{N_T} o_{kj}\leq o_k\cdot oc_i \tag{6.16}$$

$$\forall k\in[1,\ldots,\quad N_T],\quad \sum_{i=1}^{N_T} o_{kj}\cdot mpc_{kj}\leq \sigma_k\cdot b_i \tag{6.17}$$

$$\forall k\in[1,\ldots,N_T],\sum_{i=1}^{N_T} o_{kj}\cdot mmc_{kj}\leq \sigma_k\cdot m_i \tag{6.18}$$

S_{N_s+1} is a virtual satellite, and all tasks that cannot be scheduled by satellite resources will be assigned to the satellite; Eq. (6.15) is the single execution constraint of a task, that is, each task can only be scheduled once; Eqs. (6.16)–(6.18) are the load capacity constraints of satellites, that is, the number of tasks assigned to each satellite, the total power consumption, and storage capacity required cannot exceed the load capacity of satellites under a given degree of relaxation (Xu & Wang, 2020).

Prescheduling is the first stage of multisatellite scheduling, so its optimization goal is to optimize the scheduling. When the single satellite scheduling phase is completed, the allocation matrix o will be updated, mainly because some tasks cannot be successfully scheduled in the single satellite scheduling phase and can only be assigned to the virtual satellite. The ultimate optimization goal is to maximize the total priority of tasks that can be successfully scheduled after the single satellite autonomous scheduling algorithm is executed, which can be expressed as (Xu & Wang, 2020).

$$\max\{\sum_{i=1}^{N_T}\rho_i\cdot(1-o_{i(N_S+1)})\} \tag{6.19}$$

S.t. (6.15)–(6.18)

Algorithm Description: The algorithm is proposed to deal with the multisatellite prescheduling problem. The basic idea of the algorithm is to calculate the probability of successful execution of the task on each satellite according to the current allocation situation and assign the task to the satellite with the greatest imaging possibility. It is proposed that the factor affecting the imaging probability is the coincidence degree of the available time window of the current task and that of other tasks on the same satellite (Xu & Wang, 2020). However, it is one-sided to consider this factor only, because, in the multisatellite scenario, multiple tasks with conflicting time windows are assigned to different satellite resources, and the mutual influence does not necessarily exist. As shown in Fig. 6.1, to make full use of the heuristic information of multisatellite scheduling problems, the tasks are divided into conflict tasks and nonconflict tasks. The conflict imaging probability of the conflict task is composed of potential conflict coefficient, actual conflict coefficient, and energy coefficient, while the nonconflict task only needs to consider the energy constraint, that is, the energy coefficient, to get the preallocation result. The first step of the prescheduling phase is to divide tasks into conflict tasks and nonconflict tasks according to the coincidence degree of available time windows between tasks, to simplify the solution of conflict imaging probability. $mcit_{ij}$ is the longest continuous collision free time of available time window of task j on satellite i *(Xu & Wang, 2020)*. if

$$\exists i \in \{1, 2, \ldots, N_s\}, \quad mcit_{ij} > md_j \tag{6.20}$$

In other words, on satellite i, if the longest continuously available noncollision duration of task j exceeds the minimum execution duration of the task, it is said that task j has no conflict on satellite i. If a task has no conflict on any satellite, it is a nonconflict task; otherwise, it is a conflict task. The probability of collision imaging consists of p_{ij}, r_{ij} and e_{ij} in which, δ is infinitesimal as follows (Xu & Wang, 2020).

$$c_{ij} = \frac{1}{p_{ij} + \delta} + \frac{1}{r_{ij} + \delta} + \frac{1}{e_{ij}} \tag{6.21}$$

Among them, p_{ij} is the potential conflict coefficient, which reflects the potential impact between tasks, that is, tasks that have not been assigned to satellite i have available time windows on satellite i, and may be assigned to satellite i in subsequent scheduling; the calculation method of the potential conflict coefficient of task j is to find out all the unallocated tasks with available time windows on satellite i and calculate $(tw\_start_{ij}, tw\_end_{ij})$ (Xu & Wang, 2020).

$$p_{ij} = \sum_{k \in (T - r) \cap \wedge^*} \frac{tw\_overlap_{ijk} \cdot \rho_k}{d_k} \tag{6.22}$$

In Eq. (6.22), r_{ij} is the actual conflict coefficient, which reflects the actual impact between tasks. That is, when calculating the conflict imaging probability when task j is assigned to satellite i, some tasks have been assigned to satellite i. If these tasks conflict with the execution of task j, this conflict must be minimized at this time, Each satellite maintains its own directed acyclic graph to calculate the actual collision coefficient. The graph of satellite i is G_i, and its longest weighted path is $G_i\_longest - path$.When a task j is added, the task node j is inserted into the graph to form the graph G_i', Find the longest weighted path $G_i'\_longest - path$ of node j in the graph. The total weight of all assigned tasks that cannot be executed by the task is the actual conflict coefficient R_{ij} of task j on satellite i, which can be expressed as Xu and Wang (2020).

$$r_{ij} = G_i\_longest - path + p_i - G_i'\_longest - path \tag{6.23}$$

To balance the system load and meet the energy constraints, the energy consumption of the total tasks assigned by the satellite should be considered, so the energy system is defined. The number is calculated as follows (Xu & Wang, 2020):

$$e_{ij} = \frac{B_i}{\sum_{k=1}^{N_T} mpc_{ik}} + \frac{M_i}{\sum_{k=1}^{N_T} mmc_{ik}} \tag{6.24}$$

Algorithm collision probability-based schedule (CIPBS) (Xu & Wang, 2020)

Generate scene(S,T)
Initialization parameters
Pretreatment, calculation tw,mmp,mpc,mcit
for $j \leftarrow 1$ to N_t do
for $i \leftarrow 1$ to N_{SC} do
if $MCIT_{ij} > MD_j$
$\Phi \leftarrow \Phi \cup \{T_j\}$
$\Gamma_i \leftarrow \Gamma_i \cup \{SC_l\}$
end if
end for
end for
while $\wedge^* \neq \varnothing$ do
Randomly selected tasks $T_j \in \wedge^*$
for i $\leftarrow 1$ to N_{SC} do
calculation Tw_overlap$_{ijk}$ of each task in $\wedge^*$
$p_{ij} \leftarrow \sum_{k \in (T-\Gamma) \cap \wedge^*} \frac{TW\_overlap_{ijk} \cdot \rho_k}{md_j}$
Constructing directed acyclic graph G of tasks assigned on satellite S_l
Insert task T_j into G to form G′

(Continued)

(Continued)

19: $r_{ij} \leftarrow G_i\_longestpath + p_j - G_i'\_longestpath$

20: $e_{ij} \leftarrow \frac{B_i}{\sum_{k=1}^{N_T} o_{ik} \cdot mpc_{ik}} + \frac{M_i}{\sum_{k=1}^{N_T} o_{ik} \cdot mmc_{ik}}$

21: $c_{ij} \leftarrow \frac{1}{p_{ij} + \delta} + \frac{1}{r_{ij} + \delta} + \frac{1}{e_{ij}}$

22: end for

23: If c_{ij} is the maximum of $\{c_{ik} \mid 1 < K < N_S\}$ and satisfies the constraints (3) $\sim$ (6)

24: $o_{ij} \leftarrow 1$

25: $G \leftarrow G'$

26: end if

27: $\wedge^* \leftarrow \wedge^* - \{T_j\}$

28: end while

29: Arrange Λ tasks in $\frac{p}{md}$ descending order

30: while $\Lambda \neq \varnothing$ do

31: Select task $T_j \in \Lambda$

32: for $i \leftarrow 1$ to N_{SC} do

33: if $S_i \in \Gamma_j$

34: $e_{ij} \leftarrow \frac{B_i}{\sum_{k=1}^{N_T} o_{ik} \cdot mpc_{ik}} + \frac{M_i}{\sum_{k=1}^{N_T} o_{ik} \cdot mmc_{ik}}$

35: else $e_{ij} \leftarrow \infty$

36: end if

37: end for

38: If e_{ij} is the maximum value in $\{e_{ik} \mid 1 < K < N_S\}$ and satisfies the constraints (3) $\sim$ (6)

39: $o_{ij} \leftarrow 1$

40: end if

41: $\wedge \leftarrow \wedge - \{T_j\}$

42: end while

43: return O

Among them, Φ represents a collection of all conflict-free tasks; Γ_i is the set of all collision-free tasks on satellite and S_i; $\wedge^*$ represents a collection of all conflict tasks that have not been assigned; $\wedge$ represents a collection of all unassigned conflict-free tasks (Xu & Wang, 2020).

Algorithm Analysis: The time complexity of the CIBPS algorithm is mainly composed of three parts (1) the complexity of dividing conflict and nonconflict tasks are $O\ (K \cdot N^2)$. Among them, the time complexity of calculating the longest continuous available collision-free time of a task on a satellite is O (N), The time complexity of calculating the longest continuous available collision-free time of n tasks on K satellites is O ($K \cdot N^2$); (2) The complexity of conflict task scheduling is O ($K \cdot N^3$). For a single task to calculate the imaging probability on a satellite resource, the complexity of calculating the potential conflict coefficient is O (N). Dijkstra algorithm is used to construct a directed acyclic graph and calculate its shortest single source path. The complexity of the actual conflict coefficient is O (N^2), The complexity of calculating the energy coefficient is O (N), so the complexity is O (N) + O (N^2) + O (N) = O (N^2). For the conflict task scheduling problem with task size n and satellite size k, the computational time complexity is O ($K \cdot N^3$); (3) For conflict-free tasks, only the energy coefficient needs to be calculated, and the complexity is O ($K \cdot N^2$). Therefore, the total time complexity of the CIBPS algorithm is O ($K \cdot N^2$) + O ($K \cdot N^3$) + O ($K \cdot N^2$) = O ($K \cdot N^3$) (Xu & Wang, 2020).

At the beginning of scheduling, the p_{ij} is larger and the r_{ij} is smaller. At this time, the calculation of c_{ij} mainly depends on the potential conflicts between tasks. With the progress of task allocation, after each task is allocated to satellite resources, the potential conflicts between the task and other tasks will be eliminated in the subsequent scheduling calculation, and, at the same time, the potential conflicts between the task and other tasks will be eliminated, The task will be added to the directed acyclic resource map of the satellite resource. With the increase of the assigned task information of the satellite resource, p_{ij} decreases and r_{ij} increases. At this time, the calculation of c_{ij} mainly depends on the actual conflict between tasks, It can reduce the contention of multiple tasks for the same satellite resource at the same time, to achieve the goal of balanced and reasonable allocation of resources (Xu & Wang, 2020).

According to this algorithm (Xu, Liu, He, & Chen, 2020), they carried out simulation experiments. The experimental results show that CIBPS can find a better task scheduling solution in almost every case. CIBPS algorithm is obviously better than the SA algorithm in performance. CIBPS algorithm has obvious advantages in effectiveness and stability

6.5 Future trend of algorithms and models and solutions of satellite scheduling problem

Algorithm for Single Satellite Scheduling Optimization Problem: Many scholars have initially studied the single satellite scheduling optimization problem and designed efficient solving models and algorithms. Wu, Du, Fan,

Wang, and Wang (2020) proposed an adaptive SA algorithm based on dynamic task partition, and introduced the Tabu list to solve the scheduling problem of single satellite multiorbit Earth observation. Some work is related to discretizing the task's visible time window and building a 0/1 linear programming model and constructing the task clustering graph model by clustering the point targets in the remote sensing satellite imaging task planning, and designing an improved monorail optimal clustering method. This series of work has made remarkable progress in the generation of optimal feasible solutions and the description of problem bounds for single satellite scheduling planning.

Algorithm for Multisatellite Joint Scheduling Problem: With the complexity of the space mission and the improvement of the requirements, the in-orbit operation is becoming more and more important.

With the increasing number and types of satellites, the demand of users is increasing, and it shows complex and diverse characteristics. The satellite network system composed of multiple satellites has been widely used in the fields of communication, navigation, and remote sensing. Compared with the single satellite scheduling problem, it is necessary to consider observation resource selection, execution time window selection and more execution constraints when scheduling multiple cooperative satellites under complex task constraints. Considering that the traditional task planning algorithm cannot adapt to the application characteristics of multisatellite resources, some scholars divide the multisatellite joint scheduling problem into the main problem of the task to resource allocation and the subproblem of multiple single satellite scheduling. Renjie He et al. deeply analyzed the imaging satellite task planning technology, regarded it as a parallel machine scheduling problem with time window constraints and used the Tabu search algorithm to solve the large-scale task collaborative planning problem. Salman et al. (2015) Proposed a meta heuristic random diffusion search algorithm based on hybrid differential evolution, and studied the scheduling problem of multisatellite joint broadcasting. Xu et al. (2020) proposed an ant colony algorithm based on the task priority index and studied the scheduling problem of multiple agile satellites to Earth observation with maximum revenue.

Satellite Observation Scheduling With a Novel Adaptive SA Algorithm and a Dynamic Task Clustering Strategy: Effective scheduling is very important for the efficient use of scarce satellite resources. Task clustering has been proved to be an effective strategy to improve the efficiency of satellite schedule. However, the current task clustering strategies are static, that is, they are integrated into the scheduling in a two-stage way rather than in a dynamic way, and they do not show their full potential in improving satellite scheduling performance. In this study, we propose a scheduling algorithm based on adaptive SA, which is integrated with the dynamic task clustering strategy (asa-dtc) to solve the satellite observation scheduling problem (sosp). First, Wu, Wang, Pedrycz, Li, and Wang (2017) developed a formal model for Earth observation satellite schedule. Secondly, they analyze the related constraints involved in the process of observation task clustering. Thirdly, they introduce the implementation of dynamic task clustering strategy and adaptive SA algorithm in detail. Adaptive SA algorithm is effective and contains complex mechanisms, such as adaptive temperature control, Tabu list-based short-term avoidance mechanism, and intelligent combination of neighborhood structure. We show adaptive temperature control, Tabu list-based short-term return visit avoidance mechanism, and an intelligent combination of neighborhood structure. Finally, we report an experimental simulation study to demonstrate the competitive performance of asa-dtc. We show that asa-dtc is particularly effective when the sosp contains a large number of targets or the targets are densely distributed in a certain area.

Hierarchical Iterative Algorithm for Multisatellite Observation Scheduling: Jianyin Liu et al. (2018) proposed a multisatellite observation hierarchical scheduling framework based on the divide and conquer strategy. In this framework, an ant colony optimization algorithm was used to allocate tasks to each orbit circle, and an adaptive SA algorithm was used to solve the scheduling problem of each orbit circle. According to the feedback of the scheduling results of each track cycle, the task allocation scheme is adjusted, and the above process is repeated until the termination condition of the algorithm is reached. To improve the performance of the algorithm, the domain knowledge of satellite scheduling problem should be fully considered when designing the heuristic information model of ant colony algorithm; two neighborhood structures are designed in SA algorithm, and the optimal neighborhood search structure is determined in the optimization process by using dynamic selection strategy. Simulation results show that this method can effectively reduce the complexity of problem solving, especially in solving large-scale multisatellite observation scheduling problems.

Unified Modeling and Multistrategy Collaborative Solution Method for Satellite Task Scheduling: To break the barrier of "one satellite, one system" for satellite task scheduling, the implementation of integrated management and control of the measurement, operation, and control tasks, Yonghao Du et al. (2019)

systematically combed the key events, resources, and constraints of satellite task scheduling problem, and designed a unified modeling and multistrategy collaborative solution method for satellite task scheduling. Satellite operation and control task scheduling and measurement and control task scheduling were integrated into a unified modeling method that supports the incremental system development and flexible multisatellite management and breaks the barrier of "one satellite, one system" for satellite task scheduling.

In addition, a loosely coupled and modular system architecture of satellite task scheduling algorithm strategy and scheduling model is built. The proposed unified modeling method provides a unified interface between the scheduling model and algorithm strategy, the unified modeling and multistrategy solving method proposed by them can solve the practical difficulties of satellite task scheduling at the present stage, and give full play to the satellite task scheduling (Du et al., 2019).

It is of great practical significance to improve the payload capacity, utilize the satellite ground control resources and realize the satellite mission

Multisatellite Scheduling Framework and Algorithm for Large Area Observation: Xu et al. (2020) proposed a multisatellite scheduling problem. The problem is based on the satellite capacity and customer demand, under the condition of given specific constraints for a very large area of observation. It is assumed that the profit is proportional to the coverage of the acquisition area, so the goal is to maximize the total profit of the generated observation plan. To solve the satellite scheduling problem, we first demonstrate the detailed problem description and then transform it into a problem covering multiple standards and constraints. On this basis, the mathematical model is established. To solve the multisatellite scheduling problem of large-area observation, a new solution framework is proposed. The framework consists of three phases. In the discretization stage, the regional discretization method is used to establish the evaluation system. In the target decomposition stage, the regional target is decomposed into strips, and the corresponding visible time window is calculated. In the scheduling stage, through crossover, mutation, and feasibility operators, a genetic algorithm is introduced to generate the optimal observation scheduling. The effectiveness and reliability of the proposed solution framework are verified by a large number of computational experiments on the actual problems of Chinese satellite platforms.

An Improved Differential Evolution Algorithm for Multiimaging-Satellite Scheduling: Photoreconnaissance satellite scheduling problem is a kind of complex scheduling problem with mixed time and resource allocation, and it has been proved to be a kind of nondeterministic polynomial complete, as a key supporting technology in the field of space information network construction and satellite application control, imaging reconnaissance satellite scheduling problem has become an important research content in practical application, and has been highly valued by academic and engineering circles (Peng et al., 2020).

6.6 Benchmarking and simulation platforms

The heuristic search method cannot provide a reliable basis for the optimal solution, so it is very important to evaluate its performance. Performance evaluation provides evaluation results for the combination of search methods and problems. In this case, the development of benchmark and simulation tools enables researchers and developers to carry out performance evaluation and experimental research under different parameter settings. Based on the requirement of instance benchmark, the simulator STK toolbox can be used to generate instance benchmark, which is composed of instances of different sizes. Examples are expressed in XML, which is easy to read from various programming environments. The purpose of benchmarking is to capture the real characteristics of the problem, and the benchmark can show the effectiveness of the search method for experimental evaluation (Xhafa & Ip, 2021).

Satellite Tool Kit. STK is an analytical tool developed at analytical graphics, Inc. as a general-purpose simulator for complex and integrated analysis of land, sea, air and space. STK provides an analysis engine for calculating data and can display various forms of two-dimensional maps, display satellites, and other objects such as launch vehicles, missiles, aircraft, ground vehicles, targets, etc. STK's core capabilities are to generate position and attitude data, acquisition time, and remote sensor coverage analysis (Xhafa & Ip, 2021). For specific analysis tasks, STK provides additional analysis modules that can solve communication analysis, radar analysis, coverage analysis, orbit manoevring, precise orbit determination, and real-time operation. In addition, STK also has a 3D visualization module, which provides a leading 3D display environment for STK and other additional modules. We have used STK to generate a set of static problem instances for the ground station scheduling problem, which is then used for validating the various resolution methods as well as their performance. While using STK, we have selected real satellites/spacecraft and ground stations as part of instances information (Xhafa & Ip, 2021).

6.7 Conclusions and future work

Based on the rapid development of space technology and the increasing popularity of satellites, combined with people's urgent demand for scheduling systems, this paper summarizes the satellite scheduling problem in the communication task planning between spacecraft and ground station, as well as the related formulas and models. The satellite scheduling problem is a complete-time window scheduling problem, which belongs to NP-hard problem. With the rapid development of satellite technology, people also focus on the production of low-cost small satellites to promote scientific research or educational projects of various scientific research institutions and organizations. But the number of ground stations is small, and the request for satellite task planning is increasing rapidly, which leads to the increasingly serious satellite scheduling problem. We propose various satellite scheduling problems and discuss their complexity. This kind of problem is highly complex and highly constrained. We propose a centralized solution method and summarize the algorithms, models, research strategies, and solutions used for different types of satellite scheduling problems in recent years. At the same time, we emphasize the importance of solving the optimization problems of spacecraft design, operation and satellite deployment systems. It provides further development direction for optimization, simulation, design, and satellite deployment system development integrated tools and platforms.

Acknowledgments

All authors have been contributed equally to this work. Corresponding author: Ming Gao (gm@dufe.edu.cn). This work was supported in part by the National Natural Science Foundation of China under grant nos. 71772033, 71831003, in part by the Scientific and Research Funds of the Education Department of Liaoning Province of China under grant no. LN2019Q14, in part by the Natural Science Foundation of Liaoning Province of China (joint open fund for key scientific and technological innovation bases) under grant no. 2020-KF-11-11, and in part by the Department of Industrial and Systems Engineering of the Hong Kong Polytechnic University under grant no. H-ZG3K and 4-45-35-690E (Change'e Phase 3 - Sample Return Instruments).

References

Abedinniaa, H., Glocka, C. H., Grossea, E. H., & Schneider, M. (2017). Machine scheduling problems in production: A tertiary study. *Computers & Industrial Engineering*, *111*, 403–416.

Aksin, O. Z., Armony, M., & Mehrotra, V. (2007). The modern call-center: A multidisciplinary perspective on operations management research. *Production and Operations Management*, *16*(6), 665–688.

Barbulescu, L., Howe, A. E., Watson, J. -P., & Whitley, D. (2002). Satellite range scheduling: A comparison of genetic, heuristic and local search. In *Parallel problem solving from nature – PPSN*, VII, pp. 611–620.

Barbulescu, L., Watson, J.-P., Whitley, D., & Howe, A. E. (2004). Scheduling space-ground communications for the air force satellite control network. *Journal of Scheduling*, 7(1), 7–34.

Bartak, R., Salido, M. A., & Rossi, F. (2010). Constraint satisfaction techniques in planning and scheduling. *Journal of Intelligent Manufacturing*, *21*, 5–15.

Brunner, J. O., & Edenharter, G. (2011). Long term staff scheduling of physicians with different experience levels in hospitals using column generation. *Health Care Management Science*, *14*(2), 189–202.

Burke, E., & Petrovic, S. (Eds.), (2004). *European journal of operational research, special issue on timetabling and rostering*. Amsterdam: Elsevier.

Castaing, J. (2014). Scheduling downloads for multi-satellite, multi-ground station missions. In *28th Annual AIAA/USU conference on small satellites*.

Chen, X., Reinelt, G., Dai, G., & Wang, M. (2018). Priority-based and conflict-avoidance heuristics for multisatellite scheduling. *Applied Soft Computing*, *69*, 177–191.

Chen, X., et al. (2018). Priority-based and conflict-avoidance heuristics for multisatellite scheduling. *Applied Soft Computing*, *69*. Available from https://doi.org/10.1016/j.asoc.2018.04.021.

Damiani, S., Dreihahn, J. N., Nizette, M., & Calzolari, G. P. (2007). A planning and scheduling system to allocate esa ground station network services. In *The international con ference on automated planning and scheduling*. United States.

Day, P., & Ryan, D. (1997). Flight attendant rostering for short-haulairline operations. *Operations Research*, *45*(5), 649–661.

De Bruecker, P., Van den Bergh, J., Belien, J., & Demeulemeester, E. (2015). Workforce planning incorporating skills: State of the art. *European Journal of Operational Research*, *243*(1), 1–16.

Du, Yonghao, Xing, Lining, Chen, Yingguo, et al. (2019). Unified modeling and multi-strategy collaborative optimization for satellite scheduling. *Control and Decision*, 1847–1856.

Ernst, A. T., Jiang, H., Krishnamoorthy, M., & Sier, D. (2004). Staff scheduling and rostering: A review of applications, methods and models. *European Journal of Operational Research*, *153*(1), 3–27.

Fathollahi-Fard, A. M., Hajiaghaei-Keshteli, M., & Tavakkoli-Moghaddam, R. (2018). A bi-objective green home health care routing problem. *Journal of Cleaner Production*, *200*, 423–443.

Graves, S. C. (1981). A review of production scheduling. *Operations Research*, *29*, 646–675.

Gupta, J. N. D., Koulamas, C., & Kyparisis, G. J. (2006). Performance guarantees for flowshop heuristics to minimize makespan. *European Journal of Operational Research*, *169*, 865–872.

Halsey, K., Long, D., & Foz, M. (2004). CRIKEY—A temporal plannerlooking at the integration of planning and scheduling. In: *Proceedings on the ICAPS 2004 workshop on integrating planning and scheduling* (pp. 46–52). Whistler, Canada.

Hassani, Z. I. M., Barkany, A. E., Jabri, A., & Abbassi, I. E. (2018). Models for solving integrated planning and scheduling problem: Computational comparison. *International Journal of Engineering Research in Africa*, *34*, 161–170.

Hindman, B., Konwinski, A., Zaharia, M., Ghodsi, A., Joseph, A.D., Katz, R., ...& Stoica, I. (2011). Mesos: A platform for fine-grained resource sharing in the data center. In: *Proceedings of USENIX conference on networked systems design and implementation* (pp. 295–308).

Holland, J. H. (1975). *Adaptation in natural and artificial systems*. MIT Press Karapetyan.

Jian, T., Shicong, M., Xiaoqiao, M., & Li, Z. (2013). Improving reduce task data locality for sequential MapReduce jobs. In: *Proceedings of IEEE INFOCOM* (pp. 1627–1635).

Jorn, V. D. B., Jeroen, B., Philippe, D. B., Erik, D., & Liesje, D. B. (2013). Personnel scheduling: A literature review. *European Journal of Operational Research, 226*(3), 367–385.

Karapetyan., Daniel., Mitrovic, Minic, Snezana., Malladi., et al. (2015). Satellite downlink scheduling problem: A case study. *Omega*, 115–123.

Kis, T., & Pesch, E. (2005). A review of exact solution methods for the nonpreemptive multiprocessor flowshop problem. *European Journal of Operational Research, 164*, 592–608.

Ko, Sui-man, & Yung, Kai-Leung (2006). Function Deployment Model for Continuous and Discontinuous Innovation Product Development. *International Journal of Innovation and Technology Management, 3*, 107–128.

Koeleman, P. M., Bhulai, S., & van Meersbergen, M. (2012). Optimal patient and personnel scheduling policies for care-at-home service facilities. *European Journal of Operational Research, 219*(3), 557–563.

Kurazumi, S., Tsumura, T., Saito, S., & Matsuo, H. (2012). Dynamic processing slots scheduling for I/O intensive jobs of Hadoop MapReduce. In *Proceedings of international conference on networking and computing* (pp. 288–292).

Lee, K., Lee, Y., Choi, H., Chung, Y., & Moon, B. (2012). Parallel data processing with MapReduce: A survey. *SIGMOD Rec, 40*(4), 11–20.

Leung, J. Y. T. (2004). *Handbook of scheduling: Algorithms, models,and performance analysis*. Boca Raton, FL: Chapman & Hall.

Levner, E., Kats, V., Alcaide Lopez de Pablo, D., & Cheng, T. C. E. (2010). Complexity of cyclic scheduling problems: A state-of-the-art survey. *Computers & Industrial Engineering, 59*, 352–361.

Lin, C. C., & Wang, P. C. (2013). A greedy genetic algorithm for the TDMA broadcast scheduling problem. *IEICE Transactions on Information and Systems, E96-D*(1), 102–110.

Liu, J., Xueqing, J., & Zhongwei, W. (2018). Hierarchical iteration algorithm for multisatellite observation scheduling. *Journal of National University of Defense Technology*, 183–190.

Liu, N., Yang, X., Sun, X. H., Jenkins, J., & Ross, R. (2015). YARNsim: simulating Hadoop YARN. In *Proceedings of IEEE/ACM international symposium on cluster, cloud and grid computing* (pp. 637–646).

Luo, K., Wang, H., Li, Y., & Li, Q. (2017). High-performance technique for satellite range scheduling. *Computers & Operations Research, 85*, 12–21.

Maenhout, B., & Vanhoucke, M. (2013). An integrated nurse staffing and scheduling analysis for longer-term nursing staff allocation problems. *Omega, 41*(2), 485–499.

McGann, C., Py, F., Rajan, K., Ryan, J., & Henthorn, R. (2008). Adaptive control for autonomous underwater vehicles. In *Proceedings of AAAI' 08* (pp. 1319–1324). Chicago, IL.

Monch, L., Fowler, J. W., Dauzere-Peres, S., Mason, S. J., & Rose, O. (2011). A survey of problems, solution techniques, and future challenges in scheduling semiconductor manufacturing operations. *Journal of Scheduling, 14*, 583–599.

Nickel, S., Schroder, M., & Steeg, J. (2012). Mid-term and short-term planning support for home health care services. *European Journal of Operational Research, 219*(3), 574–587.

Ouelhadj, D., & Petrovic, S. (2009). A survey of dynamic scheduling in manufacturing systems. *Journal of Scheduling, 12*, 417–431.

Pemberton, J. C., & Galiber, III, F. (2000). A constraint-based approach to satellite scheduling. In *DIMACS workshop on constraint programming and large scale discrete opti- mization* (pp. 101–114).
Peng, B., Hosseini, M., Hong, Z., Farivar, R., & Campbell, R. (2015). R-Storm: Resource-aware scheduling in storm. In *Proceedings of annual middleware conference* (pp. 149–161).
Peng, P., Bai, Y., Chen, C., et al. (2020). *An improved differential evolution algorithm for multi-imageing-satellite scheduling*. Aerospace Shanghai (Chinese & English). https://doi.org/10.19328/j.cnki.1006-1630.2020.01.004.
Petrovic, S., & Berghe, G. V. (Eds.), (2007). *Annals of operations research.* , Special issue on personnel scheduling and planning. Netherlands: Springer.
POLYU. (2016). Media Releases. *Microsatellite platform revolutionises space research* [Online], March. Available at: https://www.polyu.edu.hk/cpa/milestones/en/201603/research_innovation/technology/microsatellite_platform_revolutionises_space_resea/index.html.
POLYU. (2020). Media releases. *PolyU contributes to the Nation's first Mars mission with the Mars camera* [Online], 23th July 2020. Available at: https://www.polyu.edu.hk/media/media-releases/2020/0723_polyu-contributes-to-the-nations-first-mars/.
Qian, Z., Bi, Z., Cao, Q., Ju, W., Teng, H., Zheng, Y., & Zheng, S. (2017). Expert-guided evolutionary algorithm for layout design of complex space sta- tions. *Enterprise Information Systems, 11*(7), 1078–1093.
Ruml, W., Do, M. B., & Fromherz, M. (2005). On-line planning and scheduling for high-speed manufacturing. In *Proceedings of ICAPS' 05* (pp. 30–39). Monterey, CA.
Salman, A. A., Ahmad, I., & Omran, M. G. H. (2015). A metaheuristic algorithm to solve satellite broadcast scheduling problem. *Information Sciences, 322*, 72–91.
Sarin, S. C., Varadarajan, A., & Wang, L. (2011). A survey of dispatching rules for operational control in wafer fabrication. *Production Planning & Control, 22*, 4–24.
Segal, M. (1974). The operator-scheduling problem: A network-flow approach. *Operations Research, 22*(4), 808–823.
Shi, Y., Boudouh, T., & Grunder, O. (2017). A hybrid genetic algorithm for a home health care routing problem with time window and fuzzy demand. *Expert Systems With Applications, 72*, 160–176.
Soualhia, M., Khomh, F., & Tahar, S. (2017). Task scheduling in big data platforms: A systematic literature review. *The Journal of Systems and Software, 134*, 170–189.
Waluyo, B. A., Rahayu, W., Taniar, D., & Srinivasan, B. (2011). A novel structure and access mechanism for mobile broadcast data in digital ecosystems. *IEEE Transactions on Industrial Electronics, 58*(6), 2173–2182.
Weiss, P., & Yung, K. L. (2009). Mission architecture decision support system for robotic lunar exploration. *Planetary and Space Science, 57*(12), 1434–1445.
Weiss, Peter, & Yung, Kai Leung (2009). Mission architecture decision support system for robotic lunar exploration. *Planetary and Space Science*, 1434–1445.
Wu, G., Du, X., Fan, M., Wang, J., & Wang, X. (2020). *Ensemble of heuristic and exact algorithm based on the divide and conquer framework for multisatellite observation scheduling*.
Wu, G., Wang, H., Pedrycz, W., Li, H., & Wang, L. (2017). *Computers & Industrial Engineering, 113*, 576–588.
Xhafa, F., Barolli, A., & Takizawa, M. (2013a). Steady state genetic algorithm for ground station scheduling problem. In *IEEE 27th international conference on advanced information networking and applications (AINA-2013)* (pp. 153–160). IEEE CPS.
Xhafa, F., Herrero, X., Barolli, A., & Takizawa, M. (2013b). Using STK toolkit for evaluating a GA base algorithm for ground station scheduling. In *CISIS 2013* (pp. 265–273). IEEE CPS.
Xhafa, F., Sun, J., Barolli, A., Biberaj, A., & Barolli, L. (2012). Genetic algorithms for satellite scheduling problems. *Mobile Information Systems, 8*(4), 351–377.

Xhafa, F., & Ip, A. (2019). Optimisation problems and resolution methods in satellite scheduling and space-craft operation: A survey. *Enterprise Information Systems* (1), 1–24.

Xhafa, Fatos, & Ip, Andrew W. H. (2021). Optimisation problems and resolution methods in satellite scheduling and space-craft operation:a survey. *Enterprise information systems*, *15*, 1022–1045.

Xu, Mingming, & Wang, Junfeng (2020). A collision probablity-based algorithm for multi-satellities task pre-scheduling problem. *Journal of Sichuan University(Natural Science Edition)*, 894–902.

Xu, Y., Liu, X., He, R., & Chen, Y. (2020). Multi-satellite scheduling framework and algorithm for very large area observation. *Acta Astronautica*, *167*, 93–107.

Zaharia, M., Chowdhury, M., Franklin, M. J., Shenker, S., & Stoica, I. (2010). Spark: Cluster computing with working sets. In: *Proceedings of USENIX conference on hot topics in cloud computing* (pp. 1–7).

Zulch, G., Rottinger, S., & Vollstedt, T. (2004). A simulation approach for planning and re-assigning of personnel in manufacturing. *International Journal of Production Economics*, *90*(2), 265–277.

Zufferey, N., Amstutz, P., & Giaccari, P. (2008). Graph colouring approaches for a satellite range scheduling problem. *Journal of Scheduling*, *11*(4), 263–277.

CHAPTER 7

Colored Petri net modeling of the manufacturing processes of space instruments

Ang Li[1], Bo Li[1], Ming Gao[1], K.L. Yung[2] and Andrew W.H. Ip[3,4]
[1]School of Management Science and Engineering, Key Laboratory of Big Data Management Optimization and Decision of Liaoning Province, Dongbei University of Finance and Economics, Dalian, P.R. China
[2]Department of Industrial and Systems Engineering, The Hong Kong Polytechnic University, Hong Kong, P.R. China
[3]College of Engineering, University of Saskatchewan, Saskatoon, SK, Canada
[4]Department of Industrial and Systems Engineering, The Hong Kong Polytechnic University, Hong Kong SAR, P.R. China

7.1 Introduction

In this part, we will introduce the development, classification, and application of Colored Petri net (CPN).

7.1.1 Development of Petri net

Petri net (PN) was first introduced to describe the asynchronous communication of computer system by Dr. Carl Adam Petri in his doctoral paper in 1963 (Yung, Gao, Liu, Hung Ip, & Jiang, 2020). Petri theory and PN application has made great progress in the past years.

The aim of PN research in the 1960s was to create techniques and application methods according to the isolated net. PN was called special net theory in this period.

In the 1970s, PN developed as general net theory. The classification and the relationship between nets were developed by scientists. Concurrency theory, synchronization theory, network logic and topology were created based on PN.

PN entered the stage of comprehensive development in 1980. Different theories and computers were used to solve complex issues. Level, time, color was also added to enrich PN theory.

7.1.2 Classification of Petri net

This section will concentrate on the explanation of PN from classical to some extensions.

IoT and Spacecraft Informatics
DOI: https://doi.org/10.1016/B978-0-12-821051-2.00005-2

7.1.2.1 Classical Petri net

The classical PN, a modeling language, which is used to discrete event-based distributed systems (Zahid and Tauseef, 2020). It is graphically a directed graph that includes three different elements: places, transitions, and functions. Mathematically, a classical PN can be represented as a tuple $(P, T, \text{and} F)$ (Yung et al., 2020)

There are brief explanations for these three elements $(P, T, \text{and } F)$:

Places (*P*): $P = (p_1, p_2, \ldots, p_n)$ (Wang, Fei, Chang, & Li, 2019) stands for a finite set of places. These are the first type nodes, and they are represented by circles in a PN. Moreover, places represent resources in manufacturing processes. For example, p_n is the resource condition for place *n*.

Transitions (*T*): $T = (t_1, t_2, \ldots, t_n)$ stands for a finite set of transitions. For each transition $t \in T$, they can be given one or two from two concepts: input arc or/and output arc which can be expressed as equations:

$$\rightarrow t = p \in P\{\text{input}(t, p) > 0\}, t \rightarrow\ = p \in P\{\text{output}(p, t) > 0\}$$

Transitions are another type of node, and they are represented by rectangular boxes in a PN. In practical problems, transitions represent transformation processes between places.

Functions (*F*): *F* stands for a finite set of functions, and these functions involve the logical relationships between places and transitions. In addition, tokens with arcs in a PN can represent the states of the system.

Tokens: Tokens are represented by dots in the places, and they are regarded as marks to show the presence of actual objects or indicators of conditions in a quantitative sense in the article (Zahid & Tauseef, 2020). In other words, the number of tokens is used to represent the status of the place.

Arcs: Arcs are directed arrows connecting places and transitions which are immediate accessibility instead of channels which resources can flow. Moreover, they are labeled with weights (positive integers) which is the number of arcs.

A triple $N = (P, T, F)$ is a PN that meets the following conditions (Liang, 2010):

$$P \cup T \neq \varnothing$$

$$P \cap T = \varnothing$$

$F \subseteq (P \times T) \cup (T \times P)$ indicates the function between places and transitions

$\text{dom}(F) \cup \text{cod}(F) = P \cup T$ shows PN has no isolated nodes. In a PN, you can find equation as follows: $\text{dom}(F) = \{x|, \exists y:(x, y) \in F\}$; $\text{cod}(F) = \{x | \exists y:(y, x) \in F\}$.

A set $X = P \cup T$ is a collection of net elements.

PN model builds a network to simulate paralleled, conflicting, and mutual exclusions activities from the graphical perspective (Nabi & Aized, 2019) in manufacturing systems and has capability to solve quantitative issues (supply of materials, throughput) and qualitative issues (quality control) from a mathematical perspective. In addition, PN can model a large system's performance in a dynamic vision by connecting three elements.

Although classical PNs were applied in many practical problems such as performance evaluation, process optimization, and simulation. Some drawbacks of the classical PN are as follows:

Firstly, the structure of PNs is so complex which is hard to distinguish tokens,

Secondly, lack of consideration in time aspect cannot be used to solve time-related questions,

Thirdly, the computational power of PN is strictly weaker than that of Turing machines, making them insufficient for modeling certain real-world systems such as chronological systems.

PN theory has been investigated in the past years and many different formulaic representations of PN have been created. To facilitate understanding and methodology, we used a 5-tuple $\text{PN} = (P, T, F, W, M_0)$ (Zahid & Tauseef, 2020) and summarized the explanation of extensions base on this 5-tuple PN. Moreover, we made a list of the explanations of elements in Table 7.1.

7.1.2.2 Timed Petri net

In this section, we will introduce the definitions, theories, and methods of Timed PNs.

First, we need to explain "firing" to help understand to follow-up contents. Firing means the action of launching in dictionary, so the firing of a transition can be understood as the occurrence of an event when all conditions are satisfied in a PN.

Table 7.1 Petri net element explanation.

Petri net element	Explanation
P	$P=(p_1, p_2, \ldots p_n)$, which stands for a finite set of places
T	$T=(t_1, t_2, \ldots t_n)$, which stands fora finite set of transitions
F	a finite set of functions
W	$\{0, 1 \ldots \text{n}\}$, a nonnegative finite set of weighting of arcs
M_0	$P \cap T = \emptyset P \cup T \neq \emptyset$, initial marking
C	A colored function in PN
f	firing time $f(t)$, which assigns the nonnegative value for each transition t
E	E function returned a value and allocated it to each token in a place
R	Priority index for different products in process
V_i	The set of variables in NHTCPN

Timed PN was developed from classical PN. The difference between PN and Timed PN is the transitions in Timed PN take "real-time" to fire (Zahid & Tauseef, 2020).

A timed PN can be represented with a 6-tuple $TPN = (P, T, F, W, M_0, f)$ (Yung et al., 2020)

M in this tuple is the marking which assigns to each places a nonnegative integer $\{0, 1 \ldots, n\}$. For example, M_p, $p \in P$ is the state of tokens in place p. For the initial marking M_0, which has some transitions and the choice of if fire or not. When firing of transition takes place, M_0 will transit to M_1. Repeat the process until M_0 transit to M_n. (Xi, 2009)

f in this tuple represents firing time function, which assigns the nonnegative (average) firing time $f(t)$ for each transition t,that is, $f(t) = a$ means that the occurrence of transition t requires a units time to finish.

For the "firing time" we mentioned in the last paragraph means the duration of transition's firing (Zuberek & Kubiak, 1999). So Timed PN has the information of time which enhanced the capability in performance evaluation related to "firing-time," such as leading time, delays, throughput times in real-world systems (Jensen, 2007).

7.1.2.3 Colored Petri net

In this section, we will introduce the CPN.

The most significant difference between CPN and classical one is that CPN allows tokens using colors for distinguishing. You can also consider that tokens carry multiple layers of information (numbers and colors) in CPN.

There is a 5-tuple $CPN = (P, T, F, M_0, E)$ presenting Colored PN.

E function returned a value and allocated it to each token in places, so the PN is embedded with the capability to express the firing of transitions which can remove the number and color of tokens from input places and new tokens are created to put into output places (Lu et al., 2021).

Based on above explanations, CPN can carry different tokens in one place which can reduce the complexity of PN. In a dynamical manufacturing system, using many colors in one token to represent multiple types or sequences of products is a good way to monitor the supply of original materials and manufacturing process.

7.1.2.4 Timed colored Petri net

The Timed CPN(TCPN) can be represented as a 6-tuple $N = (P, T, F, W, M_0, f, R)$ (Yung et al., 2020). TCPN developed from classical PN, TPN, and CPN. So, it has assembled all symbols of these five existing PN. Moreover, a new symbol R was added into the tuple. R represents the priority index for tokens in a place.

In real-world system, different products at the same production line will adjust the process sequence based on the current priority, such as the delivery date and importance index of customers. The same principle can apply to manage multiple production lines.

7.1.2.5 Hierarchical Petri net

Hierarchical PN is another extension of PN theory. There are two mechanisms proposing in literature (Figat & Zieliński, 2020) for composing the hierarchy:

Transition refinement, and Subnet abstraction. The mechanism of transition refinement and subnet abstraction are used for two systems. Transition refinement is used for top-down system which detailed tasks were involved in transitions. On the contrary, tasks were removed in form of subnet and replaced by a transition in subnet abstraction.

Hierarchical in literature (Wang & Wei, 2009) provide an extended way to combine CPN with hierarchy. So Hierarchical TCPN (HTCPN) was created to dynamically evolution a manufacturing system (Yung et al., 2020). We provide the details of HTCPN as follows:

HTCPN subnets: $\mathrm{HTCPNsubnet} = \{\mathrm{HTCPN}_1, \mathrm{HTCPN}_{2,\ldots}, \mathrm{HTCPN}_n\}$ stands for a finite set of subnets. In addition, $\mathrm{HTCPN}_n \in \mathrm{HTCPN}$, n here is the number of types in the model.

HTCPN = ($HTCPN_i, M_{0i}, V_i$), Vi in this tuple is used to represent a set of variables for each subnet of HTCPN.

In a complex system, there are many alternative transitions and each transition is associated with a subnet in complex HTCPN (Yung et al., 2020). And each place is allowed to contain more detailed information of activities. During the replacement, the same place could be abstracted by different unique HTCPN. From another view of point, there are two places which have their own tokens, but they are mixed with each other at the same place. Fusing places enables the connection of nets (Figat & Zieliński, 2020). That is, p_a in $HTCPN_a$ and p_b in $HTCPN_b$ are single places. There is a p_{mixed} represent the same place which contains both p_a and p_b. So, the complexity of net can be reduced and the structure of the net can be well understood.

7.1.3 Petri net properties

PN could simulate the actual manufacturing system, and the properties with this tool are as follows: accessibility, activity, and fairness.

7.1.3.1 Accessibility

As the most basic behavior characteristic of PN, other properties of PN can be deduced from the definition of accessibility.

Reachability of key process: it indicates that final state can be reached from the initial state in a given PN by triggering a series of migrations. For example, M_0 as an identifier can arrive at the only execution token at the last place. The sequence of transitions in this process is limited.

Reachability of subprocess: the arrival of M in the final place indicates the ending of all tokens. So, the subprocesses are allowed to execute in multithreaded mode, this subprocess can simulate multiuser or multitask process.

7.1.3.2 Activity

In a computer operating system, a deadlock occurs when limited resources are allocated based on an unreasonable policy. Deadlock problems are very important for PNs, which reflect the activities of PNs.

7.1.3.3 Fairness

Each place in a PN has the opportunity to receive an execution token after limited sequential transitions.

7.1.4 Modeling with TCPN

In this part, we will introduce the steps to build models with HTCPN.

The steps are as follows:

Step1: Build the overview structure of the discrete event-based net which aims to build up a first-level PN.

Step2: Further define the subnet for the first-level PN.

Step3: Explicitly check the rules of definition and evaluation. If the terms are satisfied, transit to step 5; otherwise, transit to step 4.

Step4: Modify the structure of PN to meet the definitions and evaluation rules.

Step5: Repeat step 2 until further development is prohibited.

Step6: Ensure all links are consistent through the entire HTCPN.

7.1.5 Application of Petri net

7.1.5.1 Modeling workflow

Workflow consisting of tasks, resources, and cases is a kind of process that can be fully or partially automated. For example, documents, information can be transited and executed between different participants according to rules.

In addition, the workflow includes the following four properties:

1. *Effectiveness*: Testing whether the workflow operates in accordance with the requirements.
2. *Case driven*: The workflow changes based on a different case.
3. *Correctness*: The workflow should be error-free.
4. *Performance analysis*: Evaluating whether the design can meet the requirements in terms of throughput, response time, resource utilization, etc.

Overall, the research focus of workflow is on the description and verification of changes in a process. Modeling workflow with PN is an effective method to optimize workflows.

7.1.5.2 Supply chain

Using PN theory and simulation analysis in the supply chain process modeling and simulation optimization was one of the applications. The wonderful research achievements are as follows:

(Che & Gu, 2005) discussed based on the generalized stochastic PNs on the supply chain of each enterprise problem between the working process. The main research is the generalized stochastic PN modeling and analysis methods in delayed production strategy of supply chain management. To be specific, using generalized stochastic PNs and homogeneous Markov chain relationships established a performance evaluation model.

(Zhang, 2005) analyzed the working process of large construction enterprises in dealings with cooperative units, modeled and optimized these working processes with PN, then proposed the system structure of supply chain and introduced the design method of supply chain with cost minimization as the guiding principle. Finally, the author gave Specific software implementation algorithms.

(Wang & Bai, 2006) used the modeling method combining discrete variables and continuous variables to the supply chain with hybrid PN simulate and proved that it was superior to discrete PN simulation.

(Fang, Ye, & Li, 2006) used time PN to model and simulate supply chain business processes, analyzed the impact of order completion amount on supply chain inventory and order response time, and determined the best product differential rate. The results show that less order completion is beneficial to inventory control and quick response of the supply chain. The existence of differential parts is conducive to rapid product customization and basically does not affect the inventory of the whole supply chain.

We can predict the research of supply chain will be further in the future. The research trend in the future will be closer links between the specific industries and PN.

7.1.5.3 Flexible manufacturing system

The application of PN in the FMS system is typical. In general, stochastic PN is used to model FMS. PN has an important advantage that they provide a unified modeling tool from the first step of the design to the implementation of the system. PN model will provide:

1. An accurate formal description of a figure, which prepare for an in-depth dialogue between the designer, owner, and user about the behavior of the system;
2. A well-defined theory that can verify the properties of the model (activity, fairness, boundedness, etc.);
3. Theory and method of performance evaluation (referring to random networks);
4. System implementation technology, including technology that generates code of the software which can be controlled in real-time.

Because FMS is a complex and large system, when using random PNs to simulate such a system, the method of refinement and consolidation can be adopted. The Elaborate method of the simulation can build a hierarchical network model. The merging of submodels can be simple subsystems or complex entire systems.

7.1.5.4 Database system

With the rapid development of the Internet, the database technology based on the Web has penetrated into various fields. With the increase of information content, static Web pages cannot meet the users' requirements of dynamic, real-time, and interactive information services. It is necessary to integrate Web technology with database technology to realize the interactive application of Web databases. As a mathematical graphical tool, PN can accurately describe the dependent and nondependent relationships among the events in the system from the perspective of organizational structure, control, and management. Therefore, PN can be applied to the system fault tolerance, concurrency control, and deadlock monitoring of the database.

7.1.6 Optimization tools

This part of the article will concentrate on the introduction of some important concepts. These concepts can help understand the process of manufacturing the soil preparation instruments.

7.1.6.1 Random simulation with colored Petri net tool

Simulation is regarded as an effective tool to evaluate performance in the manufacturing process. These changes can be monitored and controlled dynamically in terms of process (Kasemset, Pinmanee, & Umarin, 2014). In above part, a manufacturing system has been modeled by HTCPN theory which connects classical PN theory with time, color, and hierarchy for verification. Then output data from random simulation which we can use in the following models (Yung et al., 2020). First, we should analyze the problem within the limited parameters and build a model based on this problem by using Timed CPN tools. Next, CPN tools can be used to stimulate the process and output some data, followed by checking the reliability of the process and the requirements. If the process needs to be improved, we can use 6 sigma model and key time analysis to measure the simulation data, and then we use elimination, combination, rearrangement and simplification (ECRS) method and constraint theory to acquire

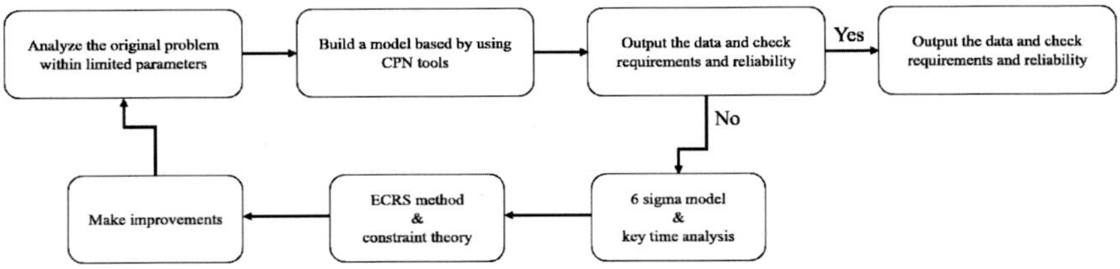

Figure 7.1 Modeling process.

the improvements. The framework is shown in Fig. 7.1 (Yung et al., 2020).

So, the random simulation is an effective and indispensable process to obtain improvements.

7.1.6.2 Six sigma system

The six sigma was created as a metric for finding defects and improving quality by Bill Smith in the mid-1980s at Motorola Inc. (Byrne, McDermott, & Noonan, 2021). In late 1990s, six sigma system was introduced to the General Electric as a technique of business management by Jack Welch (CEO of GE) and Larry Bossidy (CEO of AlliedSignal) (Pheng & Hui, 2004). Since then, six sigma has received public attention in manufacturing and writing books and journal papers.

Six sigma has integrated robust continuous quality management and business process improvement initiatives to achieve aims.

Six Sigma as a management philosophy or a business strategy was widely utilized to enhance enterprise performance and competitiveness in terms of customers, such as increasing customer satisfaction or saving cost.

Sigma σ is used to represent the standard deviation in statistics. Six Sigma is, therefore, to be defined as a statistical measure of processes and products. In real-world manufacturing, process control plays an important role in facilitating the product quality or reducing variation. Manufacturing department needs to monitor and control carefully each process to meet the requisites. So, six sigma can be regarded as a technique in the quality management field.

In this paper, we pay more attention to the research that the six sigma system was used to analyze the output data from random simulation with CPN models.

There are two limits: upper specification limits (USL) and lower specification limits (LSL) in six sigma system. X represents the output data and that X follows a normal distribution. μ is the arithmetic mean value of

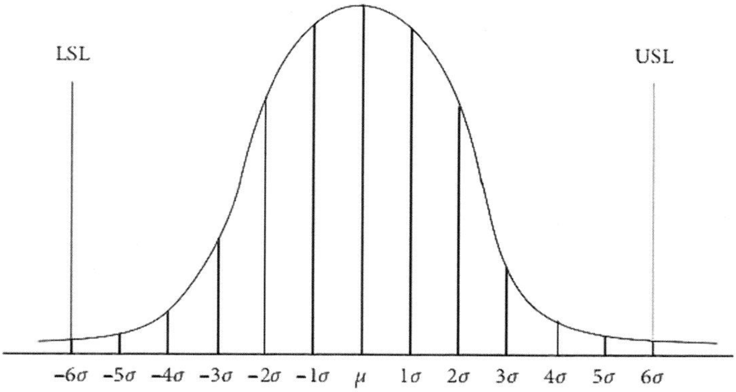

Figure 7.2 six sigma system.

data X and σ represents the arithmetic square root of μ. The distribution of X (Chen, Chang, & Guo, 2020) can be represented on equation:

$X \sim N(\mu_x, \sigma_x^2)$, and the system is shown in Fig. 7.2 (Yung et al., 2020).

The distribution figure is within acceptable range of plus and minus six sigma. The results greater than $\mu + 6\sigma$ and less than $\mu - 6\sigma$ are defectives account for 0.00034%.

1. Defects Per Million Opportunities (DPMO)

It is necessary to introduce DPMO to facilitate understanding of six sigma systems.

DPMO represents the number of defectives in one million opportunities. There are two equations:

$$\text{DPO (Defects Per Opportunities)} = \frac{\text{defective numbers}}{\text{product numbers} * \text{opportunities}} * 100\%$$

$$\text{DPMO} = \frac{\text{defective numbers} * 10^6}{\text{product numbers} * \text{opportunities}} * 100\%$$

As shown in Table 7.2 (Yung et al., 2020), if DMPO is 0.00034%, the percentage of products without defects is 99.99966%.

2. Define, Measure, Analyze, Improve, Control (DAMIC)

To improve the detailed operations from simulation to implementation of the quality control system (Al Khamisi, Khan, & Munive-Hernandez, 2018; Mkhaimer, Arafeh, & Sakhrieh, 2017). DAMIC as a systematic and fact-based methodology provides best results when the

Table 7.2 Defects per million opportunities.

Sigma levels	Percentage of products without defects(%)	Defects per million opportunities (DPMO)
1	30.9	6,90,000
2	69.2	3,08,000
3	93.3	66,800
4	99.4	6210
5	99.98	230
6	99.99966	3.4

process is flexible (Bhargava & Gaur, 2021). It was introduced as a five-phase improvement cycle to enhance the quality of processes in manufacturing industries. The cycle includes five steps:

1. *Define*: Define issues and objectives of a certain process which are in terms of the actual requisites.
2. *Measure*: Identify the key points of the issues and narrow the scope of problems from collecting data.
3. *Analyze*: Determine the root causes of problems by using tools to analyze collected data. Such as observational survey, Interview Survey, logical analysis.
4. *Improve*: Generate some solutions and plot them on a small scale to determine if they positively improve the process performance. Then the more successful methods can be implemented on a wider scale (Pheng & Hui, 2004).
5. *Control*: Improve and solve problems due to predetermined standards to remain the product quality at a desired level (Pheng & Hui, 2004).

3. Advantages and problems

The clear explanations of six sigma is demonstrated the effectiveness and methodology for enhancing the quality and controlling process in industries. But it is also necessary to provide the advantages of six sigma system and reveal the problems in using six sigma System:

Advantages:

1. *Effective management frame*: Six sigma management is driven by data and facts which can be implemented in manufacturing or service. Such as General Electric Company and Motorola Inc.
2. *Saving cost*: The defects created in the process of six sigma system help reduce the percentage of defects so that the cost of maintaining or redo can be reduced.

3. *Satisfy the requisites of clients*: Understand and meet customers' needs to reach the goal of maximum profit between each link and achieve a virtuous cycle
Problems may meet:
1. *Lack of scientific plan and implement*: Lack of statistics and management knowledge
2. *Weak basic management*: Obsolete management mode and lax managers
3. *Direct copy*: Copy the six sigma model from other companies directly and ignore the reality situation of their own company.

7.1.6.3 Critical time analysis

It is necessary to introduce some explanations to help understand critical time analysis. The related explanations are as follow:

1. Theoretically Earliest Start Date (TESD): Each task is assigned to the theoretically earliest start date which is the earliest time that process can begin without considering resource constraints (Yung et al., 2020). To facilitate methodology, we use TESD_n to present the earliest start time of task. Because the task $n-1$ must be finished before task, So the TESD requests the start time of each task must be later than or equal to TESD_n.
2. LT (Lead Time): LT can be understood as the duration between when a process is placed and the process is finished. It can also be considered as the execution of task n like LT_n.

   To build the connection between TESD and LT, we introduce such equation:

$$\text{TESD}_n = \text{TESD}_{n-1} + \text{LT}_{n-1}$$

3. Latest Start Date (LSD):LSD is another important concept to analyze the performance of manufacturing. The LSD of each assignment is calculated forward according to the deadline for each entire process (Yung et al., 2020). Compared with the concept of TESD, LSD represents that each process cannot be completed if the start point is later than LSD. The LSD can also be calculated by the following equation:

$$\text{LSD}_n = \text{LSD}_{n+1} - \text{LT}_n$$

Through the TESD and LSD of each assignment by using the simulation output data, we can conclude the real-time window between tasks, which time window_n equals to $\text{LSD}_\text{n} - \text{TESD}_n$. Then sum all time windows of each task up, we will have the entire time window of this process. Moreover, if the simulative results cannot be completed within the entire time window, the task of the process is regarded as a key object to analyze (Yung et al., 2020).

So critical time analysis is considered an efficient way to figure out the key points during the process, we may pay more attention to key points to improve overall performance.

7.1.6.4 ECRS Method

The ECRS method is a fundamental processing tool in industrial engineering (Kasemset et al., 2014). There are four elements analyzing for the possibility of ECRS application.

1. *Elimination*: Redundant inspection of the process was applied for removal (Miranda, 2011) in a manufacturing process.
2. *Combination*: Two or more objects can be merged into one. Combine the operation and inspection function together.
3. *Rearrangement*: An efficient sequence arrangement is vital to achieve smooth manufacturing flow. For instance, layout in industry can be transferred from a scatter type to cluster type to reduce walk time and reduce machine assignments.
4. *Simplification*: Make a deeper and further research to lessen the load of the process.

The ECRS method has been used in designing lead screws and face plate (Gnanavel, Saravanan, Chandrasekaran, & Pugazhenthi, 2017), ice cream manufacturing (Miranda, 2011), electronic manufacturing industries. The ECRS method is the basic principle for lifting efficiency. ECRS has been demonstrated as a technique theory for optimized production, focusing on identifying the constraints and smoothing out the process flow (Liu, Huang, Cheng, & Wu, 2012). The theory of constraints list eight rules for the manufacturing process:

1. Making a balance the supply and usage instead of making balance in capacity;
2. Using system constraints to control the noncritical resource utilization;
3. High resource utilization does not necessarily reflect the effectiveness of work;
4. Avoiding delay in critical paths of this system;
5. Critical paths have the most significant influence on the entire system;
6. Noncritical paths are not as vital as you think;
7. Batch size should vary according to the actual situation;

8. Batch sequence should be determined based on the system constraints.

The criterions and critical paths we mentioned above was the core we need to follow and improve. ECRS can be an efficient method to maintain and enhance capability in industries.

7.2 Case study

7.2.1 Case modeling and simulation

7.2.1.1 Case description

Phobos is a moon orbiting Mars (Phobos-Grunt's sad return, 2012). Its samples can be used for scientific research and providing illustration for the retrieval of future space missions (Zuberek & Kubiak, 1999). So the exploration plan is to make TePhobos -Grunt land at Phobos. Unfortunately, the mission failed because the spacecraft did not manage to leave the earth's orbit. The second PHOBOSGRUNT mission is tentatively scheduled for 2025, and some changes will be made in order to make the research interpretive and useful. The architecture of the Phobos-Grunt land preparation system (SOPSYS) logged onto Phobos is as shown in Fig. 7.3 (Yung et al., 2020). The actuator *E* will analyze particle samples from inside the spacecraft before sending the debris outside the spacecraft into space. The research focuses on the manufacturing process of part *E*.

The convey unit *E* is composed of five parts. The codes of five parts in the manufacturing process are as follow: *PART*0030(conveying motor

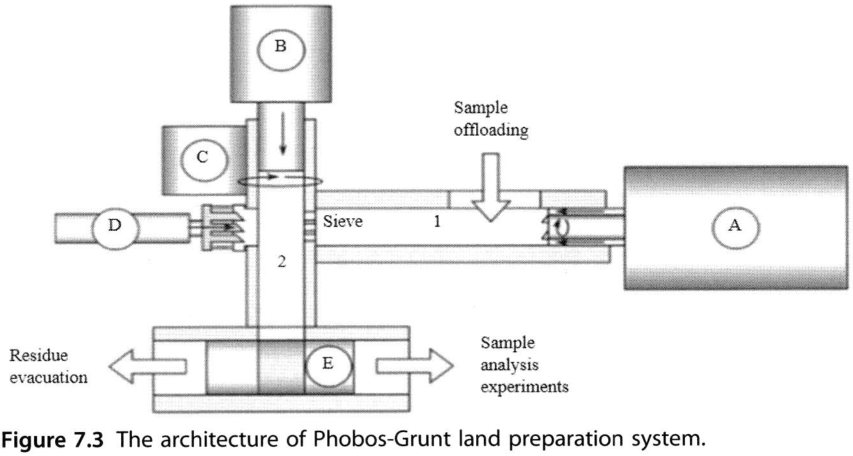

Figure 7.3 The architecture of Phobos-Grunt land preparation system.

with encoder), *PART*0031(screen encoder PCB), *PART*0032(Screen encoder receiver), *PART*0033(Screen encoder LED) and *PART*0034 (conveying chamber). A subpart of *CPN*0012(conveying encoder) is assembled with *PART*0031*and PART*0032. The convey units are assigned to make the Sample particles are sent to spacecraft while debris are sent into space.

In this case, there are five parts that should be manufactured or purchased; Fig. 7.4 (Yung et al., 2020) shows the sequence of the manufacturing process. Five engineers, four operators, and one inspector are required for this process. The beginning of the whole process is the preliminary design of CPN0005(Phobos-Grunt SOPSYS convey unit). After the design is completed, the other four tasks can be started at the same time: design, budgeting, and test equipment design for PART0034, the detailed specifications of PART0030 and PART0033, the design and budgeting for CPN0012. After the detailed specification, PART0030 and PART0033 should be purchased, and buying PART0031 and PART0032 never comes before designing CPN0012.

The only part which should be produced in-house is PART0034. At the end of the design step, the manufacturing of PART0034 will begin immediately. For every purchasing, manufacturing, and assembly process, an inspection is an inevitable process to ensure its quality. Therefore, they need to be checked for the subsequent assembly process after receiving parts PART0030, PART0031, PART0032, and PART0033 and finished PART0034. Then assembly of CPN0012 need to be tested after checking

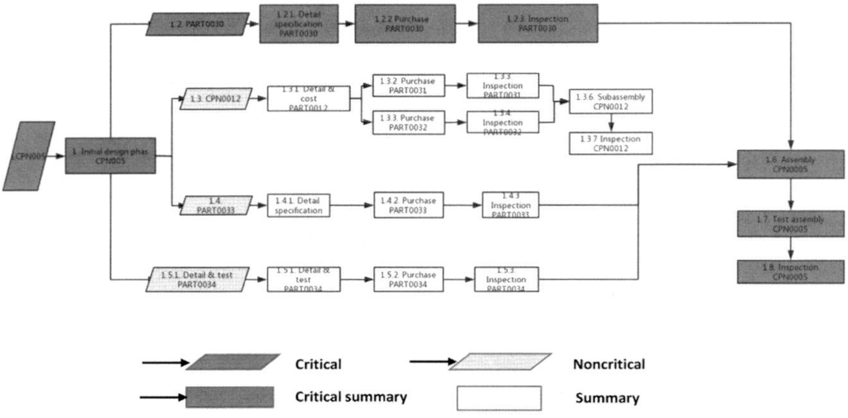

Figure 7.4 Manufacturing process.

PART0031 and PART0032. Finally, PART0030, CPN0012, PART0033, and PART0034 will be meticulously assembled into CPN0005, which will be inspected and checked again. The network flow and required resources are shown in Table 7.3 (Yung et al., 2020).

Table 7.3 Manufacturing process and utilized resources.

ID	Task name	Duration (days)	Predecessor	Resource name (number)
1	1. CPN0005	100		
2	1. 1. Initial design phase CPN0005	40		Designer (2)
3	1. 2. PART0030	38		
4	1. 2.1. Detail specification PART0030	5	2	Designer (2)
5	1. 2. 2. Purchase PART0030	30	4	
6	1. 2. 3. Inspection PART0030	3	5	Inspector (1)
7	3. CPN0012	34		
8	1. 3. 1. Design and Consume CPN0012	7	2	Designer (2)
9	1. 3. 2. Buy PART0031	21	8	
10	1. 3. 3. Buy PART003	5	8	
11	1. 3. 4. Inspection PART0031	1	9	Inspector (1)
12	1. 3. 5. Inspection PART0032	3	10	Inspector (1)
13	1. 3. 6. subassembly CPN0012	2	11, 12	Operator (1)
14	1. 3. 7. Inspection CPN0012	3	13	Inspector (1)
15	1. 4. PART0033	23		
16	1. 4.1. Detail specification PART0033	5	2	Designer (2)
17	1. 4. 2. Purchase PART0033	15	16	
18	1. 4. 3 Inspection PART0033	3	7	Inspector (1)
19	1.5. PART0034	21		
20	1. 5. 1. Design and test CPN0034	8	2	Designer (2)
21	1. 5. 2. Manufacturing PART0034	10	20	Operator (2)
22	1. 5. 3. Inspection PART0034	3	21	Inspector
23	1. 6. Assembly CPN0005	8	16, 14, 18, 22	Operator (4)
24	1. 7. Test CPN0005	7	23	Inspector (4)
26	1. 8. Inspection CPN0005	7	24	Inspector

7.2.1.2 Mapping workflow elements into colored Petri net

In this section, we will introduce the basic structure of workflow of HCPN (Wang et al., 2019) to formalize the SOPSYS process model. The purpose of the first stage is to map the Phobos-Grunt SOPSYS transport cell manufacturing process case to CPN, which is shown in Table 7.4 (Yung et al., 2020).

We use $P = (p_1, p_2, \ldots, p_n)$ to represent a finite set of positions in methodology. In this part, you can consider the places in the CPN as a warehouses among different work which stores materials along with the manufacturing process, and the employee reserve will provide humanity resources. Two distinct places represent two different physical locations, because the character of each place is determined by the token it hold.

$T = (t_1, t_2, \ldots, t_n)$ stands for a finite set of transitions. T stands for the dynamic process of advancing from one node to the next node (provided certain conditions or rules are observed). We can model resources and products as tokens in the CPN by defining a task as a transition. Otherwise, we can also use tokens in CPN to stand for the resources and products costed and produced along with the PN. For instance, for task $PART0032$, the check requires an inspector and an unchecked $PART0032$ to produce a checked $PART0032$. Two tokens are prerequisites for this transition, one of them is an inspector and another one is an uninspected $PART0032$. During the firring of this transition, these two tokens are costed and two new tokens are generated. Therefore, a token standing for the checked $PART0032$ will be prepared for its next transition, and another token standing for the uninspected will be produced for its next transition.

In the SOPSYS system, distinct colors are used to distinguish things; so, the colors can represent real things. There are four colors in this model. Each color stands for a resource or a token. The three colors are color engineer, color operator and color checker.

Table 7.4 Mapping between case and colored Petri net.

Workflow element	*CPN element*
Tasks	Transitions
Resources and products	Tokens
Types of resources and products	Colors
Warehouse and staff pool	Places

$f = (REFT:T \rightarrow IN, RLFT:T \rightarrow IN \cup \infty)$ The above equation could be used to describes the relative latest and relative earliest firing times of the transitions. For every $t \in T$, $REFT(t) \leq RLFT(t)$, and IN is a natural set (containing 0). The gaps between the relative earliest and relative latest firing times of the transitions are the domains for enabled transitions. The definitions of REFT and RLFT are different in the colored transitions (Jensen, 2007).

$HTCPN = (A_i, HTCPN_i, V_i, M_{0_i})$ is a HCPN model, which represents the hierarchy model of SOPSYS. In this paper, a bottom-up approach will be applied to build a CPN model. The subprocess will be modeled as a subnet which will be plotted as a transformation in the CPN overview (Liang, 2010). In this case, HCPN is inappropriate to apply the resources in CPN modeling, so it is also uneasy to find a single port subnet. However, introducing a bottom-up way in the modeling process could help improve the readability of CPN. The first process is to build an easy CPN for a distinct subprocess.

A subprocess totally contains five parts. Two parts have a similar process and could be introduced in Fig. 7.5 (Yung et al., 2020).

As shown in Fig. 7.5 (Yung et al., 2020), $PART0030$ and $PART0033$ start from place $P1$ (disticnt positions in the last CPN). The first transition is specification which costs tokens from $P1$ (tokens here stand for the status where or is ready to be specified) and engineer (tokens here stands for the feasible state of engineers for the tasks).

After the transition specified is fired, new tokens will be produced into $P2$ and Engineer. The new generated tokens in $P2$ have similar color set as the token from $P1$. This color set represents the state of the products in workflow. The new generated tokens in Engineer have the same color set of tokens, which represents the available engineers. Similarly, transition examinations use the product status color tokens from $P3$ inspector color token from place Inspector and generate the product status color token into $P4$, and returns the Inspector color token to the place Inspector.

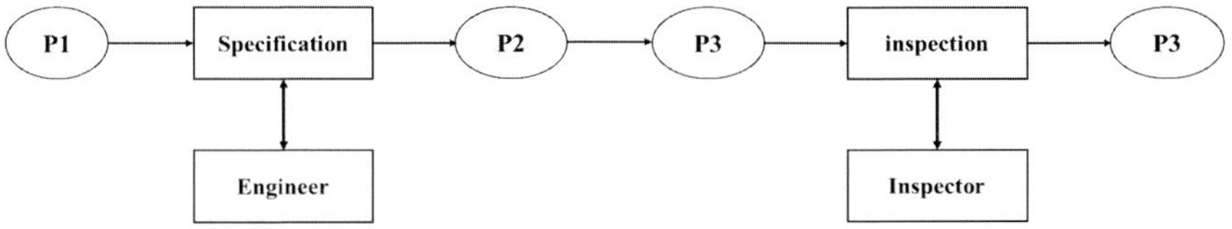

Figure 7.5 Subprocess of PART0030 and PART0033.

As shown in Fig. 7.6 (Yung et al., 2020), two other parts also share the same process.

Part0031 and Part0032 begin immediately after the design of *CPN*0012, so there is no specification transition compared with other processes.

As shown in Fig. 7.7 (Yung et al., 2020), the special task *PART*0034 is in-house manufactured which needs operator in the manufacture process. The operator color token will be removed from the place Operator, and the product status color token will be costed from *P*2. The new operator color token is going to be returned back to the place Operator, and the new product status color token will be produced into *P*3 for further handling.

The process of *CPN*0012 can be modeled as follows.

Fig. 7.8 (Yung et al., 2020) shows the subnet for the CPN0012 process. The two aspects in the subprocess of CPN0012 are as follow: First, P2 provides tokens for PART0031 and PART0032, and both transitions can be triggered at the same time. Therefore, there must be two tokens in place for triggering both transformations. The next aspect is that two places of Inspectors in Fig. 7.7 (Yung et al., 2020) which are the same and

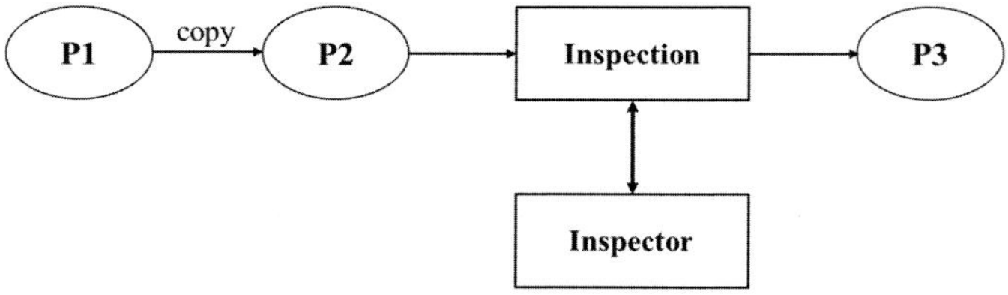

Figure 7.6 Subprocess colored Petri net for PART0031 and PART0032.

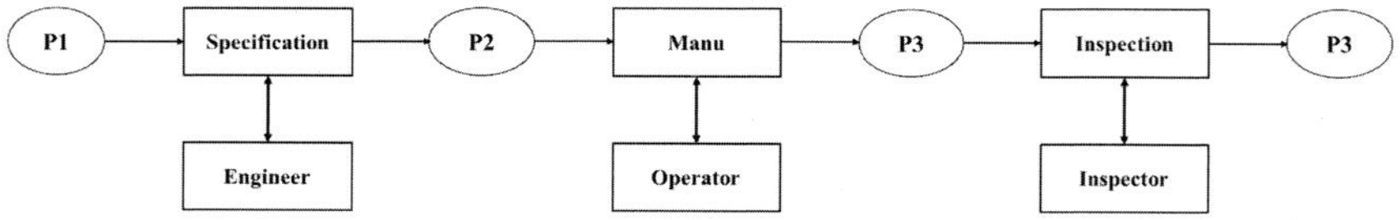

Figure 7.7 Subprocess colored Petri net of PART0034.

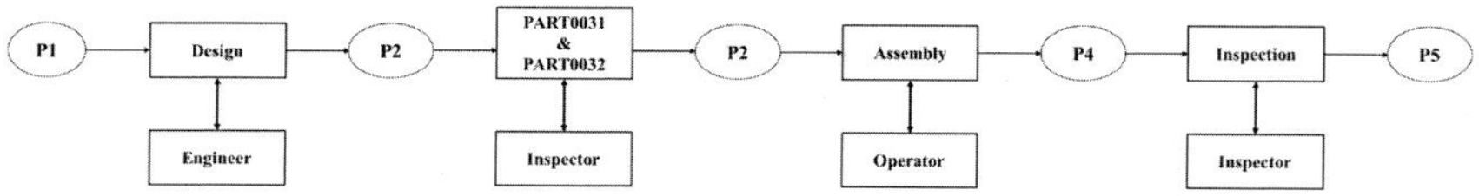

Figure 7.8 Subprocess colored Petri net of CPN0012.

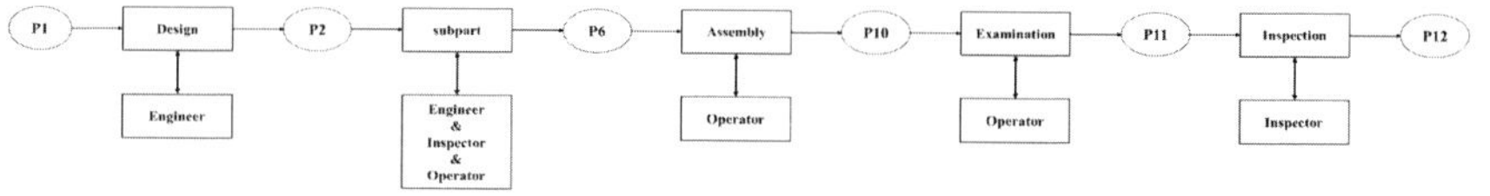

Figure 7.9 Colored Petri net model for the whole process.

only a limited number of inspectors are included. In this case, there is only one available examiner.

At the beginning of the middle reaches, there are four places to distinguish the four processes. Otherwise, one of the transitions may be triggered multiple times, while the other transitions may not be triggered at all. As shown in Fig. 7.9 (Yung et al., 2020), if the tokens which represent the state of the product can be divided into four types that can stand for four distinct subparts, then a more readable CPN could be created. The transition subpart is a complex subnet that contains all processes of these four parts. The aim of transition is to facilitate the readability of CPN, but it is not a unique port subnet that is used to simulate in this manner.

Modeling process

1. CPN tools modeling

   As a software tool for constructing, simulating, and analyzing CPNS (CPN Tools, 2018a), CPN tools can also simulates model performance in real-time. The data type of the CPN model (Wang et al., 2013) is defined by the declarations in CPN Tools. The declaration for the CPN tools is shown in Fig. 7.10 (Yung et al., 2020). The structure of the network must be modified when using the CPN model tools. It is necessary to use computer logic and the following parameters will be included in structure modification: setting place color set, building the output transition, the real-time element, and the rework structure.

2. Real-time elements.

   As shown in Fig. 7.10 (Yung et al., 2020), by adding time delays to the transitions, real-time elements are integrated into the CPN model.

   The syntax of implementing real-time is @ + (*time*).

   As shown in Figs. 7.4–7.10 (Yung et al., 2020) transition inspection will require seven units to fire in real-time. If the timed token cost in this transition is (element In Color set) @*t*, then the newly produced token will be (element in color set) @($t + 7$).

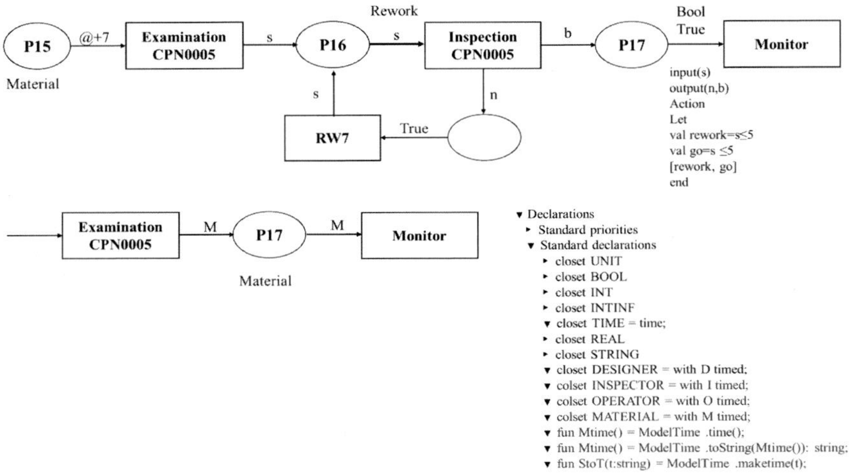

Figure 7.10 Rework structure and output transition.

3. Define color groups

Only certain types of tokens can be stored in the place. As shown in Fig. 7.10 (Yung et al., 2020), the text "Inspector" indicates its color settings. If there are multiple types of tags in a place, their color set should be explicitly defined as a list of tag types. Similarly, if a place has multiple types of tokens, its color set need to be particularly defined as a list of token types. Therefore, we can think of a list as a tuple containing variables from distinct color sets. For example, if a place stands for a staff pool of all three types of workers (operators, engineers, and supervisors), the elements in the color set here could be expressed as (e, O, and I).

As shown in Fig. 7.10 (Yung et al., 2020), the task before inspection will be randomly given an integer token within a range from 1 to 100. The model assumes a 5% scrap or 5% rework rate of all these tasks. Therefore, the code segments that inspect the transition will use a value of 5 as a marker to distinguish between good and bad product. If the value is not greater than 5, a flag with a color bool value of false will be generated at the next location. This won't fire anything in the future. At the same time, generate another mark with the color value of true at the rework point. The token will initiate a rework transformation that will send the M token back before inspecting the task. If the randomly generated value is greater than 5, the processing advances to the next phase.

We must build an additional transition in the CPN to export the performance data. The transition is shown in Fig. 7.10 (Yung et al., 2020). The conveying unit is inspected and confirmed qualified (if CPN0005 is finished). The token at the last place indicates a completed delivery unit, which will be used in another conversion called "monitor." The total LT can be defined as the consuming token and time stamp in token for this monitor transition.

Overall, we modeled the main process as CPN and described the rework structure, the inspection has been modeled as normal transitions which will push the process forward. In addition, the model satisfies all the official verification rules. The complete model is shown in Fig. 7.11 (Yung et al., 2020).

7.2.2 Simulation result and analysis

7.2.2.1 Simulation result

The data types had been recommended in the previous section, and we will run the model ten thousand simulations to ensure that all possible conditions occur. The rework rate of each work before the inspection is 5%. These types of tasks occur no more than three times at most in a workflow and three tasks we mentioned in the CPN0012 workflow should be inspected. Therefore ,if the rework is evenly distributed, the probability of three consecutive reworks is $20 \times 20 \times 20 = 8000$. Theoretically, $10,000$ times simulations with CPN model can generate reasonable results.

After $10,000$ simulations, the average LT was 118.06 days. The best possible LT is 109 days, which resources are completely contributed and all tasks have been approved by inspection for one time (see Tables 7.5 and 7.6 (Yung et al., 2020) for specific metrics). But the deadline for the program is 100 days. Thus, the result cannot meet the requirements which mean the process need to be modified.

7.2.2.2 Result analysis

1. Analysis of critical time

According to the simulation results, we summarized the data of the start time as shown in Table 7.5 (Yung et al., 2020). The three metrics should be compared before visualizing all summary data: Actual Early Start Date ($AESD$),Theoretical Early Start Date ($TESD$), and Latest Start Date (LSD). The $DELAY = AESD - TESD$ indicate how much the actual start time is later than $TESD$, which is not the earliest theoretically

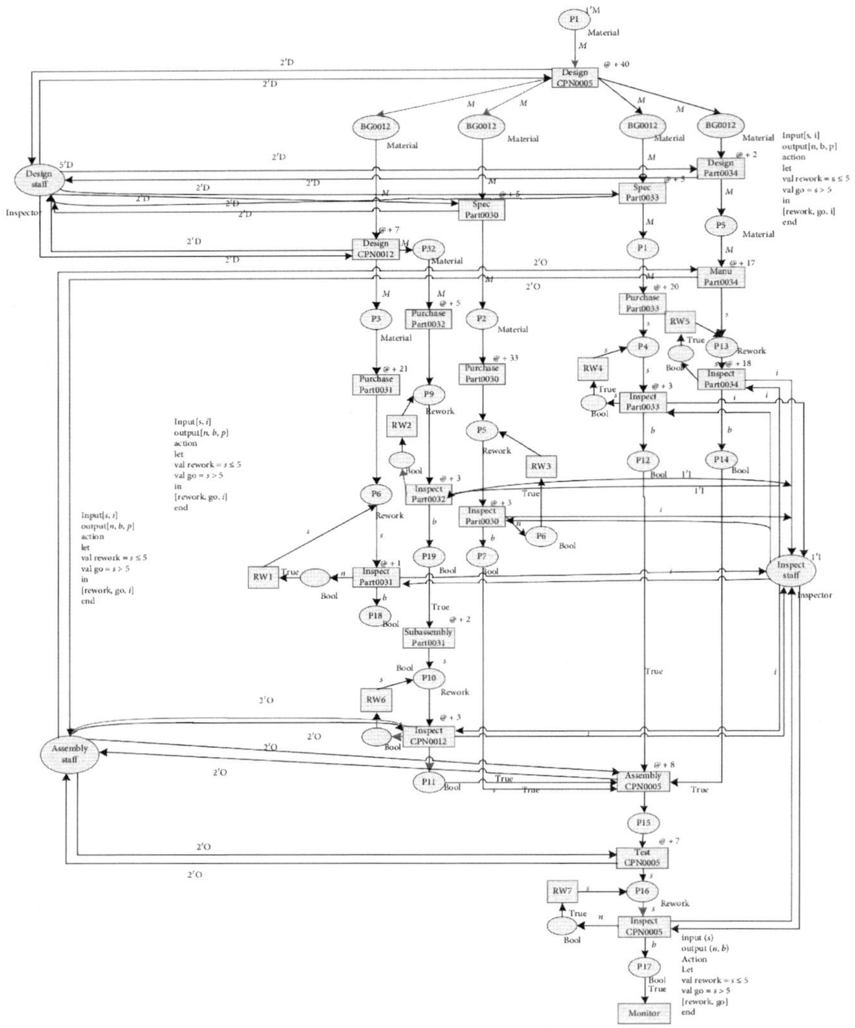

Figure 7.11 Original colored Petri net model.

according to resource limited. The $FAIL = AESD - LSD$. When the FAIL value is greater than 0(marked in red), it means that the real start time of the task is greater than the latest start time of the task, which necessarily causes the delayof completion for all systems. The $WINDOW = LSE - TESD$, which stands for the buffer time of the task (i.e., the task starts on any day within the time window, and the complete

Table 7.5 Data summary.

Task	AESD	TESD	LSD	DELAY	FAIL	WINDOW
D0012	40	40	44	0	−4	4
D0034	40	40	57	0	−17	17
S0030	40	40	40	0	0	0
S0033	40	40	55	0	−15	15
A0005	87	78	78	9	9	0
A0012	69	69	73	0	−4	4
M0034	42	42	65	0	−23	23
T0005	95	86	86	9	9	0
I0005	102	93	93	9	9	0
I0012	81	71	75	10	6	4
I0030	77	75	75	2	2	0
I0031	68	68	72	0	−4	4
I0032	52	52	70	0	−18	18
I0033	65	65	75	0	−10	10
I0034	59	59	75	0	−16	16

Table 7.6 Reliability data.

Task	AVG	LSD	FAIL	STD	Six sigma
D0012	42.01	44	−1.99	2.24	4.73
D0034	42.8	57	−14.2	2.89	−5.53
S0030	42.38	40	2.38	2.76	10.66
S0033	42.3	55	−12.7	2.75	−4.45
A0005	95.28	78	17.28	10.56	48.96
A0012	84.55	73	11.55	7.06	32.73
M0034	45.69	65	−19.31	5.03	−4.22
T0005	103.67	86	17.67	10.68	49.71
I0005	110.67	93	17.67	10.68	49.71
I0012	88.23	75	13.23	5.97	31.14
I0030	83.72	75	8.72	5.1	24.02
I0031	82.7	72	10.7	5.36	26.78
I0032	54.58	70	−15.42	3.54	−4.8
I0033	80.85	75	5.85	6.39	25.02
I0034	63.06	75	−11.94	5.12	3.42

manufacturing system will not exceed the prescribed completion time). If the *DELAY* value is greater than *WINDOW*, the manufacturing process cannot finished on time which is the same one as the judgment using *FAIL* value.

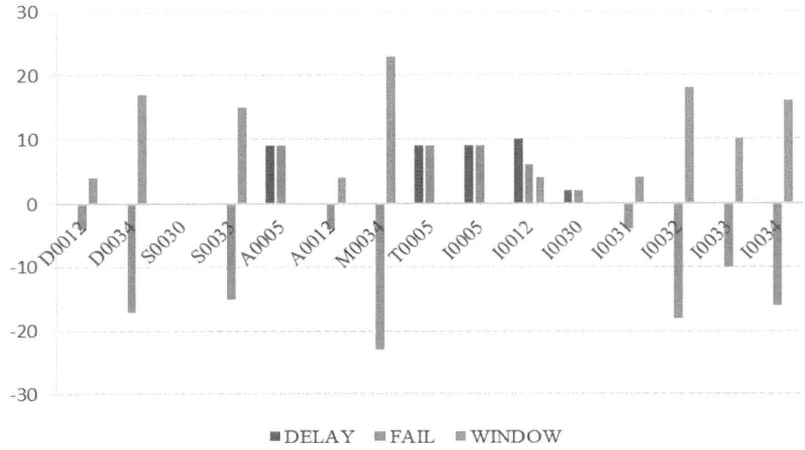

Figure 7.12 Earliest start date consumption.

After collating and listing all of the data, we will perform a visual analysis of them, as shown in Fig. 7.12 (Yung et al., 2020).

Because of a simple one-stream process for last three tasks, they have the similar *FAIL* value. Therefore, it is the previous task that causes the positive *FAIL* value.

*I*0030 is an interesting task. The data shows that this task could only be executed one day after being scheduled (*I*0033 has a *FAIL* value of 1). The task has a preceding resource-required *S*0030. Also, the FAIL value of *S*0030 is 0, which indicates that the delay of *I*0030 is not caused by *S*0030*s*. Therefore the reason for the delay in *I*0033 should be limited supervisor resources.

*I*0012 is another failed task. Analyzing task *I*0012 has the same logic as for *I*0033. The previous resource-required task has a *FAIL* value of − four, so the shortage of inspectors for task *I*0012 should be responsible for the failure. Therefore one of the key problems is the lack of inspectors.

Five tasks have no Windows at all. This is not acceptable, especially when there are two tasks may need to be reworked. If any of the on-window tasks are reworked, the total service level will at least drop to 95%, which indicates the service level of the task.

We have observed that most tasks have delay time more than their windows. There could be a lot of reasons for this, Think about the logic of CPN: there are only three possible factors for this: rework, limited

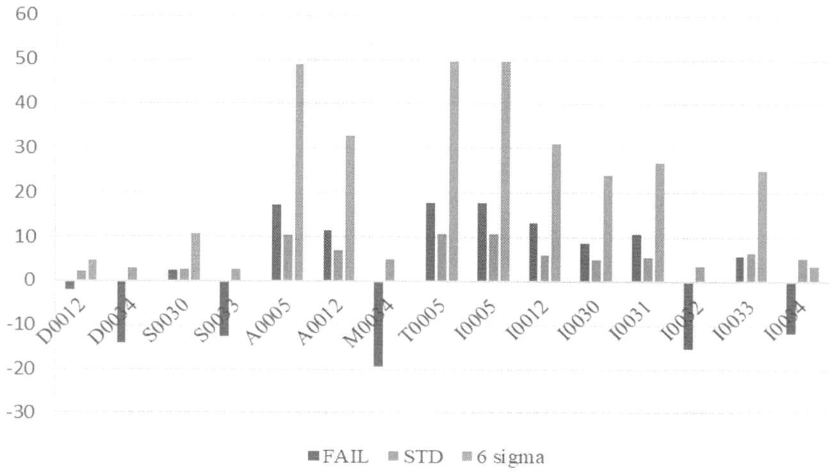

Figure 7.13 Six sigma analysis.

resources, and poor workflow configuration design. However, if we examine the task more carefully, tasks involving engineers generally have a delay time similar to or less than the window. Most of the delays occur for the inspectors. We can therefore conclude that the shortage of inspectors contributing to this critical problem: resource shortage.

2. Analysis about reliability

In order to analyze dependability, we have finished the following data, as shown in Table 7.6 (Yung et al., 2020). We sum up the standard difference (STD) and mean initiate time (AVG) of the simulation results, in which FALT is the difference between AVG and LSD, and FALT + 3 × STD is calculated based on the six sigma system.

Fig. 7.13 (Yung et al., 2020) lists several key issues. The six sigma system has been described previously. If the ceremony level is set to 99.865%, the value of Fail + 3 × STD should be negative. Unluckily, only four tasks can meet the criteria.

We have observed that almost all examination tasks with large standard deviations except *I*0034 and *I*0032. As the inspection work is reasonably short, the previous inspection tasks for distinctive parts maybe not be meanwhile finished, and the lack of inspectors will not lead a large standard deviation for all inspection tasks, which may cause potential rework. The standard deviation of the *I*0032 is much smaller because the purchase approach is very fast. *I*0034 is a similar task.

7.2.3 Improvement strategy

Here, we will use the ECRS method and constraint theory introduced in the methodology section to propose new scheduling strategies.

7.2.3.1 Workflow structure

According to constraint theory, the crucial path needs to be identified at first, and then analyze the whole system step by step. The stream of $PART0030$ is the crucial path which all tasks in it and has no window at all. The Stream of $CPN0012$ also has a reasonably smaller window that can be concentrated at the back of the $PART0030$ stream. Because $PART0033$ and $PART0034$ have a relatively big window, this means that their properties are stored in the final place.

The workflow structure of $PART0030$ must be changed because the rework ratio will straightly result in the entire process dependability being less than 95%. This can be solved through buffer times using the ECRS way, and one of that is to change the strategic objective in sourcing and, therefore modify the recent suppliers. Suppliers who can deliver faster are prior. Some models, such as Kraljic's model (CPN Group, 2018b) and the Strategic SRM model (Oosterhuis & Severens, 2018), which can Sustain this process. However, in many cases, shorter lead time (LT) may result in worse quality, which will improve rework rate. Another way to shorten purchasing time is to provide technical encouragement, such as through engineers or mechanisms for suppliers. Studies have shown that if technical and resource support is provided, it is possible to improve the performance of suppliers in terms of quality and delivery time (CPN Group, 2018b). Therefore engineers could be sent to suppliers to help to develop and manufacture $PART0030$. This approach could also dispose of inspection tasks because engineers helping with the procurement process can help ensure manufacturing quality.

7.2.3.2 Assemble, rework, and inspection

The appropriate strategy for disposing of rework was described in the previous section. Finding ways to eliminate the negative effects of rework is one of the most fantastic methods to mediate the negative effects of rework. The best way to do this is to send engineers to suppliers to help with the production process. Because this strategy has three theoretical active effects that we can make the following decisions based on ECRS method.

Incoming products should not be inspected, since engineers helping with the manufacturing process is able to take the place of inspection. As the entire production process is being checked, so another positive effect is reducing the rework risk.

One alternative is to acquire inspectors checking along with the gathering and test tasks. These two tasks may take more time to complete because of the quality control, but the risk of wholesale rework is eliminated. The assembly and test of tasks will take longer and more variable time to complete.

According to the proposed improvement strategy, the new model is then simulated, and the execution time of each task and the number of varieties of types of personnel are randomly produced. In this process, the six sigma system is still used as a neighborhood search rule to produce higher quality candidate solutions to determine the manufacturing time and number of people for each task. Finally, the simulation results of the new fabricating process are compared with the original model, and the outcome is as follows.

7.2.3.3 Result comparison

According to the improvement suggestion, we set up a new scheduling scheme. The model is placed in Fig. 7.14 (Yung et al., 2020) and the data is summarized in Tables 7.7 and 7.8 (Yung et al., 2020).

As shown in Table 7.7 (Yung et al., 2020), the NTLT performs much more superior in all aspects. The real earliest start time improved by 38%. The DELAY value of NTLT is less than that of OTLT, indicates that if the sufficient resources are provided, the running efficiency of the process will be increased. The FAIL value of NTLT is now negative, indicates that the task can now be completed on time. The NTLT has a window too, which means it has free time to deal with emergencies, such as rework. The WDELAYs for OTLT are significantly longer than NTLTIs, so the old workflow will perform much worse than the new workflow.

As the tables show, the average LT is significantly reduced in the new workflow. The FAIL value is now negative, so the process is more possible to succeed. The standard deviation of the new workflow is also slightly decreased; As a result, the new process is much better in risk control. The six sigma value drops from over fifty to six, which indicates the reliability of this process. The new workflow is within the range of plus or minus two sigmas, so the whole process has a 95% successful rate.

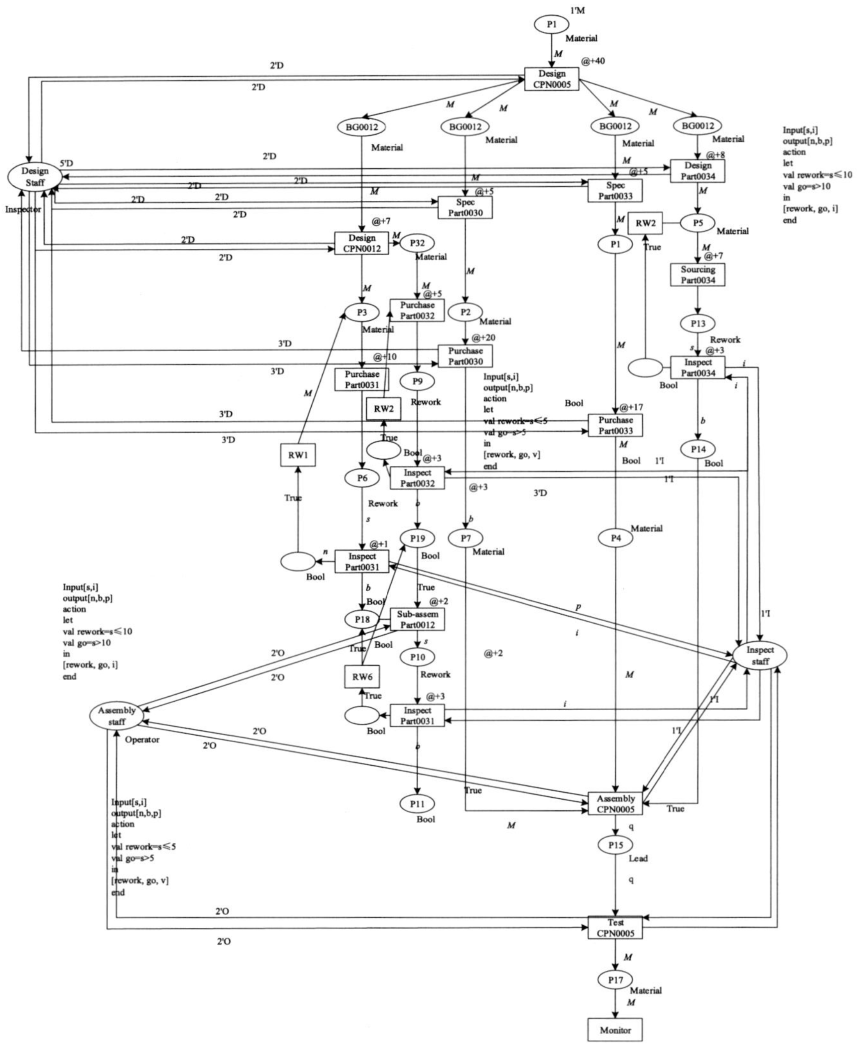

Figure 7.14 Optimized colored Petri net model proposed.

Table 7.7 Earliest start date comparison.

Task	AESD	TESD	LSD	DELAY	FAIL	WINDOW	ALSD	WDELAY
OTLT	109	100	100	9	9	0	202	102
NTLT	79	76	100	3	−21	24	131	31

Table 7.8 Reliability comparison.

Task	AVG	LSD	FAIL	STD	Six sigma
OTLT	118.06	100	18.06	11.06	51.24
NTLT	86.7	100	−13.3	6.67	6.71

7.3 Fault diagnosis of Rocket engine starting process

The fault diagnosis of partial observable Petri nets in rocket engine start-up process was proposed, and the application of petri nets in Aeronautics and Astronautics is further illustrated (Liu et al., 2017). Dr. Liu and his team came up with a new online fault diagnosis method of observable Petri nets to analyze the observable part of the system, so the unobservable part can be speculated based on the experience of observable part.

7.3.1 Online fault diagnosis method of observable Petri net

7.3.1.1 Observable Petri nets

At the above part of this chapter, we have introduced the Petri nets clearly. We will introduce a new concept of observable Petri Nets which was used in online fault diagnosis of rocket engine start process (Liu et al., 2017).

Partially overserved Petri nets could be presented as a tuple $POPN = (P, T, F, L, H)$; L here represents events array and H here represents signal array.

7.3.1.2 Partial observable Petri net online fault diagnosis method

For the specific method of partial observable Petri net fault diagnosis, two theorems are introduced firstly. One is related to the length of basic ignition sequence, which specifies the maximum length of ignition sequence and avoids the existence of infinite basic ignition sequences. Another theorem identifies the system, models the discrete event system, and expounds the change of ignition sequence. The third theorem judges the existence of the fault transition set in the ignition sequence by the ignition number vector corresponding to the ignition sequence in the observation sequence set. The third theorem gives the sufficient and unnecessary conditions of fault diagnosis and uses integer linear programming to solve the problem of fault diagnosis. The later theorem mainly focuses on the reliability of the fault set in the observation sequence.

7.3.1.3 Partial observable Petri nets for LOX/CH4 expansion cycle engine analysis of fault diagnosis results

The principle of mutual inspection is adopted in the diagnosis process. When the fault is caused by a node fault in the multinode diagnosis process, the system is considered to have a fault at that node. In order to verify the effectiveness of the fault diagnosis algorithm of partial observable Petri net for fault diagnosis, multiple and multiclass fault simulation experiments are carried out on the system.

Different fault types are set at different locations, and the optimal parameter k is selected according to the fault category. Statistical results: in 1000 experiments, 800 faults and 200 no faults are set. The actual algorithm diagnoses 793 faults, 201 no faults, and 6 uncertain faults. The reliability of algorithm diagnosis is 99.3%. According to the above data, it is proved that the fault diagnosis algorithm proposed in this chapter can meet the requirements of the practical application.

7.3.1.4 Example analysis and verification

Institute 101 of the sixth Academy of Aerospace Science and technology of China has developed a 60 ton liquid hydrogen methane rocket engine based on China's new-generation liquid oxygen kerosene rocket engine yf-77 through various extrusion tests, linkage tests, ignition tests, and so on. The main analysis and simulation experiments are LOX/CH4 expansion cycle engine starting process and LOX/CH4 expansion cycle engine fault diagnosis. Through the analysis of some observable Petri net fault diagnosis results of LOX/CH4 expansion cycle engine.

7.3.1.5 Conclusion

An important conclusion is that in the future research work, it is necessary to further study the constraint conditions and optimization calculation method for parameter K selection and fuse the state structure information of Petri net with the fault diagnosis algorithm. Due to the limited space, this part is briefly introduced. If you want to know more about this part, you can refer to the literature (Liu et al., 2017).

7.4 Conclusion

This paperhas presented a research approach for analyzing and improving the workflow of manufacturing Phobos-Grunt SOPSYS delivery unit via CPN.

First, we explained the development history, types, and applications of CPN. The performance of CPN implementation in workflow optimization is robust and efficient compared with the traditional method. Otherwise, some extensions of CPN are introduced to construct models of the complex system in the real world.

Second, we proposed a framework to simulate the process with the CPN tool, then using evaluation tools generate improvement proposals based on the simulation data.

Third, we introduced the CPN application in the Phobos-Grunt SOPSYS delivery unit manufacturing process and fault diagnosis during rocket engine start-up. It is the first attempt to apply CPN and random simulation in space project workflow optimization.

The first implication of this article is explaining the CPN model and introducing two actual cases to propose the capability of CPN and the feasibility in reducing LT in space projects. The results of the first case have demonstrated its achievement which helps avoid the potential failure in the SOPSYS delivery unit manufacturing process. The search in the second case demonstrates the feasibility of fault diagnosis.

The second implication of this article is the extened value for all the other time-sensitive single-unit manufacturing process. Because the logic of the process is similar to most other time-sensitive single-unit manufacturing processes, it can be used for workflow optimization in various space missions to the moon, Mars, and other planets.

According to the data related to resource and time obtained in this research and the requirement of precision and customization, the resource can be accurately positioned in the manufacturing system. In addition, it is necessary to build a long-term, stable, and resource-intensive manufacturing system according to this study's result, and the framework and techniques can be used in this system.

Overall, the research provides some explanations and experiences for applying CPN in various space missions in the future.

Acknowledgments

All authors have been contributed equally to this work. Corresponding author: Ming Gao (gm@dufe.edu.cn). This work was supported in part by the National Natural Science Foundation of China under grant nos. 71772033, 71831003, in part by the Scientific and Research Funds of the Education Department of Liaoning Province of China under grant no. LN2019Q14, in part by the Natural Science Foundation of Liaoning Province of China (joint open fund for key scientific and technological innovation bases) under grant no. 2020-KF-11-11, and in part by the Department of Industrial and Systems Engineering of the Hong Kong Polytechnic University under grant nos. H-ZG3K and 4-45-35-690E (Change'e Phase 3—Sample Return Instruments).

References

Al Khamisi, Y. N., Khan, M. K., & Munive-Hernandez, J. E. (2018). Assessing quality management system at a tertiary hospital in Oman using a hybrid knowledge-based system. *International Journal of Engineering Business Management, 10*, 1–13.

Bhargava, M., & Gaur, S. (2021). *Process improvement using six-sigma (DMAIC Process) in bearing manufacturing industry: A case study, . In IOP conference series: Materials science and engineering* (Vol. 1017, p. 012034). IOP Publishing, No. 1.

Byrne, B., McDermott, O., & Noonan, J. (2021). Applying lean six sigma methodology to a pharmaceutical manufacturing facility: A case study. *Processes, 9*(3), 550.

Che, J., & Gu, S. (2005). Supply chain management modeling and analysis based on Generalized Stochastic Petri Net. *Journal of Fujian Computer, 21*(08).

Chen, K. S., Chang, T. C., & Guo, Y. Y. (2020). Selecting an optimal contractor for production outsourcing: A case study of gear grinding. *Journal of the Chinese Institute of Engineers, 43*(5), 415–424.

Fang, Z., Ye, F., & Li, Y. (2006). A petri-net modeling and analysis of flexible supply chain. *China Mechanical Engineering*.

Figat, M., & Zieliński, C. (2020). Robotic system specification methodology based on hierarchical Petri nets. *IEEE Access, 8*, 71617–71627.

Gnanavel, C., Saravanan, R., Chandrasekaran, M., & Pugazhenthi, R. (2017). Case study of cycle time reduction by mechanization in manufacturing environment. *In IOP Conference Series: Materials Science and Engineering, 183*, 012023.

Jensen, K. (2007). Special section on coloured Petri nets. *International Journal on Software Tools for Technology Transfer, 9*(3–4), 2–34, 209–411.

Kasemset, C., Pinmanee, P., & Umarin, P. (2014, October). Application of ECRS and simulation techniques in bottleneck identification and improvement: A paper package factory. In *Proceedings of the Asia Pacific industrial engineering & management systems conference*.

Liang, W. (2010). Modeling and analysis of manufacturing supply chain business process based on Petri Net. *China Metalforming Equipment & Manufacturing Technology, 4*, 94–98. Available from https://doi.org/10.16316/j.issn.1672-0121.2010.04.031.

Liu, J. F., Sun, Y., Yu, J., Liu, W. Y., & Liu, H. Y. (2017). Fault diagnosis of partial observable Petri nets in rocket engine starting process. *Journal of Harbin Institute of Technology, 3*, 15–21.

Liu, Y. J., Huang, Q., Cheng, Y. Q., & Wu, J. J. (2012). Fault diagnosis method for liquid-propellant rocket engines based on the dynamic cloud-BP neural network. *Journal of Aerospace Power, 27*(12), 2842–2849.

Lu, J., Ou, C., Liao, C., Zhang, Z., Chen, K., & Liao, X. (2021). Formal modelling of a sheet metal smart manufacturing system by using Petri nets and first-order predicate logic. *Journal of Intelligent Manufacturing*, 1–21.

Miranda, F.A.A. (2011, June). Application of work sampling and ECRS (eliminate, combine, re-lay out, and simplify) principles of improvement at TO1 assembly. In *Proceedings of the twenty-first ASEMEP* national technical symposium (pp. 1-7).

Mkhaimer, L. G., Arafeh, M., & Sakhrieh, A. H. (2017). Effective implementation of ISO 50001 energy management system: Applying lean six sigma approach. *International Journal of Engineering Business Management, 9*, 1–12.

Nabi, H. Z., & Aized, T. (2019). Modeling and analysis of carousel-based mixed-model flexible manufacturing system using colored Petri net. *Advances in Mechanical Engineering*, *11*(12), 1687814019889740.

Oosterhuis, W. P., & Severens, M. J. (2018). Performance specifications and six sigma theory: Clinical chemistry and industry compared. *Clinical Biochemistry*, *57*, 12−17.

Pheng, L. S., & Hui, M. S. (2004). Implementing and applying six sigma in construction. *Journal of Construction Engineering & Management*, *130*(4), 482−489. Available from https://doi.org/10.1061/(ASCE)0733-9364(2004)130:4(482).

Phobos-Grunt's sad return. (2012). *Astronomy & Geophysics*, *53*, 10−11.

Wang, J., Fei, Z., Chang, Q., & Li, S. (2019). Energy saving operation of manufacturing system based on dynamic adaptive fuzzy reasoning Petri net. *Energies*, *12*(11), 2216.

Wang, J. W., Ip, W. H., Muddada, R. R., & Zhang, W. J. (2013). On Petri net implementation of proactive resilient holistic supply chain networks. *The International Journal of Advanced Manufacturing Technology*, 1−4. Available from https://doi.org/10.1007/s00170-013-5022-x.

Wang, Z., & Wei, D. (2009, October). Subnet abstract and transition refinement in Petri nets model. In *Proceedings of the second* international workshop on computer science and engineering (Vol. 2, pp. 418-421). IEEE.

Wang, W., & Bai, S. (2006). Research on supply chain hybrid petri nets simulation based on discrete-continuous combined modeling. *Science Technology and Engineering*.

Xi. S. (2009). *Research on knowledge representation based on Petri net*. Master's thesis, Shandong University. <https://kns.cnki.net/KCMS/detail/detail.aspx?dbname = CMFD2011&filename = 2010067257.nh>.

Yung, K. L., Gao, M., Liu, A., Hung Ip, W., & Jiang, S. (2020). Colored Petri net-based verification and improvement of time-sensitive single-unit manufacturing for the Soil preparation instrument of space missions. *Discrete Dynamics in Nature and Society*, *2020*.

Zahid, N. H., & Tauseef, A. (2020). Performance evaluation of a carousel configured multiple products flexible manufacturing system using Petri net. *Operations Management Research Through Theory*, 7.

Zhang, Y. (2005). The research on the mechanism and the methods of constructing supply chains for large construction enterprises. *Huazhong University of Science and Technology*.

Zuberek, W. M., & Kubiak, W. (1999). Timed Petri nets in modeling and analysis of simple schedules for manufacturing cells. *Computers & Mathematics with Applications*, *37*(11-12), 191−206.

http://cpntools.org/2018/01/12/monitors/, 2018a.

http://cpntools.org/, 2018b.

CHAPTER 8

Product performance model for product innovation, reliability and development in high-tech industries and a case study on the space instrument industry

Yuk Ming Tang[1,3], Andrew W.H. Ip[2,4], Yim Shan Au[1,3] and K.L. Yung[1]

[1]Department of Industrial and Systems Engineering, The Hong Kong Polytechnic University, Hong Kong, P.R. China
[2]Department of Mechanical Engineering, University of Saskatchewan, Saskatoon, SK, Canada
[3]Laboratory for Artificial Intelligence in Design, Hong Kong Science Park, New Territories, Hong Kong SAR, P.R. China
[4]Department of Industrial and Systems Engineering, The Hong Kong Polytechnic University, Hong Kong SAR, P.R. China

8.1 Introduction

8.1.1 Project background

According to Wolf and Terrell (2016), high-tech industries can be divided into two groups: high-tech manufacturing industries with a subset of the goods-producing industry and the high-tech service industry. However, there are very few studies that have directly addressed the relationship of innovation, product reliability, and product development in high-tech industries. Most of the studies pointed out that the above three dimensions are strongly related to product success or organization performance but most of them had limitations when applying their theories to the high-tech industries, which is a significant field for the future development of a lot of organizations and even for the future global economic environment. In addition, these three dimensions, innovation, product reliability, and product development are also affected by a series of factors. As suggested by Zhang, Lettice, and Zhao (2015), product innovation capability is influenced by funding, such as social capital, and individual creativity (Sok & O'Cass, 2015). In addition, it can be driven by product design through creating needs (Guo et al., 2016). Moreover, product reliability represents the quality standard of high-tech products and has a great impact on

IoT and Spacecraft Informatics
DOI: https://doi.org/10.1016/B978-0-12-821051-2.00006-4

high-tech industries. It can be dramatically improved by employee training and during the development process of the product and in the early design phase of product architecture decision making (Salonen, Holtta-Otto, & Otto, 2008). It is likely that the perspectives and challenges of product reliability can be predicted in the product development process (Yung, Tang, Ip, & Kuo, 2021). The design process is helpful for risk management, which can strongly improve the level of product reliability. Chau, Tang, Liu, Ip, and Tao (2021) investigated critical success factors for improving supply chain quality management in manufacturing. However, there is a lack of research studies investigating the relationship of innovation, product reliability and product development in high-tech industries, the correlation between those three dimensions can be observed by the subfactors which positively affect these aspects. The investigation is important not only to determine the product performance model, but also essential for the development of space instruments that involves a large of different processes such as inventory classification, component replenishment, etc. (Yung, Ho, Tang, & Ip, 2021). So, in this chapter, we seek to examine the relationship between the three dimensions and product performance in high-tech industries and the correlation among these three dimensions through analyzing their influence factors for improving the development of high-tech industries. In particular, we also describe a case study of the design and development of a space instrument to illustrate our discussion.

8.1.2 Project objectives

To achieve the research aim, a high-tech product and its research and development team will be analyzed in this paper. The chosen high-tech product is the soil preparation system (SOPSYS), which is a mission-critical space tool designed to extract the soil of Mars' innermost moon, Phobos, looking for signs of life. This experiment was initiated by the Russian Federal Space Agency to collect soil samples on the Martian moon Phobos under a vacuum and low gravity environment. The research and development process of SOPSYS is analyzed as a case study to discuss the relationship of innovation, product reliability and product development in high-tech industries.

8.2 Literature review

8.2.1 Definition of innovation

Production innovation has significant meaning for not only the firm and organizations directly involved but also the whole society by improving our daily life. Ferràs (2008) stated that innovation is an important tool for the organization and a vital process for its strategic organization management. In addition, as pointed out by Zuñiga-Collazos et al. (2015), there are four different types of innovation which are described by the Organization for Economic Co-operation and Development (OECD) and Eurostat as given in Table 8.1.

In high-tech industries, innovation mainly focuses on products and processes, as only those two dimensions of innovation can truly improve the product itself and create new things. Furthermore, although an innovative product can bring a relatively high payoff and great benefit for the organization, the development and research process of such products can be risky and costly (Evanschitzky, Eisend, Calantone, & Jiang, 2012). Hence, finding out the effectiveness and success factors for new product innovations is essential due to the high failure rate and high cost of innovation.

8.2.2 Factors affecting innovations

Summarizing the results of the relevant literature, there are four main effective factors of innovation, especially for product innovation. The first is due to the "individual." Azarmi (2016) put forward that this factor refers to the characteristics of an individual who takes part in the overall scheme or process of product innovation. It involves three dimensions, social skills and teamwork, education and diversity.

The second factor affecting innovation is social capital. As mentioned above, powerful financial support is essential for product innovation, especially in the high-tech industries. However, social capital is not only about funding but also is related to different kinds of potential social resources.

Table 8.1 Four types of innovation.

Types of innovation	Description
Product innovation	Referring to a significant change in the characteristics of goods and services, meaning new products and improved existing products.
Process innovation	Referring to significant changes in the methods of production and distribution
Organizational innovations	Referring to the implementation of new methods in an organization;
Marketing innovation	Referring to all practices or developing new marketing processes, and selling products or services

The third factor affecting innovation is management. The management concerns about different aspects which refer to the significant influence of the whole process of innovative activities. According to Azarmi (2016), as an important influence factor of innovation, management primarily includes five different aspects: knowledge management, intellectual property (IP) management, human resource (HR) management, risk management, and financial management. All of them concern the resource allocation and distribution to different but important aspects, which are also closely related to the performance of innovative activities.

The fourth-factor affects innovation in product design. As a vital part of the innovative activity process, product design determines innovation performance. The ultimate purpose of innovative products is to fulfill the needs of customers and improve their satisfaction level in using those products.

8.2.3 Definition of product reliability

The most common understanding of product reliability is about reducing faults and ensuring smooth operation. It can be regarded as a kind of ability to operate its design and multiple functions. Mackelprang, Habermann, and Swink (2015) suggested that product reliability is an important indicator for measuring product performance based on whether it can match its specified functionality. Moreover, it reveals the ability to operate its specified function under preset conditions. They also suggested that product reliability is likely to fail because of product design limitations and production limitations, which would have a great impact on the ultimate product performance, resulting in a loss of profit. Reliability failures would have a great influence not only on the customers but also the manufacturers. Although a manufacturer cannot avoid product failures completely, product reliability can be controlled and monitored during preproduction decision making and in avoiding product failures as far as possible (Murthy, Rausand, & Virtanen, 2009). Hence, finding out the factors affecting product reliability can greatly help manufacturers to improve product reliability.

8.2.4 Factors affecting product reliability

To some extent, product quality is the core value of a new product and quality control against various uncertainties can be enabled to increase the level of product reliability (Mrugalska & Tytyk, 2015). Quality is determined by its core technologies, hence, the first and the most important factor affecting product reliability is technology and knowledge support.

The second factor affecting product reliability is funding. As mentioned before, developing an innovative new product, especially for high-tech industries, is costly and risky. As suggested by Murthy et al. (2009), adequate funding is the initial and vital condition to guarantee product quality and new product research and development. The estimates of increased reliability always come with increasing development costs (Levin & Kalal, 2003). Adequate funding is essential to support and affect these factors during the development process in enhancing product reliability.

The third factor affecting product reliability is product design. During the product design process, the designers and manufacturers need to make decisions on what kinds of components, technologies, and materials need to be used for designing and producing products. These decisions made in the design process would have a great impact on various aspects of the product, which are reflected in the product appearance, functionality, complexity, quality, and especially reliability. Identifying the linkage between product reliability and decision-making in design will bring a great benefit for the later development process (Salonen et al., 2008).

8.2.5 Definition of new product development

Generally, a new product can be classified into two different aspects, according to the newness to the market and the firm. As Crawford and Di Benedetto (2008) stated, there are six categories of new products, new-to-the-world products, new-to-the-firm products, additions to existing products, improvements and revisions to existing products, repositioning, and cost reductions products. Crawford and Di Benedetto (2008) also outlined the basic NPD process which is composed of opportunity identification and selection, concept generation, concept or project evaluation, development, and testing and launch. Atilgan-Inan, Buyukkupcu, and Akinci (2010) stated that "as NPD is a vital and risky process due to the hundreds of millions of dollars it can cost in case of a failure, its determinants should be carefully analyzed" (p. 97). Therefore various reports in the literature refer to defining different factors affecting new product development (NPD) and its processes (Brown & Eisenhardt, 1995). From the previous research, three main factors affect NPD.

8.2.6 Factor affecting new product development

The first factor affecting NPD is human resource management (HRM). Atilgan-Inan et al. (2010) pointed out that the new NPD process is an important stage for cross-functional integration. Based on the multifunctional tasks, the way to design and manage an NPD team is significant since NPD teams can direct and control the development process (Soderquist & Kostopoulos, 2012). Therefore excellent management of human resources (HR) will bring great benefits for NPD.

The second factor affecting NPD is product design. Roper, Micheli, Love, and Vahter (2016) pointed out that investment in design can make a significant contribution to NPD as it involves designers, who are the functional specialists, at different stages of the NPD process and can have a positive influence on product performance.

The third factor affecting NPD is government support and adequate investment. Government support can make a great contribution to the success of product development in the form of direct investment, tax policy, risk fund, industry policy, and so on (Zhang et al., 2016).

8.2.7 Product relationships in high-tech industries

Nowadays, product innovation, which is affected by several factors resulting in various innovation outcomes, has become an important competence measure and driving force of a firm or organization (Rocheska, Nikoloski, Angeleski, & Mancheski, 2017). To be specific, much previous research illustrated that product innovation is a key indicator of the ultimate new product performance (Evanschitzky et al., 2012; Guo et al., 2016). However, there is no direct research referring to the relationship between the subfactors of product innovation, reliability, product development and new product performance in high-tech industries. Identifying the determinants of those three dimensions and their relationship toward new product performance can provide a helpful and valuable reference for project managers, manufacturers, and designers when considering developing new products. By analyzing the subfactors of these three aspects of new product performance in high-tech industries, the project team can more easily manage the resources and risks, avoiding failure and maximizing the product performance. As mentioned above, from the previous literature, product innovation, reliability and product development are the four affecting factors respectively, which are listed in Table 8.2.

Table 8.2 Affecting factors of innovation, product reliability, and product development.

Subfactors dimensions	Individual	Adequate funding	Product design	Technology & knowledge support	Government support	Human resource management
Innovation	✓	✓	✓			✓
Product reliability		✓	✓	✓		✓
Product development		✓	✓		✓	✓

Table 8.2 shows that the three aspects of product performance have several of the same determinants and their subfactors. Therefore according to their subfactors and their relationship toward innovation, reliability and product development, six hypothesizes can be put forward as discussed in Chapter 3.

8.3 Methodology

8.3.1 Research framework

From the previous research results, it can be summarized up that product performance is highly related to three dimensions: product innovation activities, product reliability, and product development. However, there are no studies that directly and systematically investigate that how these three dimensions affect product performance. Some previous research mentioned a series of subfactors that play an important role in those three elements, with four factors affecting product reliability: adequate funding, product design, high-end technology and knowledge and HRM. In addition, product development is also influenced by adequate funding, product design, government support and HRM.

The research framework is divided into three parts in this study as shown in Fig. 8.1.

8.3.2 Research hypothesis

The hypothesis of this study suggests that those subfactors of product innovation, product reliability and product development are also strongly relevant to the ultimate product performance. Table 8.3 shows the research hypothesis for investigation of the relationship between three research dimensions: product innovation, product reliability and product

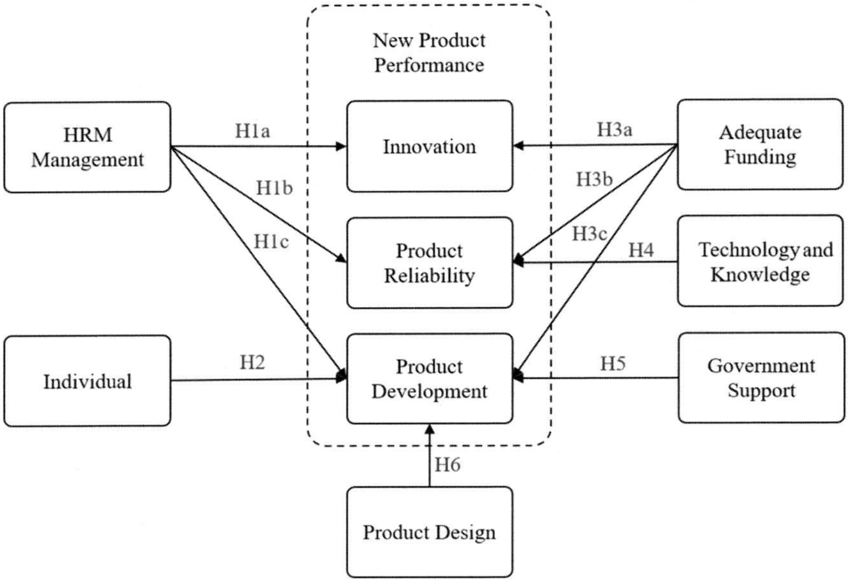

Figure 8.1 Information system model for new product performance and the individual hypothesis.

Table 8.3 Research hypothesis.

Hypothesis	Description
H1a	Positive relationship between HRM management and innovation
H1b	Positive relationship between HRM management and product reliability
H1c	Positive relationship between HRM management and product development
H2	Positive relationship between Individual and product development
H3a	Positive relationship between adequate funding and innovation
H3b	Positive relationship between adequate funding and product reliability
H3c	Positive relationship between adequate funding and product development
H4	Positive relationship between technology & knowledge and product reliability
H5	Positive relationship between government support and product development
H6	Positive relationship between product design and product development

development, and product performance, to seek how the three dimensions affect the product performance.

8.3.3 Data collection

In this research, data is obtained from a focused group survey with the research and development team of the SOPSYS, and 40 other people who are interested in or work in the hi-tech industry. The survey instrument is a questionnaire with two versions. One is distributed to the project team of SOPSYS and the other one is distributed to the other 40 people. Mostly, the questions in both two questionnaires are the same, while the version distributed to the project team has three more questions aiming at the specific product of SOPSYS. The instrument is based upon the literature in the subfactors of innovation, product reliability and product development and their relationship with product performance in the hi-tech industry, aiming to obtain a comprehensive overview of the issue. The questionnaire used in this study is summarized in Table 8.4.

8.3.4 Case study of the soil preparation system

The SOPSYS was used for the Phobos Grunt mission to Mars in November 2011. As a typical hi-tech product, SOPSYS was a prizing-awarding aerospace instrument for the internationally joint project. The ultimate purpose of the SOPSYS was to collect soil samples from Phobos, a satellite of Mars, and have primary treatment of the soil samples. Then, the samples were brought back to Earth for comprehensive scientific research into Phobos, with Mars and Martian space the final objective.

This study aims to investigate the relationship between innovation, product reliability, and product development in the hi-tech industry by identifying how these three dimensions affect the ultimate hi-tech product performance. As a successful hi-tech product model, SOPSYS and its project team provide valuable sources to be researched and studied.

8.4 Methodology

8.4.1 Respondents

In this study, samples were collected by a questionnaire that was distributed to the SOPSYS project team. Around 40 samples were collected by the questionnaire which was distributed online or by the social media platform. The respondents were divided into two groups. The first group

Table 8.4 The questionnaire summary.

#	Impact of individual ability on the ultimate performance of the hi-tech industry
1a	Individual organizing ability
1b	Individual communicating ability
1c	Individual innovating ability
1d	Individual learning ability
1e	Individual leadership
1f	Individual academic ability in specific field
#	Impact of adequate funding on the ultimate performance of the hi-tech industry
2a	Innovation
2b	Product reliability
2c	Product development
#	Impact of elements of product design on the performance of the hi-tech industry
3a	Appearance design
3b	Function design
3c	Internal operation system design
#	Impact of human resource management of the R&D team on the ultimate performance of the hi-tech industry
4a	Innovation
4b	Product reliability
4c	Product development
5	Impact of technology and knowledge support on the ultimate performance of the hi-tech industry
6	Impact of government support on the ultimate performance of the hi-tech industry
#	Impact of the element of government support on the ultimate performance of the hi-tech industry
7a	Individual organizing ability
7b	Individual communicating ability
7c	Individual innovating ability
7f	Individual academic ability in specific field

consisted of the 30 samples collected from the SOPSYS project team. The second group was the 40 samples collected from people who were interested in or work in the hi-tech industry.

8.4.2 Results of questionnaire

The questionnaire was distributed to investigate the impact factor of innovation, product reliability, and product development in the hi-tech industry. It was distributed online and sought respondents who were interested in, work in or have some experience in the hi-tech industries. After distributing for one week online and by social media platform, a total of 40 useful samples was collected.

8.4.2.1 Results of individual factors

According to the survey, most of the respondents thought that individual ability would have an impact on the ultimate performance of hi-tech products, and three of them held the opposite view. Even though the 37 respondents affirmed the impact of individual ability on hi-tech product performance, they had different opinions for the impact extent of different elements of individual ability, as shown in Table 8.5.

Table 8.5 shows that the respondents had different opinions on the impact extent of different elements of individual ability, with most considering that individual academic ability in the specific areas and individual ability to innovate were more important than other elements, represented by a higher score.

8.4.2.2 Results of product factors

According to the survey, 24 respondents (60%) regarded adequate funding had a great impact on the hi-tech product performance, while 14 respondents thought that the impact extent was moderate, and the rest of them considered that the impact of adequate funding was relatively slight. At the same time, through the investigation of how adequate funding impacts the ultimate performance of the hi-tech product in the three dimensions, innovation, product reliability, and product development, the respondents held similar views, as shown in Table 8.6.

Table 8.5 The statistical result of the question regarding the impact of individual ability on the ultimate performance of the hi-tech industry.

#	Impact of individual ability on the ultimate performance of the hi-tech industry	Mean	SD
1a	Individual organizing ability	3.55	3.17
1b	Individual communicating ability	3.825	3.42
1c	Individual innovating ability	4.275	3.83
1d	Individual learning ability	4.025	3.59
1e	Individual leadership	3.5	3.11
1f	Individual academic ability in specific field	4.325	3.89

Table 8.6 The statistical result of the question regarding the impact of adequate funding on the ultimate performance of the hi-tech industry.

#	Impact of adequate funding on the ultimate performance of hi-tech industry	Mean	SD
2a	Innovation	3.92	3.55
2b	Product reliability	4.10	3.64
2c	Product development	4.3	3.83

Table 8.7 The statistical result of the question regarding the impact of elements of product design on the performance of the hi-tech industry.

#	Impact of elements of product design on the performance of the hi-tech industry	Mean	SD
3a	Appearance design	3.52	3.08
3b	Function design	4.35	3.89
3c	Internal operation system design	4.22	3.80

8.4.2.3 Results of product design factor

According to the survey, all respondents were convinced that product design would have a certain impact on the final performance of the hi-tech product. However, different aspects of product design surely have different impact extent on the hi-tech product performance according to the results from the survey. The respondents were asked to rank the extent of impact level of each of the aspects of the product design from "Almost no impact" (equal 1) to "Strong impact" (equal 5). As shown in the following Table 8.7, the appearance of the product got the lowest score, only 141 in total, while the product function design and internal operation system design got a similar and higher score around 170.

8.4.2.4 Results of human resource management factor

According to the survey, all the 40 respondents considered that the HRM in a project team certainly has an impact on the hi-tech product performance. More than half of the 40 respondents regarded the impact extent of HRM as moderate, while another 11 people thought that HRM only had a slight impact on the hi-tech product performance, and the rest were convinced that HRM was strongly affected the hi-tech product performance (Table 8.8).

Table 8.8 The statistical result of the question regarding the impact of human resource management of the R&D team on the ultimate performance of the hi-tech industry.

#	Impact of human resource management of the R&D team on the ultimate performance of the hi-tech industry	Mean	SD
4a	Innovation	2.93	2.46
4b	Product reliability	3.63	3.31
4c	Product development	3.80	3.42

Table 8.9 The statistical result of the question regarding the impact of technology and knowledge support on the ultimate performance of the hi-tech industry.

5	Impact of technology and knowledge support on the ultimate performance of the hi-tech industry	Mean	SD
		3.28	2.66

Table 8.10 The statistical result of the question regarding the impact of government support on the ultimate performance of the hi-tech industry.

6	Impact of government support on the ultimate performance of the hi-tech industry	Mean	SD
		3.35	2.72

8.4.2.5 Results of technology and knowledge support factor

According to the survey, most of the respondents considered that the hi-technology and knowledge support have a moderate or strong impact on hi-tech product performance. Approximately about 12.5% of the respondents held the opinion that the hi-technology and knowledge support only had a slight impact on the hi-tech product performance (Table 8.9).

8.4.2.6 Results of government support factor

According to the survey, all the 40 respondents were convinced that government support would have a certain impact on the final performance of the hi-tech product. However, the government provides different support in various aspects. Table 8.10 shows that most people regard that government policy support as playing the most important role in hi-tech product performance. Government financial and technical support got a similar score according to their importance level and relatively less than the

Table 8.11 The statistical result of the question regarding the impact of the element of government support on the ultimate performance of the hi-tech industry.

#	Impact of the element of government support on the ultimate performance of the hi-tech industry	Mean	SD
7a	Individual organizing ability	3.78	3.36
7b	Individual communicating ability	4.08	3.67
7c	Individual innovating ability	3.73	3.33
7f	Individual academic ability in specific Field	3.48	3.12

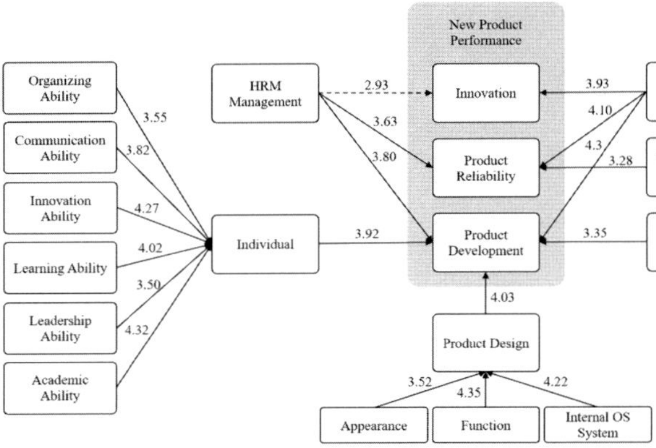

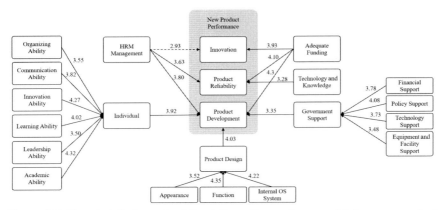

Figure 8.2 The statistical result of the information system model for new product performance proposed regarding the hi-tech industry.

government policy support factor. Most people considered that government equipment and facility support was the least important (Table 8.11).

Fig. 8.2 shows the statistical result of the proposed information system model for new product performance regarding the hi-tech industry. A value of 4.0 represents the factor being strongly positive-related with new product performance. The value of 3.0 represents the factor being slightly positively related to new product performance. The value of 2.0 represents the factor being negatively related to new product performance.

8.4.3 Result of case study on soil preparation system compared to hi-tech industry

As mentioned previously, according to the analysis of the results of the questionnaires, the SOPSYS project team agreed that individual ability has an impact on the ultimate performance of a hi-tech product, while three of the respondents who were interested in or work in the hi-tech industry held the

Table 8.12 The statistical result of question regarding the impact of individual ability on the ultimate performance of hi-tech industry and soil preparation system.

1	Impact of individual ability on the ultimate performance of hi-tech industry and SOPSYS	Hi-tech industry		SOPSYS	
		Mean	SD	Mean	SD
1a	Individual organizing ability	3.55	3.17	3.23	2.77
1b	Individual communicating ability	3.82	3.42	4.10	3.59
1c	Individual innovating ability	4.27	3.83	3.67	3.20
1d	Individual learning ability	4.02	3.59	4.57	4.07
1e	Individual leadership	3.50	3.11	3.17	2.67
1f	Individual academic ability in specific field	4.32	3.89	4.80	4.31

opposite opinion. As suggested by Walker (2002), the individual ability has a certain impact on group performance by affecting the working process step by step. There is no doubt that, from researching and developing to launching a hi-tech product, there is a need to go through a series of processes for a relatively long time, and also, the project involves working in a group or a team. Su (2006) also put forward that the individuals in each group would have an influence on the overall learning ability of the group and finally play an important role in the group performance.

In addition, the survey also investigated the impact extent of different individual ability aspects on the performance of a hi-tech product. Comparing the results from the two questionnaires, as shown in Table 8.12, it can be seen that people from the two research groups have different opinions for some aspects of individual ability.

It is clearly shown that overall, the standard deviation of different aspects of individual ability from Questionnaire 1 is relatively smaller than the results of Questionnaire 2. It means that people from the same project team may have similar opinions rather than people who have different hi-tech industry backgrounds. Furthermore, comparing the results collected from the two different research groups, it can be observed that respondents from the SOPSYS project consider individual communication is more important than the respondents from the other groups.

8.4.3.1 Adequate funding factors analysis

As mentioned in Chapter 4, according to the analysis of the results in Questionnaires 1 and 2, almost 60% of the respondents from the SOPSYS project team considered that adequate funding had a moderate impact on the ultimate performance of the product. Nevertheless, 60% of the respondents who were interested or work in the hi-tech industry thought that adequate funding would have a strong impact on the final performance of a hi-tech product. Szylar (2013) put forward that funding is quite important for a wide range of industries and it plays an important role in the risk management in different processes of the project. Even though the results from the two research groups had small differences between the impact extent of adequate funding, there is no doubt that it does have an important meaning for the hi-tech product performance.

Regarding the question of how adequate funding affects hi-tech products, the two research groups also provided different opinions in the three dimensions: innovation, product reliability, and product development.

Table 8.13 reveals that the respondents from the SOPSYS project team considered that the impact of adequate funding is mainly focused on product reliability and product development. However, most of the respondents from the other research groups considered that the impact extent of adequate funding on innovation and product reliability was similar and relatively lower than its impact on product development. Despite that the respondents from the two research groups having different opinions on the impact extent of adequate funding on the three dimensions of the product, it certainly has a visible impact on the hi-tech product performance.

8.4.3.2 Product design factors analysis

According to the analysis of the results of questionnaires, all the respondents from the two research groups agree that product design surely has

Table 8.13 The statistical result of the question regarding the impact of adequate funding on the ultimate performance of the hi-tech industry and soil preparation system.

2	Impact of adequate funding on the ultimate performance of hi-tech industry and SOPSYS	Hi-tech Industry		SOPSYS	
		Mean	SD	Mean	SD
2a	Innovation	3.92	3.55	4.47	3.97
2b	Product reliability	4.10	3.64	4.67	4.16
2c	Product development	4.30	3.83	3.47	3.02

an important influence on hi-tech product performance. Cooper (2019) suggested that product design is a crucial factor for product success. It is the first connection point between the product and the customers. Its values are not only created by its "form" or "function," but also by a self-expressive dimension (social and altruistic value) (Kumar & Noble, 2016).

Product design is related to many aspects. The survey aimed to investigate three aspects of product design, product appearance design, function design, and internal operating system design, and their impact on hi-tech product performance. The respondents were asked to rate the extent of the impact level of each of the aspects of the product design from "Almost no impact" (equal 1) to "Strong impact" (equal 5). Data collected from the two questionnaires are summed up in Table 8.14.

As the data shown in Table 8.14, the appearance design factor had a lower average score from research group one than from research group two. This means that most of the SOPSYS researchers considered that appearance design had a relatively slight impact on the product performance while the majority of the respondents held opposite opinions, most likely due to different hi-tech products. But overall, data from both groups showed that function design had the highest average score and lowest standard deviation, which means that most of the respondents were convinced that function design had a significant impact on hi-tech product performance. Finally, internal operation system design also had a relatively high average score in both groups showing its importance.

8.4.3.3 Human resource management factor analysis

The data distribution of the impact extent for the HRM factor from both research groups was very similar. More than half agreed that HRM has a moderate impact on hi-tech product performance (Table 8.15). Around

Table 8.14 The statistical result of the question regarding the impact of elements of product design on the performance of the hi-tech industry and soil preparation system.

3	Impact of elements of product design on the performance of hi-tech industry and SOPSYS	Hi-tech industry		SOPSYS	
		Mean	SD	Mean	SD
3a	Appearance design	3.52	3.08	3.10	2.61
3b	Function design	4.35	3.89	4.97	4.44
3c	Internal operation system design	4.22	3.80	4.90	4.39

Table 8.15 The statistical result of the question regarding the impact of human resource management of the R&D team on the ultimate performance of the hi-tech industry and soil preparation system.

4	Impact of human resource management of the R&D team on the ultimate performance of hi-tech industry & SOPSYS	Hi-tech industry		SOPSYS	
		Mean	SD	Mean	SD
4a	Innovation	2.93	2.46	3.07	2.56
4b	Product reliability	3.63	3.31	4.03	3.61
4c	Product Development	3.80	3.42	4.23	3.79

20% of them considered that this impact was relatively slight and the rest of the respondents held the opposite opinion, regarding HRM having a strong impact on hi-tech product performance. More and more emphasis was put on the fact that HRM is a crucial factor for product success, as well as the success of the company or organization. Kehoe, Collins, and Chen (2017) also emphasized the impact of HRM on the performance of knowledge-intensive work. They suggested that a "relationship-oriented HR system contributes to unit performance through its positive effects on employees' collective access to knowledge by fostering a social context and interpersonal exchange conditions which support employees' ongoing access to knowledge flows within and outside their unit and broader organization" (p. 1222). An R&D project for a hi-tech product involves typical knowledge-intensive work. Hence, according to the previous research and data collected from the survey, HRM has a significant impact on hi-tech product performance. In addition, by investigating how the HRM affects hi-tech product performance, the survey data showed that most of the respondents from both two research groups were convinced that HRM mainly affects product development, then product reliability, and last but least innovation.

8.4.3.4 Technology and knowledge support factor analysis

The respondents from the two research groups had different opinions on the impact extent of technology and knowledge support factors on hi-tech product performance. Data collected from two surveys are shown in the following Table 8.16.

Table 8.16 indicates that almost all of the respondents from research group one agreed that technology and knowledge support have a strong

Table 8.16 The statistical result of the question regarding the impact of technology and knowledge support on the ultimate performance of hi-tech industry and soil preparation system.

5	Impact of technology and knowledge support on the ultimate performance of hi-tech industry and SOPSYS	Hi-tech industry		SOPSYS	
		Mean 3.28	SD 2.66	Mean 3.90	SD 3.11

impact on the hi-tech product performance, except one person holding a different opinion. However, in terms of the results from research group two, more people were convinced that the impact extent of technology and knowledge was moderate and people who thought that the impact extent was strong were fewer. Also, only 12.5% of respondents from the research group regard a slight impact of technology and knowledge support. As mentioned before, there is no doubt that an R&D project for a hi-tech product involves typical knowledge-intensive work. Meeting the expectation of this study, all the respondents affirmed the impact of technology and knowledge support on hi-tech product performance. Merminod and Rowe (2012) suggested that technology and knowledge support play an important role in NPD, which increasingly takes place across organizations. The competition of hi-technology and knowledge support and transfer from universities are becoming more and more intensive, and are essential for product success (Kesting & Wurth, 2015).

8.4.3.5 Government support factor analysis

The results of the impact extent of government support on hi-tech product performance from the two research groups were very similar. Results from research group one showed that almost 60% of respondents regarded government support as having a moderate impact on SOPSYS product performance. Respondents who considered the impact extent of this factor as slight or strong amounted to 25% and 16%, respectively. In terms of research group two, most of the people also agreed that government support has a moderate impact on hi-tech product performance.

In this study, the survey mainly chose four different aspects of government support, government financial support, government policy support, government technology support, and government equipment and facility support, and investigated their different impact extents on hi-tech product performance. The respondents were asked to rate the extent of the impact

Table 8.17 The statistical result of the question regarding the impact of the element of government support on the ultimate performance of the hi-tech industry and soil preparation system.

7	**Impact of the element of government support on the ultimate performance of hi-tech industry and SOPSYS**	**Hi-tech industry**		**SOPSYS**	
		Mean	**SD**	**Mean**	**SD**
7a	Individual organizing ability	3.78	3.36	3.53	3.07
7b	Individual communicating ability	4.08	3.67	2.93	2.50
7c	Individual innovating ability	3.73	3.33	3.87	3.38
7f	Individual academic ability in specific field	3.48	3.12	3.27	2.80

Table 8.18 The statistical result of the question regarding the impact of government support on the ultimate performance of the hi-tech industry and soil preparation system.

6	Impact of government support on the ultimate performance of hi-tech industry and SOPSYS	Hi-tech Industry		SOPSYS	
		Mean	SD	Mean	SD
		3.35	2.72	3.30	2.68

level of each of the aspects of the government support from "Almost no impact" (equal 1) to "Strong impact" (equal 5). As shown in Table 8.17, the results collected from the two research groups showed some differences in the impact extent of the four different aspects of government support.

As shown in the data in Table 8.18, the average scores of four different aspects of government support collected from questionnaire one were very close. All of them were with a range of 3–4, which means that the respondents from the SOPSYS project team considered the four aspects of government have a similar impact extent on the hi-tech product ultimate performance, which is moderate. In terms of research group two, respondents think more highly of government policy support. This may be due to the impact extent of government support varying in different specific hi-tech projects. Despite that, some previous studies also placed great emphasis on the fact that government support can make a great contribution to the success of product development in the forms of direct

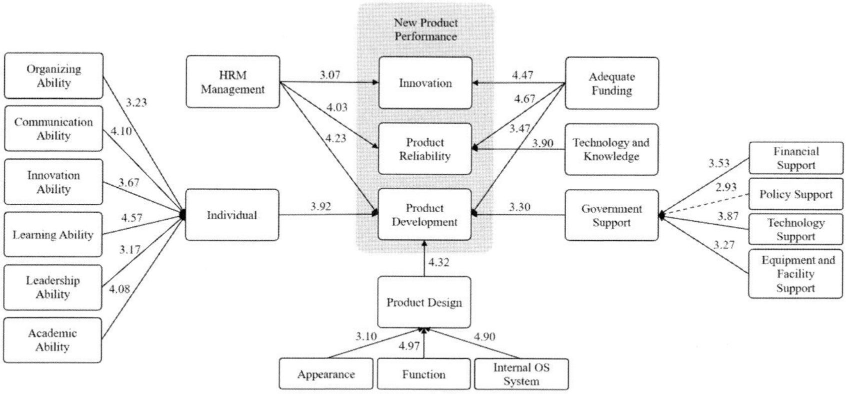

Figure 8.3 The statistical result of the information system model for new product performance proposed regarding the soil preparation system team.

investment, tax policy, risk fund, industry policy, and so on (Zhang, Li, & Liu, 2016).

Fig. 8.3 shows the statistical result of the information system model for the new product performance proposed regarding the SOPSYS team. A value above 4.0 indicates the factor is strongly positive-related with new product performance. A value above 3.0 indicates the factor is slightly positively related to new product performance. A value above 2.0 indicates the factor is negatively related to new product performance.

8.5 Discussion

8.5.1 Relationship of innovation, reliability and product development in high-tech industries

There is no doubt that product reliability is another vital element to measure and evaluate the success of NPD. Product reliability is closely related to product quality and safety, which directly influence product performance and success (Levin & Kalal, 2003). It is a crucial indicator for monitoring the performance of a product, especially in hi-tech industries.

As a risky and vital process, NPD has become an important competitive force for companies or organizations in hi-tech industries (Atilgan-Inan et al., 2010). Hi-tech product performance is strongly influenced by NPD processes (Crawford & Di Benedetto, 2008).

Therefore according to these previous literature reviews, it was shown that innovation, product reliability, and product development certainly influence hi-tech product performance. To define their relationships in hi-tech industries, six kinds of subfactors of those three dimensions are summed up from previous research and are investigated in this study. The hypotheses of the positive relationship between those six subfactors and hi-tech product performance were put forward.

8.5.1.1 Relationship between individual factor and hi-tech product performance

According to Pratoom and Savatsomboon (2012), individuals can make different contributions to a team project according to their different individual abilities. As suggested by Walker (2002), individual ability has a certain impact on group performance by affecting the working process step by step. However, there is no direct evidence proving the relationship between individual ability factors and hi-tech product performance. To investigate and verify the hypothesis of a positive relationship between the individual factor and new product performance in hi-tech industries, a questionnaire survey was distributed to two research groups.

The survey results indicated that people no matter from a specific hi-tech project team or who were interested in or work in hi-tech industries were convinced that individual ability has an impact on the hi-tech product performance, even though they expressed different opinions toward the impact extent of different individual ability aspects. Moreover, respondents from a specific project team think more highly of the individual learning ability than the respondents from the other research group. Thus the assumption that a positive relationship between the individual ability factor and hi-tech product performance is substantiated.

8.5.1.2 Relationship between adequate funding and hi-tech product performance

There is no doubt that a NPD project in a hi-tech industry is always costly and risky. The estimates of increased reliability always show the increasing development costs (Levin & Kalal, 2003). Szylar (2013) put forward that funding is quite important for a wide range of industries and it plays an important role in the risk management in different processes of the project. On the contrary, inadequate funding may cause a series of serious defects in product quality, which would require another large amount of funding to settle. Besides, adequate investment acquisition in different ways and social resources are essential conditions for supporting the research and development of a new product. Based on the previous research, the assumption that a positive relationship between adequate funding and hi-tech product performance was put forward and investigated using the survey. By analyzing the data collected from the survey in two research groups, it was found that all of the respondents were convinced that adequate funding plays a significant role in affecting hi-tech product performance, achieved by influencing product reliability and product development. Therefore the assumption that a positive relationship between adequate funding and hi-tech product performance is substantiated.

8.5.1.3 Relationship between product design and hi-tech product performance

Decision-making in the product design process has a great impact on various aspects of the product. Identifying the linkage between product reliability and decision-making in design brings great benefits for the later development process (Salonen et al., 2008). Moreover, investment in design can make a significant contribution to NPD as it can positively affect product performance (Roper et al., 2016). According to the literature review of product design and its impact on product performance, this study provides insight into the relationship between product design and hi-tech product performance. Referring to the data collected from the survey, respondents from both research groups agreed that product design surely has an important influence on hi-tech product performance.

In addition, the survey also investigated the impact extent of different aspects of product design, including appearance design, function design, and internal operating system design. The results showed that function design and internal operating system design were more important than appearance design, which also had a stronger influence on hi-tech product performance. Therefore the assumption that a positive relationship between product design and hi-tech product performance is substantiated.

8.5.1.4 Relationship between human resource management and hi-tech product performance

Highly effective HRM can have a great impact on the performance of a company or organization, especially in the hi-tech industries. Levin and Kalal (2003) suggested that an excellent reliability engineer who has a full and deep understanding of the firm, the concepts, and the whole process of manufacturing would be essential for ensuring product reliability. Having a highly efficient HRM system enables assigning different tasks to different employees appropriately, maximizing the performance of the project team, resulting in a better performance of the hi-tech product.

In addition, by investigating how the HRM affects hi-tech product performance, the survey data showed that HRM mainly affected product development, then product reliability, and last but not least innovation. Thus the assumption that a positive relationship between HRM and hi-tech product performance is substantiated.

8.5.1.5 Relationship between technology and knowledge support and hi-tech product performance

As knowledge-intensive work, the research and development of a hi-tech product are always costly, with high risk, and also requires a vast amount of talent with specific hi-end technologies or knowledge in a specific field. Merminod and Rowe (2012) suggested that technology and knowledge support play an important role in NPD, which is increasingly taking place across organizations. The competition for hi-technology and knowledge support and transfer from universities are becoming more and more intensive and is essential for product success (Kesting & Wurth, 2015). In this study, the impact extent of technology and knowledge support was investigated by a survey by the two research groups. The results indicated that technology and knowledge support play an important role in affecting hi-tech product performance by mainly influencing its innovative activities and reliability.

Hence, the assumption that a positive relationship between technology and knowledge support and hi-tech product performance is substantiated.

8.5.1.6 Relationship between technology and knowledge support and hi-tech product performance

As a matter of fact, government support can make a great contribution to the success of product development in the form of direct investment, tax policy, risk fund, industry policy, and so on (Zhang et al., 2016). Hi-tech products always are of high value (Ramani, Thutupalli, & Urias, 2017), not only their commercial value and a large potential market but also presenting meaningful issues for society, such as protecting the environment, energy-saving, or even making contributions to the progress of human civilization. However, research and development of a hi-tech product is risky and costly with a high rate of failure and takes a long time until success. It means that not many companies or organizations can afford it. Based on this point, support from the government is essential. In this study, the survey mainly investigated four aspects of government support: government financial support, government policy, government technology support, and government equipment and facility support. The results illustrated that government support has an impact on hi-tech product performance to some extent, especially government financial and policy support.

Therefore the assumption that a positive relationship between government support and hi-tech product performance is substantiated.

8.6 Conclusions

8.6.1 Summary of study

In the hi-tech industry, innovation not only refers to the early process of product research and development but also runs through the whole project of hi-tech product development. Also, product reliability is closely related to product quality and safety, which directly influence product performance and success (Levin & Kalal, 2003). In addition, hi-tech product performance is strongly influenced by NPD processes (Crawford & Di Benedetto, 2008).

Hence, concluding previous research, it was found that three dimensions, innovations, product reliability, and product development certainly have an impact on the performance of hi-tech industries. In this study, the main purpose was to investigate that how these three dimensions affect hi-tech product performance and find the relationship between them in hi-tech industries. According to some previous research, a number of subfactors of the three dimensions were put forward but did not have a systematical investigation or conclusion. Therefore investigations into the relationship between six subfactors and hi-tech product performance, and their impact extent on hi-tech product performance have been undertaken in this study. The major findings are summarized as follows.

1. Six main factors have a positive relationship with hi-tech product performance by affecting innovative activities, product reliability, and product development. The six main factors are individual ability, adequate funding, product design, HRM, technology and knowledge support, and government support.
2. Different aspects of individual ability have different impact extents on hi-tech product performance. Individual communication, learning ability, and academic ability in a specific field have a stronger impact

on hi-tech product performance than other aspects of individual ability.
3. Function design and internal operating system design are were important than appearance design, in influencing product reliability and product development.
4. The impact of government support is primarily from government financial and technical support, which play an important role in affecting hi-tech product performance.
5. Adequate funding and HR have a similar impact extent on hi-tech product performance, which is moderate, while technology and knowledge support have a strong impact as a hi-tech project is a knowledge-intensive work.

8.6.2 Limitations of study

8.6.2.1 Sampling limitations

In this study, the respondents were divided into two groups. The first group consisted of staff from the SOPSYS project team. The second group involved 40 samples collected from people who were interested in or work in the hi-tech industry. The samples are just small focused group samples that may not be able to represent all the situations that occur in hi-tech industries. Moreover, in this study, the object of the case study was a space technology-based product. Thus the results of the case study may not apply to other hi-tech products in different hi-tech industry fields as different hi-tech fields can be one of the influencing factors in the research. Furthermore, due to the unequal distributions in the size.

8.6.2.2 Questionnaire limitations

The research instrument in this study was the focused group questionnaire survey. Although the direction of the questions was generally supported the literature and designed and conducted by the author. We have demonstrated the method and some preliminary results which can be further developed into a full study in the future with larger sample size and better experimental instruments.

8.6.3 Suggestions for future work

Some of the characteristics of the proposed research dimensions and methodology to be investigated in the future study are as follows:

1. As a knowledge-intensive work, the spatial technological product is an important dimension of research in the development of the high

technology industries (Sun & Liu, 2012). Even in this study, one of the spatial technological products, the SOPSYS, was selected as the research object in the case study, the result cannot apply to all product developments in hi-tech industries. Therefore more spatial technological products are suggested to be investigated to obtain more accurate results.

2. This study is focused on the relationship of innovation, product reliability, and product development in hi-tech industries and its objective is to seek the factor affecting hi-tech product performance. Based on the findings of this study, future research is suggested to have a deeper investigation of the affecting factors and find out how to increase the success rate of hi-tech products, as well as maximize ultimate performance.
3. Since there are some limitations of the case study due to the problem with small samples, future research is strongly suggested to provide more evidence to support the analysis and findings in this study, such as the camera pointing system. As presented by Ettinger and Freund (2008), the camera pointing system was controlled by real-time calculations of sound source locations from a microphone array. It uses a series of hi-end technologies to provide the positional information for the camera to drive it. Also, it requires high precision of spot detection in laser communication acquisition, tracking, and pointing system (Qian, Jia, Zhang, & Wang, 2013). As an informative and knowledge-intensive hi-tech product, a camera pointing system is worth being researched in a future study. Some other fields in hi-tech industries, such as computers and communication consumer products are also suggested to be investigated in future work.

Acknowledgment

We acknowledge the support of the Innovation and Technology Fund (Project Ref.: PRP/071/20FX) and the Laboratory for Artificial Intelligence in Design (Project Code: RP2-1), Hong Kong Special Administrative Region, in preparing this chapter.

References

Atilgan-Inan, E., Buyukkupcu, A., & Akinci, S. (2010). A content analysis of factors affecting new product development process/yeni ürün gelistirme sürecini etkileyen faktörlerin degerlendirilmesi. *Business and Economics Research Journal, 1*(3), 87–100.

Azarmi, D. (2016). Factors affecting technology innovation and its commercialization in firms. *Modern Applied Science, 10*(7), 36.

Brown, S., & Eisenhardt, K. (1995). Product development: Past research, present findings, and future directions. *Academy of Management Review, 20*(2), 343–378.
Chau, K. Y., Tang, Y. M., Liu, X. Y., Ip, Y. K., & Tao, Y. (2021). Investigation of critical success factors for improving supply chain quality management in manufacturing. *Enterprise Information Systems*. Available from https://doi.org/10.1080/17517575.2021.1880642.
Cooper, R. G. (2019). The drivers of success in new-product development. *Industrial Marketing Management, 76*, 36–47.
Crawford, C. M., & Di Benedetto, A. (2008). *New products management* (9th edn). New York: McGraw-Hill.
Ettinger, E., & Freund, Y. (2008). Coordinate-free calibration of an acoustically driven camera pointing system. In *Proceedings of the Second ACM/IEEE international conference on distributed smart cameras, (ICDSC 2008)* (pp. 1–9).
Evanschitzky, H., Eisend, M., Calantone, R., & Jiang, Y. (2012). Success factors of product innovation: An updated metaschitzky. *Journal of Product Innovation Management, 29*, 21–37.
Ferràs, X. (2008). Towards a wide definition of innovation: Product, process, organisation and marketing. Paradigmes: Economia productiva i coneixement.
Guo, J., Tan, R., Sun, J., Ren, J., Wu, S., & Qiu, Y. (2016). A needs analysis approach to product innovation driven by design. *Procedia CIRP, 39*, 39–44.
Kehoe, R., Collins, C., & Chen, G. (2017). Human resource management and unit performance in knowledge-intensive work. *Journal of Applied Psychology, 102*(8), 1222–1236.
Kesting, T., & Wurth, B. (2015). Knowledge and technology transfer support potential of intermediate organizations. In *Competitive strategies for academic entrepreneurship* (pp. 143–170).
Kumar, M., & Noble, C. H. (2016). Beyond form and function: Why do consumers value product design? *Journal of Business Research, 69*(2), 613–620.
Levin, M. A., & Kalal, T. T. (2003). *Improving product reliability: Strategies and implementation* (Vol. 1). John Wiley & Sons.
Mackelprang, A. W., Habermann, M., & Swink, M. (2015). How firm innovativeness and unexpected product reliability failures affect profitability. *Journal of Operations Management, 38*, 71–86.
Merminod, V., & Rowe, F. (2012). How does PLM technology support knowledge transfer and translation in new product development? Transparency and boundary spanners in an international context. *Information and Organization, 22*(4), 295–322.
Mrugalska, B., & Tytyk, E. (2015). Quality control methods for product reliability and safety. *Procedia Manufacturing, 3*, 5897–5904.
Murthy, D. N. P., Rausand, M., & Virtanen, S. (2009). Investment in new product reliability. *Reliability Engineering & System Safety, 94*(10), 1593–1600.
Pratoom, K., & Savatsomboon, G. (2012). Explaining factors affecting individual innovation: The case of producer group members in Thailand. *Asia Pacific Journal of Management, 29*(4), 1063–1087.
Qian, F., Jia, J., Zhang, L., & Wang, J. (2013).). Positioning accuracy of spot-detecting camera in acquisition, tracking, pointing system. *Zhongguo Jiguang (Chinese Journal of Lasers), 40*(2), 0205007.
Ramani, S. V., Thutupalli, A., & Urias, E. (2017). High-value hi-tech product introduction in emerging countries. *Qualitative Market Research: An International Journal*.
Rocheska, S., Nikoloski, D., Angeleski, M., & Mancheski, G. (2017). Factors affecting innovation and patent propensity of smes: Evidence from Macedonia. *TEM Journal, 6*(2), 407–415.

Roper, S., Micheli, P., Love, J. H., & Vahter, P. (2016). The roles and effectiveness of design in new product development: A study of Irish manufacturers. *Research Policy*, *45*(1), 319–329.

Salonen, M., Holtta-Otto, K., & Otto, K. (2008). Effecting product reliability and life cycle costs with early design phase product architecture decisions. *International Journal of Product Development*, *5*(1–2), 109–124.

Soderquist, K. E., & Kostopoulos, K. (2012). *Factors affecting the performance of new product development teams: Some European evidence. Knowledge perspectives of new product development* (pp. 29–48). New York, NY: Springer.

Sok, P., & O'Cass, A. (2015). Examining the new product innovation–performance relationship: Optimizing the role of individual-level creativity and attention-to-detail. *Industrial Marketing Management*, *47*, 156–165.

Su, A. (2006). The impact of individual ability and favorable team member scores on students' preferences of team-based learning and grading methods. *Journal of Teaching in Travel & Tourism*, *6*(3), 27–45.

Sun, Y., & Liu, F. (2012). Evolution of the spatial distribution of China's hi-tech industries: Agglomeration and spillover effects. *Issues & Studies*, *48*(1).

Szylar, C. (2013). *Risk management under UCITS III / New challenges for the fund industry. (ISTE)*. London: Wiley.

Walker, J. T. (2002). *Exploring the influence of the individual's ability to experience flow while participating in a group-dependent activity on the individual's satisfaction with the group's performance*. Clemson University.

Wolf, M., & Terrell, D. (2016). *The high-tech industry, what is it and why it matters to our economic future*.

Yung, K. L., Tang, Y. M., Ip, W. H., & Kuo, W. T. (2021). A systematic review of product design for space instrument innovation, reliability, and manufacturing. *Machines*, *9* (10)244. Available from https://doi.org/10.3390/machines9100244.

Yung, K. L., Ho, G. T. S., Tang, Y. M., & Ip, W. H. (2021). Inventory classification system in space mission component replenishment using multiattribute fuzzy ABC classification. *Industrial Management & Data Systems*, *121*(3), 637–656. Available from https://doi.org/10.1108/IMDS-09-2020-0518.

Zhang, C., Li, X., & Liu, H. (2016). Huawei case study: country-specific factors affecting new product development. In *International operations management*, (pp. 135–148). Routledge.

Zhang, M., Lettice, F., & Zhao, X. (2015). The impact of social capital on mass customisation and product innovation capabilities. *International Journal of Production Research*, *53*, 5251–5264.

Zuñiga-Collazos, A., Harrill, R., Escobar-Moreno, N. R., & Castillo-Palacio, M. (2015). Evaluation of the determinant factors of innovation in Colombia's tourist product. *Tourism Analysis*, *20*(1), 117–122.

CHAPTER 9

Monocular simultaneous localization and mapping for a space rover application

K.K. Tseng[1], Jun Li[1], Yachin Chang[1], K.L. Yung[2] and Andrew W.H. Ip[3,4]

[1]Harbin Institute of Technology (Shenzhen), Shenzhen, P.R. China

[2]Department of Industrial and Systems Engineering, The Hong Kong Polytechnic University, Hong Kong, P.R. China

[3]Department of Mechanical Engineering, University of Saskatchewan, Saskatoon, SK, Canada

[4]Department of Industrial and Systems Engineering, The Hong Kong Polytechnic University, Hong Kong SAR, P.R. China

9.1 Introduction

Simultaneous localization and mapping (SLAM) is the process of creating the map of an environment while simultaneously deciding on the position of the rover on the map. SLAM has been used and validated by many researchers and has played an important role in intelligent rover research and other research fields. Rover sensors have a large impact on the algorithm used in SLAM, which can be applied to many sensors, such as laser, sonar, or visual camera sensors. Sometimes, additional sensorial sources are used to obtain a better understanding of the rover's state and its surrounding environment (Aulinas, Petillot, Salvi, & Lladó, 2008). Compared with these sensorial sources, a camera is not only cheaper and lighter but is also more convenient in certain conditions. There has been increasing interest in using a mono camera as the single visual sensor in using SLAM. Such a method is named Monocular SLAM (Scaramuzza & Fraundorfer, 2011).

Even though it is unable to recover information on the depth of the observed landmarks directly, Monocular SLAM is favored in the literature on visual SLAM and has received a great deal of attention in recent years. It requires less computing resources than Binocular or Panoramic SLAM. Monocular SLAM has closely related to the structure-from-motion (SFM) problem for recovering relative camera positions and three-dimensional (3D) structures from a set of camera images. SFM methods are sometimes formulated as offline algorithms that require simultaneous batch processing for all

IoT and Spacecraft Informatics
DOI: https://doi.org/10.1016/B978-0-12-821051-2.00004-0

the images acquired in the sequence. To obtain a globally consistent reconstruction of the camera trajectory and scene structure, local motion estimates are refined using offline global optimization (i.e., a bundle adjustment) through the whole sequence, and the computation time for this grows with the number of images. Unlike SFM techniques, Monocular SLAM focuses on estimating the 3D motion of the camera sequentially and in real-time. As a new image arrives, a SLAM solution can quickly update the latest state of the camera. As a result, for consistent localization over long sequences in real-time, Monocular SLAM is more suitable than SFM.

For most SLAM approaches, perceiving a map of the outside world is fundamental for an autonomous rover, since such maps are required for higher-level work. As a result, the first problem that is encountered is how to extract the important features from the visual camera and build a map of the surrounding environment. Since SLAM usually leads to a map based on features, many algorithms, such as Shi-Tomasi (Davison 200), Harris (Lemaire, Berger, Jung, & Lacroix, 2007), SIFT (Karlsson et al., 2005; Se, Lowe, & Little, 2002), SURF (Wang, Hung, & Sun, 2011; Zhang, Huang, Chao, & Kang, 2008), FAST (Civera, Grasa, Davison, & Montiel, 2010) and so on, have been applied to extract the features and create their descriptions as landmarks on the map. Furthermore, after adding the features into the map of the system, the state of the camera and all the features can be updated from new information by matching features between consecutive images. The extracted feature points used as landmarks should be robust under all kinds of changes, such as scale, viewpoint, and illumination changes, or the feature matching will be done incorrectly. That is, if the observations cannot be correctly associated with the landmarks on the map, the map will be inconsistent. This case is regarded as a data association problem. Wrong matching will cause wrong associations of data (Fraundorfer, Scaramuzza, & Pollefeys, 2011). It is, therefore, crucial to choose suitable features from the images captured by the camera as reliable landmarks. Meanwhile, it is also important to design a good feature matching algorithm to improve the data association. Some new feature extraction algorithms provide good performance for SLAM.

The other problem is how to match the feature map to estimate the rover's state. Until now, two well-known algorithms have been used in SLAM: the first is filtering-based SLAM and the other is key-frame-based SLAM. The main benefits of these two algorithms have been studied in Strasdat, Montiel, and Davison (2010a,b). These methods have the advantage of using few resources but in providing high accuracy. The Extended

Kalman Filter (EKF) is applied to solve the estimation issues in SLAM. In addition, several modified filters that have better performance have been found to improve the estimation system.

Our research presents a new Monocular SLAM to estimate the state of the visual sensor; this is then extended to the work of creating the map from sparse features (Civera et al., 2010; Davison, Reid, Molton, & Stasse, 2007). Moreover, Oriented FAST and Rotated BRIEF (ORB) (Rublee, Rabaud, Konolige, & Bradski, 2011) is used to detect the features, which can greatly reduce the initialization time in the system, and a new feature detection algorithm is proposed to obtain a suitable number of landmarks. Because of the linear error caused by EKF, we apply the Modified coVariance Extended Kalman Filter (MVEKF) (Guo, Sun, & Kan, 2003) to estimate the system state recursively.

As for space exploration, lunar rovers play an important role in exploring the moon and Mars, but because the moon and Mars do not have a global positioning system (GPS) like Earth, so shelter and high-accuracy location information cannot be provided. The SLAM algorithm can be used to assist in positioning. This allows a lunar rover to remember its own trajectory and provides a functional service for the lunar rover to return to its original point. Thus, we also apply our proposed algorithm to a space rover application.

The rest of the chapter is organized as follows. The related works of Monocular SLAM are introduced in Section 9.2. In Section 9.3, the proposed system and algorithm is described step by step in a detailed manner. The design of experiments and related experimental results are shown in Section 9.4, conclusions are given in Section 9.5.

9.2 Related work

In 1980, research on predicting a rover's movement from visual data was presented by Moravec (1980). Moravec's work can be considered as the first approach to predicting movement, but his functioning modules used other motion estimation solutions. He also proposed one of the earliest corner detectors, called the Moravec corner detector, which was are the predecessor of the popular corner detector known as the Harris corner detector (Harris & Stephens, 1988). Building upon Moravec's work, both SFM and Visual SLAM have been developed over a long period.

Nowadays, there are many research reports on Monocular SLAM (Fraundorfer, Scaramuzza, & Pollefeys, 2011; Scaramuzza & Fraundorfer, 2011). Among the Monocular SLAM systems that have been reported, two systems can be seen as milestones in the history of the development, with other improved versions based on them to some degree.

The first successful application of the SLAM methodology with one single camera was developed by Davison et al. (2007). This can operate at 30FPS and is capable of coping with an agile motion for an uncontrolled camera with the standard EKF framework. In Davison's system, point features are detected automatically using Shi and Tomasi's detection operator. The second well-known system was proposed by Klein and Murray (2008). Because of the linearization errors implicit in the filtering approaches, they advocated the optimization of a sparse selection of images, called key-frames, instead of the use of a sliding window. In their system, the FAST corner detection was applied to extract the features, and global bundle adjustment was performed over the tracked features and a selected set of key-frames in the image sequence.

Based on these works, many other researchers have reported their own improved methods to solve the issues of visual SLAM. Lemaire et al. (2007) presented a robust interest point matching algorithm that can work in very diverse environments with the improved Harris corner detector. Considering the various changes of images, Karlsson et al. (2005) proposed a method to extract the features using scale-invariant feature transformation (SIFT) to provide feature extraction. Wang et al. (2011) proposed the application of SURF in the algorithm to provide a better representation of a visual SLAM system. In addition, a tracking window and a nearest-neighbor algorithm were integrated to improve the data association in SLAM. Strasdat et al. (2010a,b) presented a new architecture for visual SLAM with a keyframe optimization method. Civera, Davison, and Montiel (2008) proposed a novel method to combine one-point RANSAC into the EKF, to get a more reliable data association. In Civera's system, the FAST algorithm is used to detect the positions of the landmarks.

Our research is built on the works of Davison et al. (2007) and Civera et al. (2010). Here, we design a new Monocular SLAM system and propose some improved measures for the issues mentioned that affect current vision-based SLAM solutions. Experiments are carried out with our system to test its performance. Furthermore, we propose a new lunar rover location module that uses an improved algorithm.

9.3 Proposed system and algorithm

The system procedure for SLAM is shown in Fig. 9.1. We first set some parameters of the system to give the initialization, which is essential for the

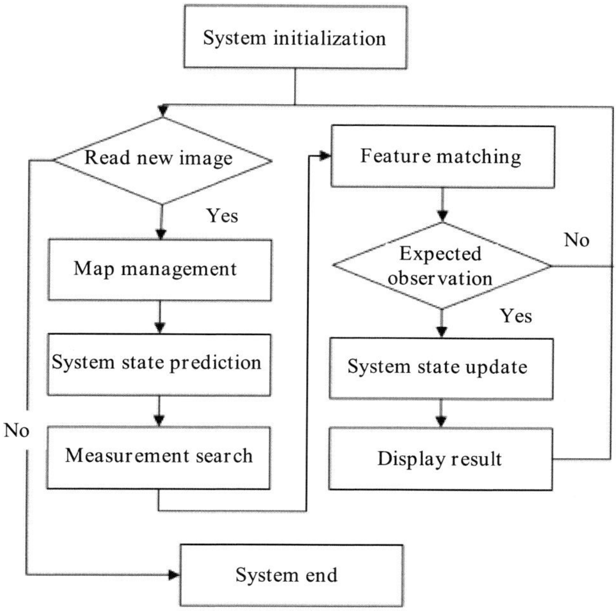

Figure 9.1 The system procedure.

system. After that, the system begins to capture new images into the loop. In the map management module, there are two important processes to be accomplished. Unstable features in the system map are removed, and new features are initialized from the new image and merged into the map. Then, the state of the system is predicted for the next time with the motion model of the camera. At this point, we load the new image into the estimation system and project all 6D features into 2D points to search the measurements in the new image plane through the measurement model of the camera. The search region of the measurements for feature landmarks can be calculated using the active search method proposed by Davison. After checking the observation data, the system state is updated with the modified filter. Finally, the new state of the camera and the positions of the sparse features in the map can be estimated. The important steps are explained below.

9.3.1 System initialization

The initialization processing phase mainly proceeds with the parameterization of the system. In our SLAM system, the system state consists of two parts: the camera state and the positions of all the feature points in the

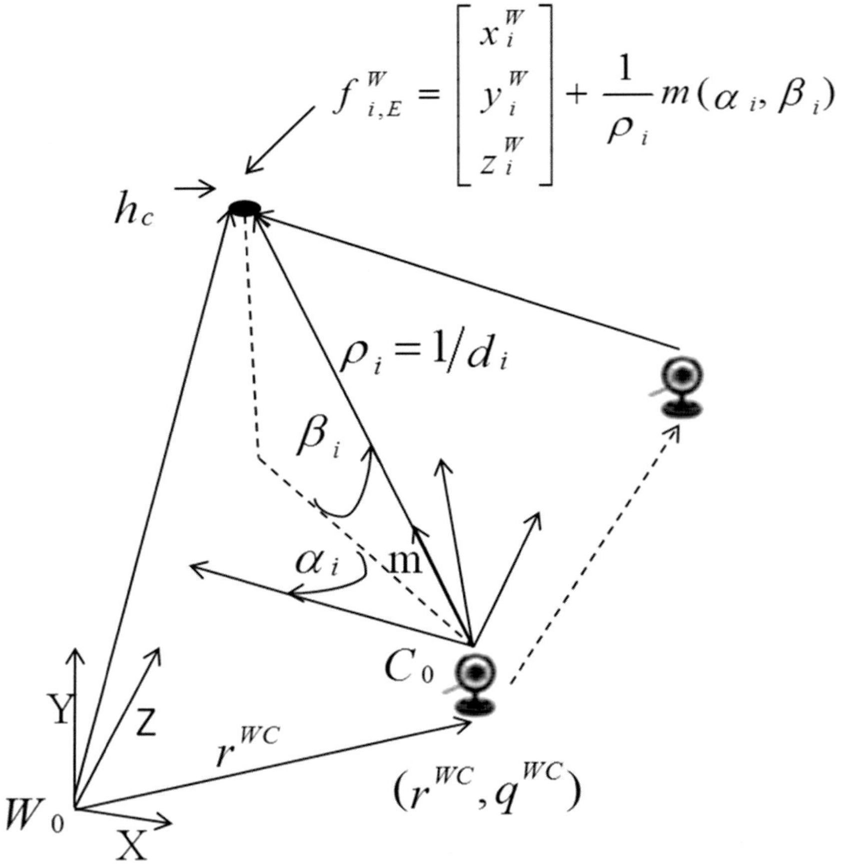

Figure 9.2 System parameterization.

map. Here we use W to specify the world coordinate frame, which we define as a static reference frame, and C stands for the visual motion coordinate frame. Fig. 9.2 illustrates the system parameterization process. The camera state can be denoted:

$$x_c = \left[r^{WC}, q^{WC}, v^{W}, w^{C}\right]^T \tag{9.1}$$

where r^{WC} means the position of the motion camera; q^{WC} stands for the camera orientation of the world frame W; and v^W and W^c are the linear and angular velocities. Therefore, we have:

$$r^{WC} = [x_r, y_r, z_r]^T \tag{9.2}$$

$$q^{WC} = \left[q_1, q_2, q_3, q_4\right]^T \tag{9.3}$$

$$v^W = \left[v_x, v_y, v_z\right]^T \tag{9.4}$$

$$w^C = \left[w_x, w_y, w_z\right]^T \tag{9.5}$$

Every feature in the map is composed of the following state vector in the form of an inverse depth parameterization (Civera et al., 2008):

$$f_i^W = \left[x_i^W, y_i^W, z_i^W, \alpha_i, \beta_i, \rho_i\right]^T \tag{9.6}$$

where α_i and β_i represent the azimuth and the elevation, respectively, for defining the unit directional vector $m(\alpha_i, \beta_i)$, and a point's depth along the ray d_i is encoded by its inverse $\rho_i = 1/d_i$. The following is the unit directional vector:

$$m(\alpha_i, \beta_i) = \left[\cos\beta_i \sin\alpha_i, \ -\sin\beta_i, \cos\beta_i \cos\alpha_i\right]^T \tag{9.7}$$

Furthermore, the above 6D feature is usually converted into a 3D Euclidean point on the map:

$$f_{i,E}^W = \begin{bmatrix} X_i^W \\ Y_i^W \\ Z_i^W \end{bmatrix} = \begin{bmatrix} x_i^W \\ y_i^W \\ z_i^W \end{bmatrix} + \frac{1}{\rho_i} m(\alpha_i, \beta_i) \tag{9.8}$$

In the next step, the system state vector includes the following parameters:

$$x = \left[x_c, f_1^W, f_2^W, f_3^W, \cdots, f_N^W\right]^T \tag{9.9}$$

Meanwhile, the covariance matrix is denoted as:

$$P = \begin{bmatrix} P_{xx}, P_{xf} \\ P_{fx}, P_{ff} \end{bmatrix}^T \tag{9.10}$$

9.3.2 Map management

Map management handles the addition of new features or the removal of unwanted features that have been found to affect the results of the SLAM system. To provide a better performance, we propose a grid-based feature detecting strategy and modify the feature deleting criterion for the map.

Fig. 9.3 shows the flowchart of the feature detection algorithm. We separate the image into M grids and apply the ORB detector to detect the minimum features in a selected grid. The features are extracted randomly until a number of features are satisfied. The following expression shows how to select a grid sub-region in an image:

$$grid_k = \begin{cases} randperm(1, 2, \cdots, M)_1 & k \leq N \\ randperm(\min_E(f_{k1}, f_{k2}, \cdots, f_{kM}))_1 & k > N \end{cases} \tag{9.11}$$

where $grid_k$ stands for the index of the grid selected at the time k; *randperm*() is a function that generates a random integer permutation (we just need to select the first value from the permutation as the grid index); and f_{kj} represents the number of features in each grid.

Davison and Civera both used a random region detection method to generate landmarks. Our method gives a better distribution of features over the whole image. The deletion criterion is defined by the ratio between the number of correct matches and the number of matching attempts. If the ratio is below a certain threshold, the landmark is deleted from the feature map. However, this criterion has two shortcomings. When occluded landmarks are removed using the ratio deletion criterion, there will be an extra cost in re them once they are observed again. Another shortcoming is obvious: if old features are removed, it is impossible to recognize previously mapped areas in the loop closing tasks, which

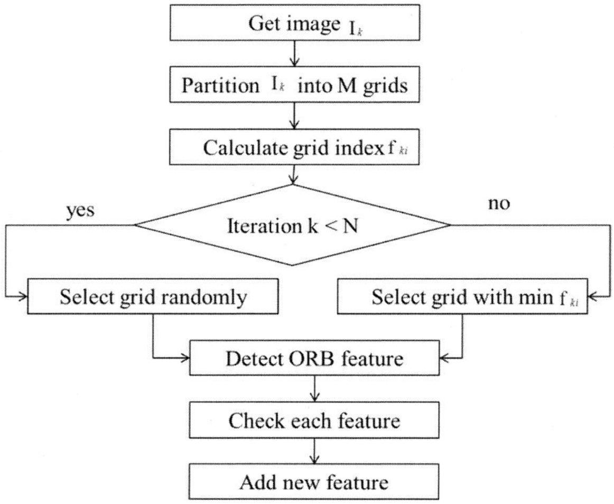

Figure 9.3 Flowchart of feature detection.

will lead to divergence in the mapping. We make some improvements to avoid these drawbacks of the ratio criterion. The result of the improvements is given in the experiments section.

The linear velocity and angular velocity in Davison et al. (2007), describes the motion of a free movement visual sensor as the following:

$$x_{k+1} = g(x_k u) = \begin{bmatrix} r_{k+1}^{WC} \\ q_{k+1}^{WC} \\ v_{k+1}^{W} \\ w_{k+1}^{C} \end{bmatrix} = \begin{bmatrix} r_k^{WC} + \left(v_k^W + m * V^W\right)\Delta t \\ q_{k+1}^{WC} + q\left(\left(w_k^C + n * W^C\right)\Delta t\right) \\ v_{k+1}^{W} + V^W \\ w_{k+1}^{C} + W^C \end{bmatrix} \tag{9.12}$$

Here V^W and W^C denote the impulsive linear and angle velocities; $q\left(\left(w_k^c + W^C\right)\Delta t\right)$ means the quaternion computed from the rotation vector $\left(w_k^c + W^C\right)\Delta t$; u denotes the changes in the linear and angular velocities; and m and n represent the dynamic parameters for the acceleration velocities. Then we have:

$$u = \begin{bmatrix} V^W \\ W^C \end{bmatrix} = \begin{bmatrix} a^W \Delta t \\ b^C \Delta t \end{bmatrix} \tag{9.13}$$

where Δt represents the sampling time interval, and a^W and b^C are Gaussian variables with zero mean.

Hence, the estimated system state for the MVEKF is modeled as:

$$\hat{x}_{k|k-1} = \begin{bmatrix} g\left(\hat{x}_{k-1|k-1}\right) \\ \hat{f}_1^W \\ \vdots \\ \hat{f}_1^W \end{bmatrix} \tag{9.14}$$

$$P_{k|k-1} = G_x P_{k-1|k-1} G_x^T + G_u Q G_u^T \tag{9.15}$$

where G_x and G_u are the derivatives of the motion model g, the camera state $\hat{x}_c$ and the control parameter u; and Q means the noise of the covariance matrix.

9.3.3 Feature search and matching

In our Monocular SLAM system, the pinhole camera model with two parameters of radial distortion is also used to take the measurements of the landmarks. With this model, we can project 6D features from the inverse depth world coordinate frame W onto the camera coordinate frame C using the following equations:

$$h_i^C = \begin{bmatrix} x_i^C \\ y_i^C \\ z_i^S \end{bmatrix} = R^{CW}\left(q_i^{WC}\right)\left(\rho\left(f_i^W - r_i^{WC}\right) + m(\alpha_i, \beta_i)\right) \tag{9.16}$$

where $R^{CW}\left(q_i^{WC}\right)$ is a rotation matrix and C is the camera frame. The following is the pinhole model:

$$h_u = \begin{bmatrix} u_u \\ v_u \end{bmatrix} = \begin{bmatrix} u_0 - \dfrac{f}{dx}\dfrac{x_i^C}{z_i^C} \\ v_0 - \dfrac{f}{dx}\dfrac{x_i^C}{z_i^C} \end{bmatrix} \tag{9.17}$$

where f is the local length of the camera, (u_0, v_0) are the central coordinates of the image and dx and dy denote the unit sizes of each pixel. An undistorted point is transformed by the radial distortion model, and its distorted pixel coordinates are:

$$h_d = \begin{bmatrix} u_d \\ v_d \end{bmatrix} = \begin{bmatrix} u_0 + \dfrac{u_u - u_0}{\sqrt{1 + k_1 r^2 + k_2 r^4}} \\ v_0 + \dfrac{v_u - u_0}{\sqrt{1 + k_1 r^2 + k_2 r^4}} \end{bmatrix} \tag{9.18}$$

$$r = \sqrt{(d_x(u_d - u_0))^2 + \left(d_y(v_d - v_0)\right)^2} \tag{9.19}$$

Using the active search principle, the predicted probability distribution for the system state, which is described in Section 3.2, can be used to find the corresponding search region automatically. From the function $h(x)$, an expected position $\hat{h}_i\left(u_i^E, v_i^E\right)$ of the projection for the 6D feature f_i is the covariance matrix:

$$\begin{aligned} \hat{h}_i &= h_i\left(\hat{x}_{k|k-1}\right) \\ S_i &= H_i P_{k|k-1} H_i^T + R_i \end{aligned} \tag{9.20}$$

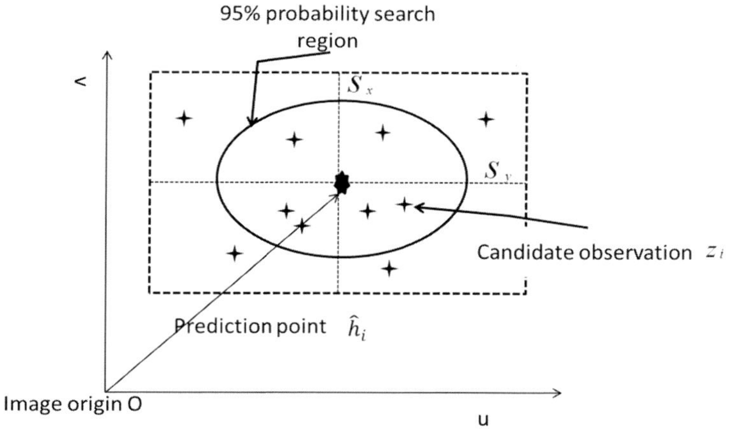

Figure 9.4 Feature search with active vision.

The boundary of the search region can then be obtained:

$$\begin{bmatrix} s_x \\ s_y \end{bmatrix} = \begin{bmatrix} 2n\sqrt{S_{i(1,1)}} \\ 2n\sqrt{S_{i(2,2)}} \end{bmatrix} \tag{9.21}$$

where n is the number of the desired search regions. The actual measurement z_i can be searched for using the 95% probability of the search region, as illustrated in Fig. 9.4.

Although the feature matching process may be robust and reasonable with the active search principle, there are still some improvements made in our experiments to enhance the robustness and efficiency of the feature matching. Fig. 9.5 shows our feature matching method.

Depending on the match between the system map features and the new observations, the system will decide on whether to use a Hamming match with the descriptors extracted by ORB. This is the main improvement over other matching methods. When the camera has a drastic move, the ratio of correct feature matches decreases greatly, and sometimes even reduces to zero. This may cause a filter failure. These measures are therefore necessary for Monocular SLAM.

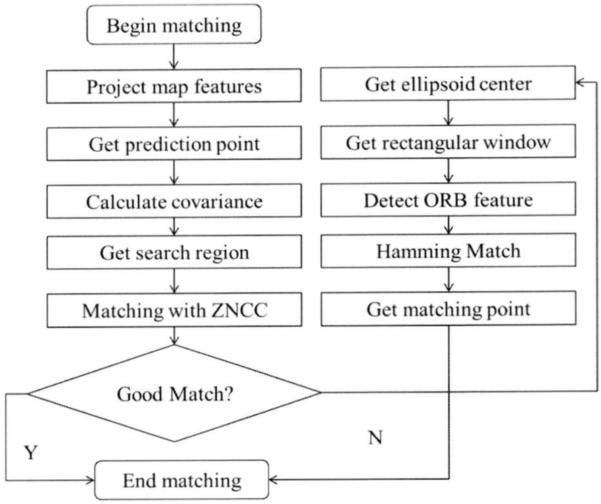

Figure 9.5 Flowchart for feature matching.

9.3.4 System state update

MVEKF (Guo et al., 2003) is a new filtering method that can be compared with the other usual filtering methods such as EKF, MGEKF, and IEKF. The idea of MVEKF is to recompute the Jacobi matrix H_k^+.

$$H_k^+ = \frac{\partial h(x)}{\partial x}\Big|_{x=\hat{x}_{k|k}} \tag{9.22}$$

The state covariance matrix is updated by

$$H_k^+ \cdot P_{k|k}^+ \approx 0 \tag{9.23}$$

When the expected observations are ready, the parameters are updated as follows:

$$K_k = P_{k|k-1} H_k^T S_i^{-1} \tag{9.24}$$

$$\hat{x}_{k|k} = \hat{x}_{k|k-1} + K_k\left(z_i - \hat{h}_i\right) \tag{9.25}$$

$$K_k = P_{k|k-1} H_k^{+T}\left[H_k^+ P_{k|k-1} H_k^{+T} + R_k\right]^{-1} \tag{9.26}$$

$$P_{k|k} = P_{k|k-1} - K_k S_i K_K^T \tag{9.27}$$

$$P_{k|k} = \left(P_{k|k} + P_{k|k}^T\right)/2 \tag{9.28}$$

The whole motion estimation algorithm is described in Listing 1:
Listing 1. Motion estimation algorithm.

Algorithm:	MVEKF State Update for Monocular SLAM
Input:	System state variable $\hat{x}_{k-1\|k-1}$
	System state covariance matrix $P_{k-1\|k-1}$
Output:	System state variable $\hat{x}_{k\|k}$
	System state covariance matrix $P_{k\|k}$
Step 1:	State prediction with MVEKF filter:
	$\hat{x}_{k\|k-1} = g\left(\hat{x}_{k-1\|k-1}\right)$
	$P_{k\|k-1} = G_x P_{k-1\|k-1} G_x^T + Q$
Step 2:	Calculate Jacobi matrix of measurement function to system state:
	$H_k = \frac{\partial h(x)}{\partial x}\Big\|_{x=\hat{x}_{k\|k-1}}$
	Calculate Kalman coefficient:
	$K_k = P_{k\|k-1} H_k^T \left[H_k P_{k\|k-1} H_k^T + R_k\right]^{-1}$
Step 3:	System state update with MVEKF filter:
	$\hat{h}_i = h_i\left(\hat{x}_{k\|k-1}\right)$
	$S_i = H_i P_{k\|k-1} H_i^T + R_i$
	$\hat{x}_{k\|k} = \hat{x}_{k\|k-1} + K_k\left(z_i - \hat{h}_i\right)$
	$P_{k\|k} = P_{k\|k-1} - K_k S_i K_k^T$
Step 4:	Modify the measurement Jacobi matrix:
	$H_k^+ = \frac{\partial h(x)}{\partial x}\Big\|_{x=\hat{x}_{k\|k}}$
Step 5:	Repeat Step 3, adjust system state matrix as a symmetric matrix:
	$K_k = P_{k\|k-1} H_k^T S_i^{-1}$
	$\hat{x}_{k\|k} = \hat{x}_{k\|k-1} + K_k\left(z_i - \hat{h}_i\right)$
	$K_k = P_{k\|k-1} H_k^{+T}\left[H_k^+ P_{k\|k-1} H_k^{+T} + R_k\right]^{-1}$
	$P_{k\|k} = P_{k\|k-1} - K_k S_i K_k^T$
	$P_{k\|k} = \left(P_{k\|k} + P_{k\|k}^T\right)/2$
Step 6:	Return new $\hat{x}_{k\|k}$ and $P_{k\|k}$

9.4 Experiments

All the experiments were run on Pentium(R) Dual-Core T4500 processors at 2.30 GHz, and we implemented three Monocular SLAM algorithms to test their performance. The first algorithm is called 1PRMSLAM (short for 1-point RANSAC for EKFSLAM), which was proposed by Civera et al. (2008). The main contribution of these authors was to propose the incorporation of one-point RANSAC, which allows the minimum sample size to be reduced to one, resulting in large computational savings without the loss of discriminative power for outlier rejection. The second algorithm is called SEKFM SLAM, and uses SIFT to detect the image features as useful landmarks; the other blocks are the same as 1PRMSLAM. The purpose of building this algorithm is to test the effect and make comparisons with other algorithms when we just change the image feature detectors. The last Monocular SLAM algorithm is the one proposed by us, and we call it MVMSLAM. This algorithm combines the above improvements. A comprehensive comparison is made with the other two SLAM algorithms.

9.4.1 System display and camera view

We use the publicly available images from Civera, which were recorded using a 320 × 240 camera capturing at 15 FPS acquired in a desktop environment. In this experiment, with the same parameters, we select 200 frames to test 1PRMSLAM, SEKFMSLAM, and our method, MVMSLAM. The result of our system running is displayed in Fig. 9.3. In the right-hand part of the image, the trajectory of the camera is shown with a black line, and each feature (landmark) in the map is a black point with a corresponding red ellipse that represents the uncertainty of its position. From Fig. 9.6, we can see that the uncertainties of most features are gradually reduced, and the red ellipses are finally collapsed to the black points. We, therefore, prove that our proposed system is feasible for solving a SLAM problem, and the whole estimation trajectory of the camera is closely approximated to the true motion of the camera.

9.4.2 Performance comparison

The public image dataset is used to test the three algorithms. From Fig. 9.7, it is obvious that the time spent by 1PRMSLAM is almost three times greater than with our system. Furthermore, we record the average time for each frame. As Table 9.1 shows, the average time for our system is 0.1799 s, while the averages for 1PRMSLAM and SEKFMSLAM are 1.5931 and 0.1901 s, respectively.

Figs. 9.8 and 9.9 show the errors in the estimation of the camera trajectory in the systems. We calculate the root mean square error to show the estimation performance for the three systems. From Table 9.2, the accuracy of our method and that of SEKFMSLAM are fairly similar to the accuracy of 1PRMSLAM. The maximum space location error is

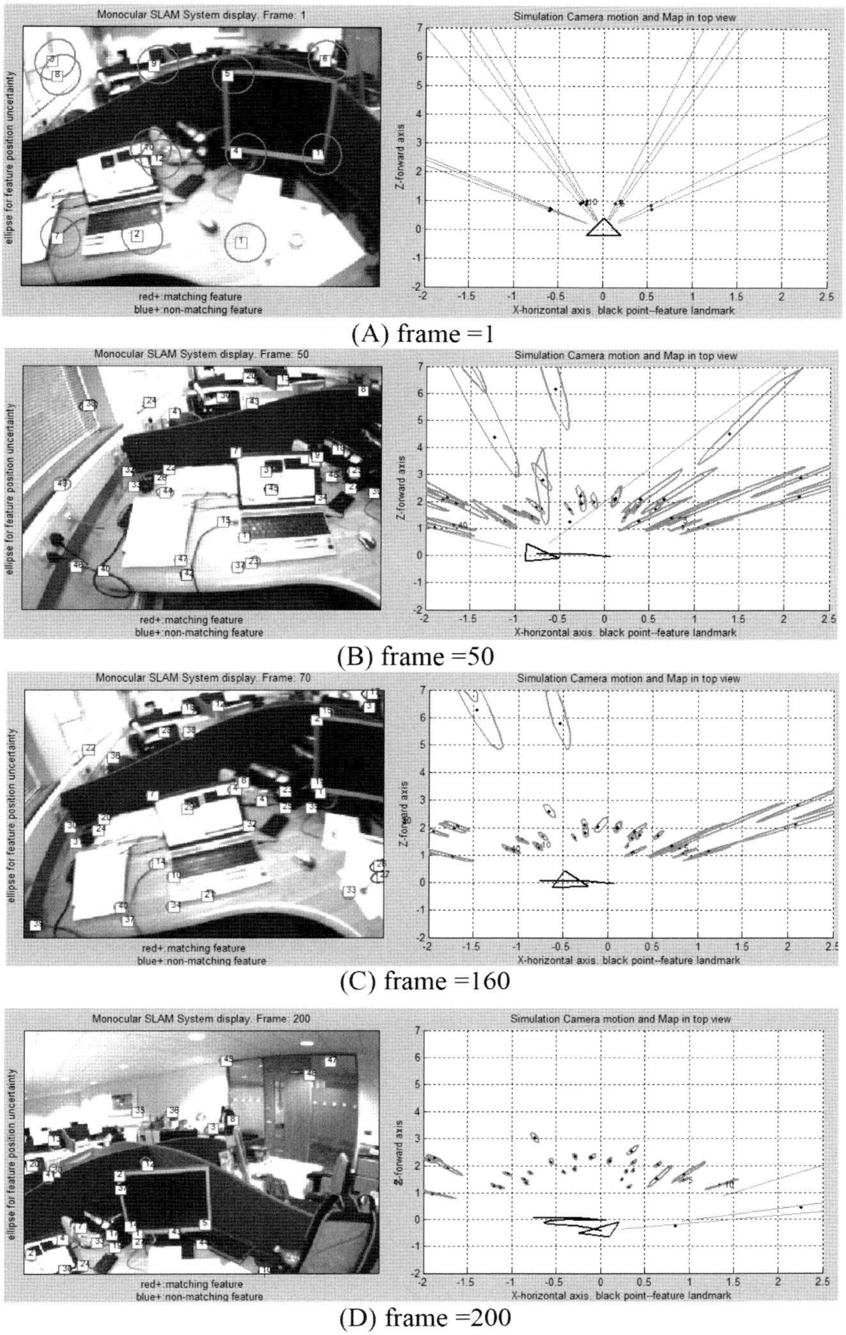

Figure 9.6 Monocular simultaneous localization and mapping system display and camera motion in top view.

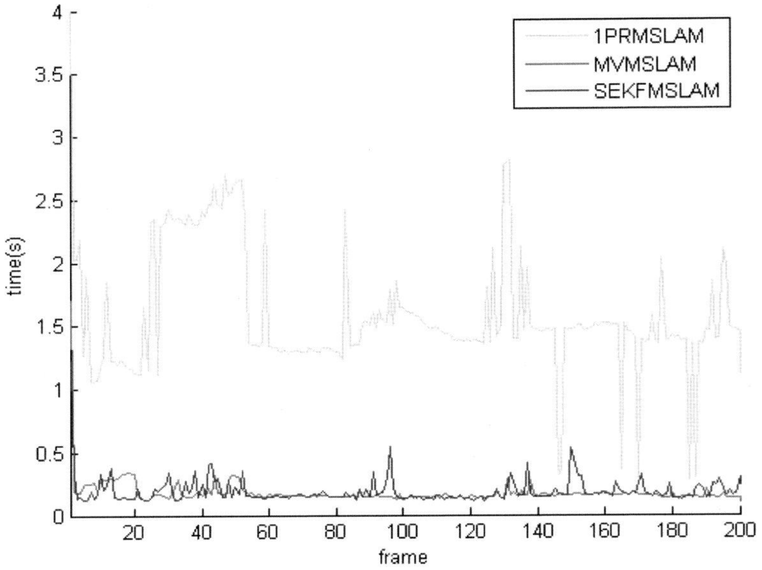

Figure 9.7 Time comparison of simultaneous localization and mapping for frame processing.

Table 9.1 Average time for each frame.

Time	1PRMSLAM	MVMSLAM	SEKFMSLAM
Frame(s)	1.5931	0.1799	0.1901

0.2005 m, while the minimum error is 0.0192 m. Meanwhile, the errors for the camera orientation are all small and these errors lie within the 95% confidence regions. This estimation proves that our algorithm, including SEKFMSLAM, is effective to track the unconstrained 6D motion. To show the performance better, we have carried out many tests with our images. We focus on the following experiment, which tests the final performance for the three methods.

9.5 Planetary rover application

Visual SLAM (Zhou et al., 2015) and planetary rovers (Fallah, Yue, Vahid-Araghi, & Khajepour, 2013) are both popular topics for vehicular technology. The lunar rover is currently the most direct tool for lunar

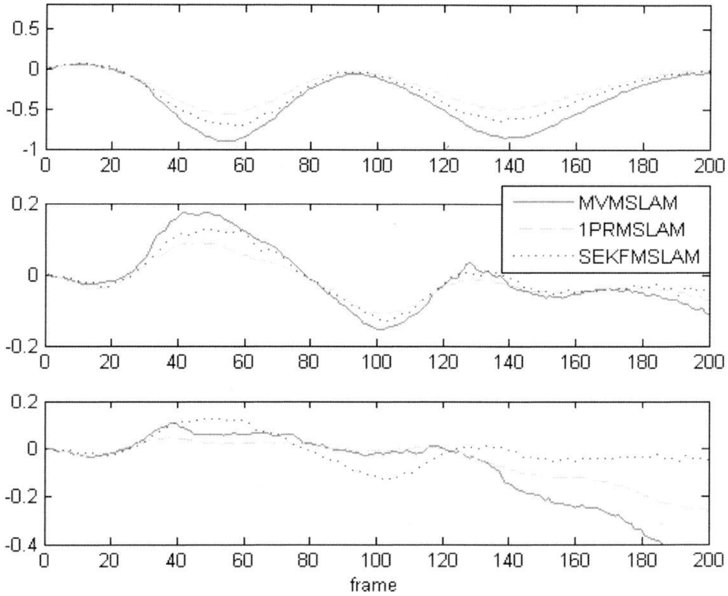

Figure 9.8 Error for camera location (XYZ).

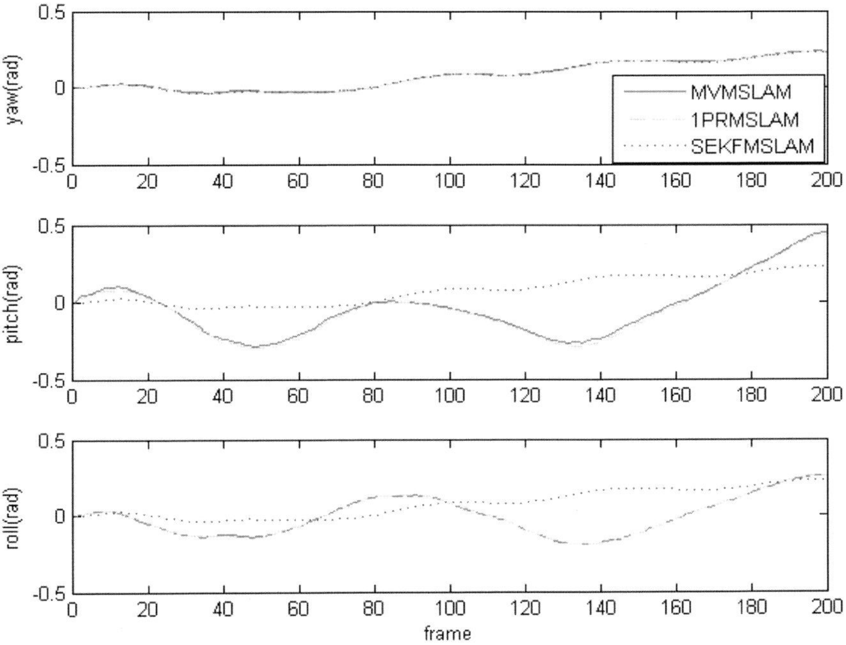

Figure 9.9 Error rate of monocular camera orientation.

Table 9.2 Error for six-dimensional pose.

Pose	SEKFMSLAM	MVMSLAM
X(m)	0.0867	0.2005
Y(m)	0.0193	0.0367
Z(m)	0.0192	0.0854
Yaw(m)	0.0038	0.0049
Pitch(m)	0.1001	0.0968
Roll(m)	0.0064	0.0043

exploration, and mainly requires location identification path planning, obstacle avoidance, and motion control modules. In the modules mentioned above, the location module is the most critical; its role is to calculate the real-time motion status and to indicate information on the position and attitude of the rover. The information on the rover's position is the premise for path planning and provides important input data for obstacle avoidance and other motion control modules. Therefore, the real-time location of the rover has important research significance for lunar exploration missions.

Currently, the well-established navigation technologies use dead reckoning, visual navigation, and radio navigation. Dead reckoning often relies on an odometer or other inertial navigation unit for location, but the fatal drawback of dead reckoning is the existence of error that gradually accumulates as time goes on. Therefore, this location method is not suitable for long-term lunar exploration missions. Besides, the specific nature of the lunar environment affects the navigation technologies: the moon has almost no atmosphere, so air media and magnetic field measuring instruments (such as ultrasonic sensors) cannot be relied on; the lunar magnetic field is weak, so a magnetic compass cannot be used for navigation; and, most importantly, there is no GPS on the moon at the moment (because of the distance, the Earth's GPS cannot provide lunar rover navigation services). Thus, the most feasible method for lunar rover location is visual navigation. Based on MVMSLAM, we propose a lunar rover locating module as shown in Fig. 9.10.

In this module, a monocular camera is used as the sensor to perceive the moon's surface environment. The real-time image sequence captured during the rover's movement is used as the input for the module to carry out real-time position and altitude information calculation for the rover in an unknown environment. Meanwhile, a good map of the environment

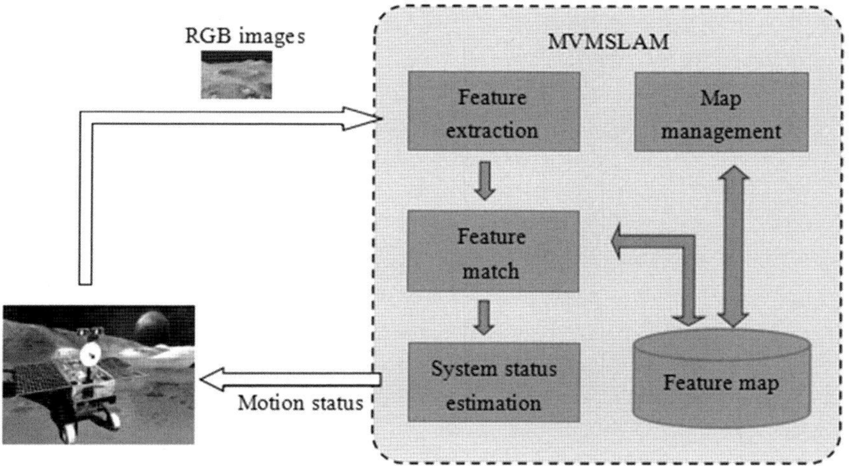

Figure 9.10 Lunar rover location module.

can be built and used in follow-up path planning and for other high-level exploration mission services. The module mainly includes the improved grid-based feature extraction, rule matching, and optimized feature map management algorithms mentioned above.

To validate the proposed module for feasibility we perform a simulated experiment for the lunar rover. Since it is difficult to obtain image sequences of the lunar surface environment, we adopt the desert landscape on earth to simulate the lunar surface environment. Desert topography consists of an open visual field and less blocking within the visual field; the surface features are mostly raised gravel and an undulating surface. These characteristics are similar to those of the lunar environment. Using a head-mounted monocular camera, we took a 36 s long monocular sequence of 1150 frames and ran it on MVMSLAM. The result is shown in Fig. 9.11.

Fig. 9.11A shows the distribution of feature points corresponding to a certain frame in the image sequence, Fig. 9.11B shows the distribution of the feature points of the constructed map, and Fig. 9.11C shows the 3D trajectory map estimated by the module.

The experiments show that the lunar rover module based on our proposed MVMSLAM can provide real-time location information for the rover and a map of the features of an unknown environment. The output of the location module mainly consists of the motion trajectory and the

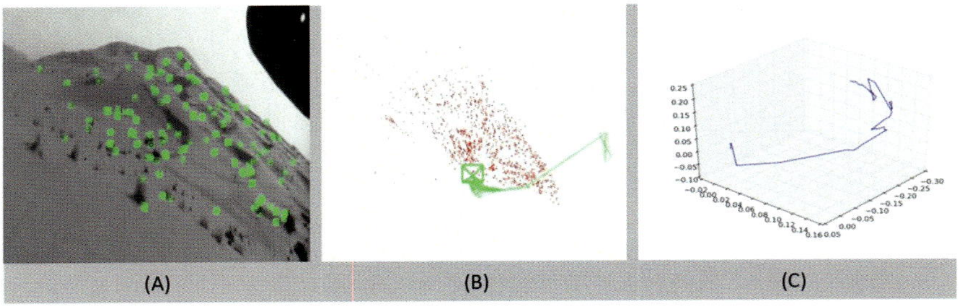

Figure 9.11 Experiment results.

feature map. The motion trajectory can be used as the motion route for the rover to return to the base station. The feature map can be used together with the path planning module to accomplish more autonomous exploration tasks.

9.6 Conclusions

This research presents a new approach, known as MVMSLAM, to the design of a visual SLAM system for the lunar rover with a single camera. In the MVMSLAM system, a modified ORB is first proposed to enhance the feature point extraction using a grid-based strategy. Furthermore, to maintain a satisfactory processing speed for the SLAM application, the ellipse search algorithm is improved, with a better feature detection approach. In addition, the MVEKF filter is optimized for the SLAM architecture. The experimental results show that our proposed new architecture is practical, with a satisfactory performance and a low error rate for a rover SLAM system.

Finally, we construct a lunar rover location system with our proposed MVMSLAM. In the simulated experiment for the lunar rover location task, MVMSLAM is shown to be capable of a lunar rover exploration task.

In the future, with the development of unmanned driving rovers, the demands for multitask (Matarić, Sukhatme, & Ostergaard, 2003) and mutlivision sensors (Potthast, Breitenmoser, Fei, & Sukhatme, 2016) will increase. The design and optimization of SLAM architecture for more advanced unmanned driving rovers will therefore be our future research focus.

References

Aulinas, J., Petillot, Y. R., Salvi, J., & Lladó, X. (2008). The slam problem: A survey. *CCIA, 184*(1), 363–371.

Civera, J., Davison, A. J., & Montiel, J. (2008). Inverse depth parametrization for monocular SLAM. *IEEE Transactions on Robotics, 24*(5), 932–945, IEEE.

Civera, J., Grasa, O. G., Davison, A. J., & Montiel, J. (2010). 1-point ransac for extended kalman filtering: Application to real-time structure from motion and visual odometry. *Journal of Field Robotics, 27*(5).

Davison, A. J., Reid, I. D., Molton, N. D., & Stasse, O. (2007). Monoslam: Real-time single camera slam. *IEEE Transactions on Pattern Analysis and Machine Intelligence, 29*(6), 1052–1067.

Fallah, S., Yue, B., Vahid-Araghi, O., & Khajepour, A. (2013). Energy management of planetary rovers using a fast feature-based path planning and hardware-in-the-loop experiments. *IEEE Transactions on Vehicular Technology, 62*(6), 2389–2401.

Fraundorfer, F., Scaramuzza, D., & Pollefeys, M. (2011). *Autonomous Systems Lab,* ETH Zurich.

Guo, F., Sun, Z., & Kan, H. (2003). A modified covariance extended Kalman filtering algorithm in passive location. In *Robotics 2003 IEEE international conference on intelligent systems and signal processing* (Vol. 1, pp. 307–311).

Harris, C. G., & Stephens, M. J. (1988). A combined corner and edge detector. In *Alvey vision conference* (pp. 147–151).

Karlsson, N., Bernardo, E. D., Ostrowski, J., Goncalves, L., Pirjanian, P., & Munich, M. E. (2005). The vSLAM algorithm for robust localization and mapping. *ICRA* (pp. 24–29).

Klein, G., & Murray, D. (2008). Parallel Tracking and Mapping for Small AR Workspaces. *IEEE & ACM International Symposium on Mixed & Augmented Reality*. ACM.

Lemaire, T., Berger, C., Jung, I. K., & Lacroix, S. (2007). Vision-based slam: Stereo and monocular approaches. *International Journal of Computer Vision, 74*(3), 343–364.

Matarić, J. M., Sukhatme, G. S., & Ostergaard, H. (2003). Multi-robot task allocation in uncertain environments. *Autonomous Robots, 14*(2–3), 255–263.

Moravec, H. P. (1980). *Obstacle avoidance and navigation in the real world by a seeing robot rover*. Stanford University.

Potthast, C., Breitenmoser, A., Fei, S., & Sukhatme, G. S. (2016). Active multi-view object recognition: A unifying view on online feature selection and view planning. *Robotics and Autonomous Systems, 84*(C), 31–47.

Rublee, E., Rabaud, V., Konolige, K., & Bradski, G. R. (2011). ORB: An efficient alternative to SIFT or SURF. *IEEE international conference on computer.*

Scaramuzza, D., & Fraundorfer, F. (2011). Visual odometry [tutorial]. *IEEE Robotics & Automation Magazine, 18*(4), 80–92.

Se, S., Lowe, D. G., & Little, J. J. (2002). Mobile robot localization and mapping with uncertainty using scale-invariant visual landmarks. *The International Journal of Robotics Research, 21*(8), 735–760.

Strasdat, H., Montiel, J., & Davison, A. J. (2010a). Scale drift-aware large scale monocular SLAM. In *Proceedings of robotics science and systems.*

Strasdat, H., Montiel, J., & Davison, A. J. (2010b). Real-time monocular SLAM: Why filter? *IEEE ICRA* (pp. 2657–2664).

Wang, Y., Hung, D., & Sun, C. (2011). Improving data association in robot slam with monocular vision. *Journal of Information Science & Engineering, 27*(6), 1823–1837.

Zhang, Z., Huang, Y., Chao, L., & Kang, Y. (2008). Monocular vision simultaneous localization and mapping using SURF. In *World Congress on Intelligent Control & Automation*. IEEE.

Zhou, H., Zou, D., Pei, L., Ying, R., Liu, P., & Yu, W. (2015). Structslam: Visual slam with building structure lines. *IEEE Transactions on Vehicular Technology, 64*(4), 1364–1375.

CHAPTER 10

Reliability and health management of spacecraft

Xilang Tang[1], K.L. Yung[2] and Bin Hu[3]
[1]Air Force Engineering University, Xi'an, P.R. China
[2]Department of Industrial and Systems Engineering, The Hong Kong Polytechnic University, Hong Kong, P.R. China
[3]Changsha Normal University, Changsha, P.R. China

10.1 Introduction

Spacecraft can be thought of as a kind of large-scale complex mechatronics equipment with high requirements for the stability and reliability of operation. For this kind of complex engineering system, faults often occur. At the same time, due to the special orbit operation environment of spacecraft, it is out of the earth's atmosphere, which is essential to the protection of the planet. This unpredictable and complex environment brings many uncertain factors to spacecraft's safe operation and presents an important hidden danger threatening the normal operation of spacecraft. There are many reasons for the failure of spacecraft, including the failure of the structure, control, power supply, propulsion, and other systems (Akin & Sullivan, 2001; Kim, Castet, & Saleh, 2012; Tafazoli, 2009). Failure will prevent the spacecraft from fulfilling its mission, bringing severe economic loss or even catastrophic consequences. For example, due to the lack of lubrication of the momentum wheel, in 2000, the attitude pointing system of the GOES-9 satellite was seriously attracted, which resulted in attitude control failure. The AO-40 satellite failed to enter the predetermined high elliptical orbit due to the internal structure failure of the motor, and some of the communication payloads on the satellite were damaged. The ERS-1 satellite failed in its attitude control on March 10, 2000, resulting in the mission failure and the end of its service. In February 2002, the battery of the MAP spacecraft was abnormal, and its voltage difference began to rise, increasing by 0.1 V every day. Finally, telemetry showed that the spacecraft lost a single battery. Among the published spacecraft failures, catastrophic failures (class I) account for about 20%, which is still a large proportion. For example, on January 28, 1986, the space shuttle Challenger exploded due to the failure of assembly joints and seals of the right solid rocket motor, killing seven astronauts.

IoT and Spacecraft Informatics
DOI: https://doi.org/10.1016/B978-0-12-821051-2.00012-X

On February 1, 2003, the space shuttle "Columbia" lost contact with the ground control center 16 m before the original landing time, and then disintegrated over central Texas, and none of the seven astronauts survived.

Therefore it is very important to take effective measures to improve the reliability of spacecraft and avoid the serious consequences of a spacecraft failure. In the past, the common method was to quickly detect and isolate the fault after it occurred, and to avoid the serious consequences of the fault through fault-tolerant design. However, due to the complex structure of spacecraft, there are many uncertain factors for the cause of a fault, and the mechanism of fault propagation is difficult to clarify. After a spacecraft breaks down, it is very difficult to find the fault in time and quickly repair the fault through reconstruction. Therefore the concept of health management was introduced into the aerospace field. Through health assessment, fault prediction, and other technologies, the occurrence of faults is predicted in advance, and predictive maintenance technology is adopted to ensure the reliable and safe operation of spacecraft (Al-Zaidy, Hussein, Sayed, & El-Sherif, 2018; Li, Shi, Li, & Liao, 2019; Yang, Li, Feng, Sun, & Zhang, 2019; Ye, Yang, Yang, Huo, & Meng, 2020).

Due to the complexity of spacecraft systems, health management is a very challenging task, and it requires deploying a large number of sensors on the spacecraft to obtain comprehensive information about the health status of the spacecraft. The Internet of Things (IoT) is an "intelligent" engineering technology system that links the "people-machine-environment" for the purpose of perception. IoT forms a close network between people and machine, machine and machine, machine and environment, people and environment, to collect relevant data, realize information interaction, mine a large amount of data and information, find the value behind the data or hidden potentially dangerous accidents in spacecraft systems (Malik, Rouf, Mazur, & Kontsos, 2019). In short, the IoT technology provides data, or the information collection and a sharing platform for spacecraft health management. This chapter introduces the basic concept, benefit, technical structure, and key techniques of health management of spacecraft based on IoT.

10.2 An introduction to "health management"

10.2.1 The basic concept of "health management"

For health management, there is no unified or clear definition, and different research teams or scholars have a different understanding of health management according to their respective research fields. NASA believes that the basic tasks of spacecraft health management include: online monitoring, diagnosis, and evaluation of the status of spacecraft in the early, middle, and later stages of the mission, and early maintenance according to its status. In addition, NASA believes that health management can prevent serious faults using reconfiguration or redundancy, start maintenance procedures when necessary, and ensure the healthy operation of spacecraft (Belcastro, 2006).

In our opinion, health management can be seen in the medical field. It is a kind of anthropomorphic technology, in which we can compare a spacecraft to the human body. People can have several different health states, such as health, subhealth, and disease. Correspondingly, a spacecraft also has many health states, such as healthy, degraded, failure and so on. People hope to monitor the health status through physical examination and adjust when the body enters a subhealth status, so as to realize the concept of "prevention is better than cure." This concept is introduced into the field of spacecraft, which evaluates or predicts the health status of a spacecraft, and to maintain the spacecraft before it is about to fail, but has not failed, so as to improve the reliability and security. Health is a description of the performance status of objects such as spacecraft, subsystems, and components, and represents the degree of performance degradation or deviation compared to the expected normal performance status (Kim, An, & Choi, 2017; Li et al., 2019). The main tasks of health management include the following three aspects: one is to collect the parameter information of the spacecraft systems in real-time based on the sensor network; the second is to evaluate and predict the health status of the spacecraft; the third is to evaluate the mission capability and safety of the equipment based on the health status, and to provide maintenance and strategies (Al-Zaidy et al., 2018; Hu, Youn, & Wang, 2019; Vogl, Weiss, & Helu, 2016).

10.2.2 Condition-based maintenance

Maintenance based on the health status of spacecraft is called "condition-based maintenance (CBM)," which is different from breakdown maintenance (BM) and Time-Based Maintenance. This kind of maintenance method monitors the status of the system in real-time and selects the most suitable maintenance time based on the health status. When it is predicted that a fault is imminent, it will be repaired immediately, which will ensure that the system will not cause a major failure. A standard called open system architecture for CBM (OSA-CBM) is specifically used to regulate this maintenance method (Lu, Liu, & Zou, 2012). The standard US Navy funding was jointly developed by machinery information management open standards alliance (MIMOSA) ARL and multiple CBM technology application companies (such as Boeing), and MIMOSA manages the standard. Although the standard was developed for the maintenance of aircraft, it has reference significance for the maintenance of spacecraft and is of great significance for understanding the concept of health management. OSA-CBM includes six levels (ISO 13374-1, 2003; ISO 13374-2, 2007), as shown in Fig. 10.1.

10.2.2.1 Data acquisition layer

To evaluate or predict the health status of a spacecraft, we need to deploy a large number of sensors on the spacecraft to collect various status data of the spacecraft. Due to the large scale of sensors, how to connect these sensors to bring data together is a question worth considering. For this

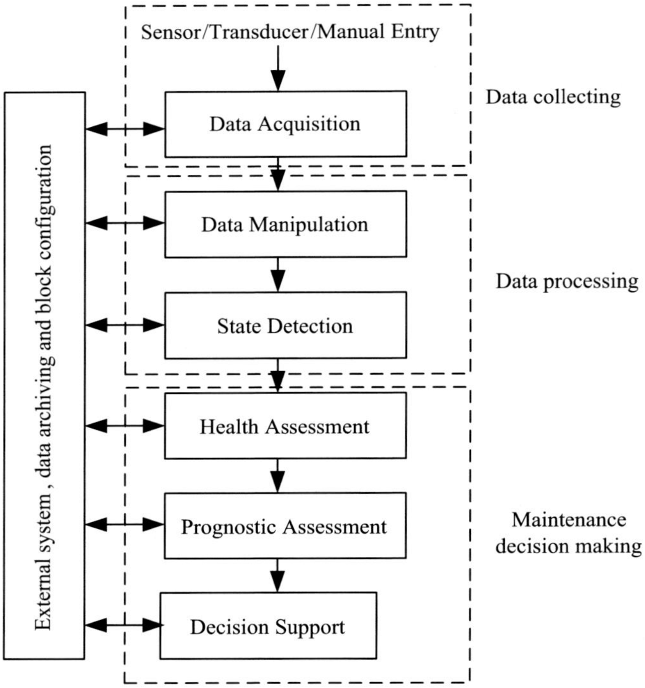

Figure 10.1 The six levels of open system architecture for CBM.

problem, the emerging IoT technology provides a solution, which is discussed later.

10.2.2.2 Data manipulation layer

This layer is responsible for signal processing (such as data filtering, Fourier transform FFT, etc.), synchronous or asynchronous averaging, executing physical models, neural networks, and other algorithms, extracting features, and finally outputting information data with time and quality indications. This information includes extracted features, waveforms from the time domain to the frequency domain, algorithm calculation results, virtual sensor data, filtered data, and time-series data (such as sampling rate).

10.2.2.3 State detection layer

This is responsible for collecting data from the first two layers and comparing these data with the standard baseline data, to obtain abnormal information of various characteristic data, including whether it exceeds the warning line and how serious it is. At the same time, State detection provides feature data, feature data status indication information, baseline data, etc. as output data to the next layer.

10.2.2.4 Health assessment layer

This accepts data from different condition monitors or other health assessment modules. When the monitored systems, subsystems, or equipment components are degraded, they determine whether they are healthy and make recommendations with a certain degree of confidence for the fault status.

10.2.2.5 Prognostic assessment layer

This layer predicts the future health status of the device based on the current health status of the device or estimates the remaining useful life (RUL) of the device under a given planned use profile.

10.2.2.6 Decision support layer

According to the output information of the previous layers, operation and maintenance decisions, capacity evaluations, recommended tasks, evidences, and explanations are generated in this layer.

To support the implementation of OSA-CBM, infrastructure, resource management, and business logic are generally required. The infrastructure architecture provides the hardware, system software, and network that support the operation of the health management system, at least including:

1. *Sensing infrastructure*. To obtain spacecraft status information and environmental information.
2. *Network transmission infrastructure*. Use various communication methods to transmit the information collected by the sensors and gather all kinds of data together.
3. *Data storage and processing infrastructure*. Responsible for the use, processing, and presentation of collected data.

As can be seen from the above, the infrastructure for spacecraft health management can be provided by the IoT technology.

10.3 The application of spacecraft health management—integrated vehicle health management

The ultimate goal of health management technology research is to develop a practical health management system and apply it to spacecraft ground tests and on-orbit flight to give full play to its effectiveness in improving spacecraft reliability and safety. In terms of spacecraft health management technology, American space research institutions have carried out a large amount of theoretical and applied research work, and are in a leading position in the world. The integrated vehicle health management (IVHM) system can be said to be the most successful and typical application of the spacecraft health management technology, which is briefly introduced in this section (Gordon & Aaseng, 2001; Scandura, 2005).

IVHM was formally proposed by NASA in its reusable launch vehicle (RLV) project. It is the integration and application of advanced software, sensors, intelligent diagnostics, digital communications, system integration, and other technologies in the spacecraft system to achieve intelligent, system-level health assessment and control, information and decision management of the spacecraft system, and help operators complete missions, reduce risks and hazards.

NASA proposed the concept of vehicle health management in the RLV project of the X-33 in 1998. In this project, the fault monitoring and function management modules of each subsystem of the vehicle are integrated and encapsulated in two LRU units to become health management computers. They are divided into three subsystems: the intelligent sensor network composed of remote health node (RHN) is used to collect the data of the aircraft structure, machinery, and environment; the communication status of six MIL-STD-1553 buses is monitored and recorded; Low-temperature oil tank monitoring by using distributed optical fiber temperature, hydrogen and stress sensors. The vehicle health management system based on the hardware composition and functional structure of X-33 has been verified and developed in the following X-34, f/A-18, DS-1, k-1, X-37 projects, and IVHM technology has been improved in the concept to system structure, function, and hardware composition. The IVHM plan has been continuously updated and improved with the change of the world environment and the development of high and new technology.

By November 2009, NASA released version 2.03 of IVHM, listing the five-year roadmap plan and detailed milestones and metrics for 2008–12. VHM 2.03 aims to develop a set of confirmed multidisciplinary integrated health management tools and technologies so that the new generation of aircraft or spacecraft can realize the automatic detection, diagnosis, prediction, and mitigation of adverse events in flight. The implementation framework of the plan is shown in the hierarchical diagram shown in Fig. 10.2, which shows the research level within IVHM and the logical relationship from basic research to planned objectives.

1. *Foundational layer.* This layer is the cornerstone of the whole IVHM program, which covers four aspects: advanced sensors and materials, modeling, advanced analysis, and complex systems, verification, and validation. This layer is particularly concerned with data acquisition, data modeling, and related verification methods of spacecraft payloads. The data about the health status of the aircraft comes from all kinds of sensors and bus data of the aircraft system. After signal modulation, a/D conversion, time synchronization, and other signal processing, it is sent to the signal processing module for processing. Sensor types in the aircraft include: temperature, pressure, displacement, strain, vibration, flow, etc., data information includes data rate, flow, bandwidth occupancy, etc., and other data include voltage, current, magnetic field strength, etc. of the instrument and equipment.
2. *Subsystem level.* This layer predicts and evaluates the health status of each subsystem, and judges whether the health status of the monitoring subsystem or components is declining generates diagnosis records to put forward possible fault status with a certain confidence and

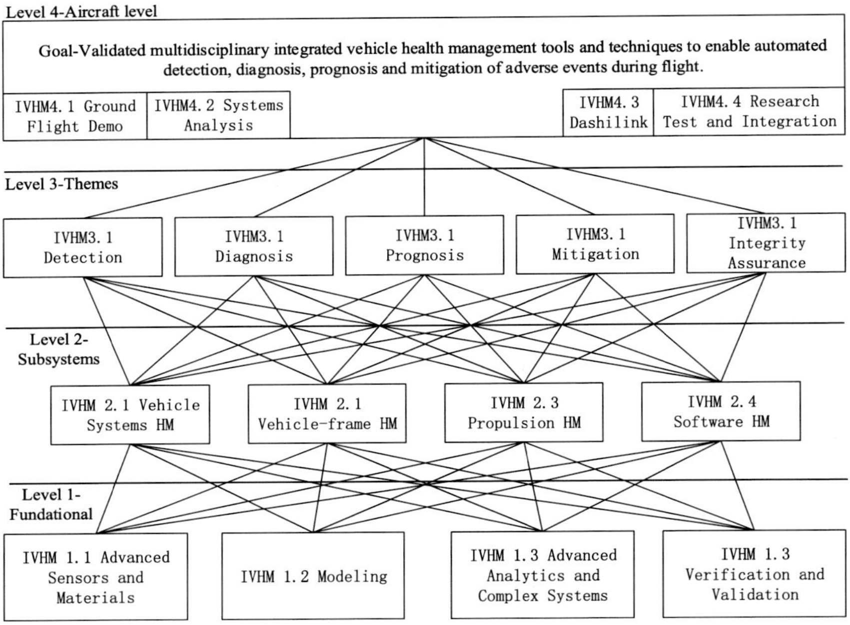

Figure 10.2 The research level within integrated vehicle health management.

integrates the information of the health history trend, operation status, and load as well as maintenance and guarantee history in the diagnosis process. These subsystems include the vehicle system, vehicle-frame, propulsion system, and software.

3. *Theme level.* The goal of the theme level is to develop an integrated toolset for the detection, diagnosis, prediction, and mitigation of adverse events in flight, as well as integrity verification.
4. *Spacecraft layer.* The spacecraft layer evaluates the health status of the whole spacecraft life cycle. IVHM is a system composed of multiple subsystems. The function division, cost-benefit analysis, mutual cooperation mechanism, data flow, data communication standard, data interface, database of each subsystem are all important parts related to the integrity, efficiency and economy of the whole IVHM system. The objectives of the spacecraft layer include a complete definition of the functions of all parts of the system, a new and efficient autonomous learning mechanism, an efficient task scheduling mechanism, an efficient and stable data communication standard, a powerful, safe and reliable database, etc.

According to the whole hierarchical structure of the IVHM system, data acquisition can be regarded as the cornerstone of the whole health management system, and the sensors used to acquire data are particularly important. With the development of sensors toward miniaturization and networking, multiple sensors can form arrays and further form networks to collect and process signals, which is the application of sensor network technology. The typical working mode of the sensor network is as follows: a large number of microsensor nodes (from hundreds to thousands) are arranged in the region of interest in the spacecraft, and the nodes form a network by self-organization. The node is not only the information collector and sender but also the information router. The collected data is sent to the gateway through multi-hop routing. The gateway (also known as sink node in some literature) is a special node, which can communicate with the ground TT & C center through mobile communication network, satellite, etc. The data collected by the whole network can be regarded as multidimensional signals, including more accurate and comprehensive information than a single sensor.

10.4 The classical structure of health management system for spacecraft

The classical structure of the health management system for a spacecraft consists of two parts: the space-borne system and the ground system.

10.4.1 The space-borne health management system

The space-borne health management system adopts the "hierarchical and distributed" structure of regional management, as shown in Fig. 10.3. It has the characteristics of "regional processing" which reduces the state information from bottom to top and extracts the key information layer by layer. It can quickly form system-level auxiliary decision-making information and realize the comprehensive application of vehicle test data, real-time detection and the diagnosis of fault information, and quick decision making in fault recovery.

The functions of a space-borne health management system include diagnosis, fault report, fault recovery management, possible maintenance, and decision support, etc. According to the characteristics of different subsystems or regions, the content of health management is quite different. For example, the health management of aircraft structures and the health management of electronic systems are as follows.

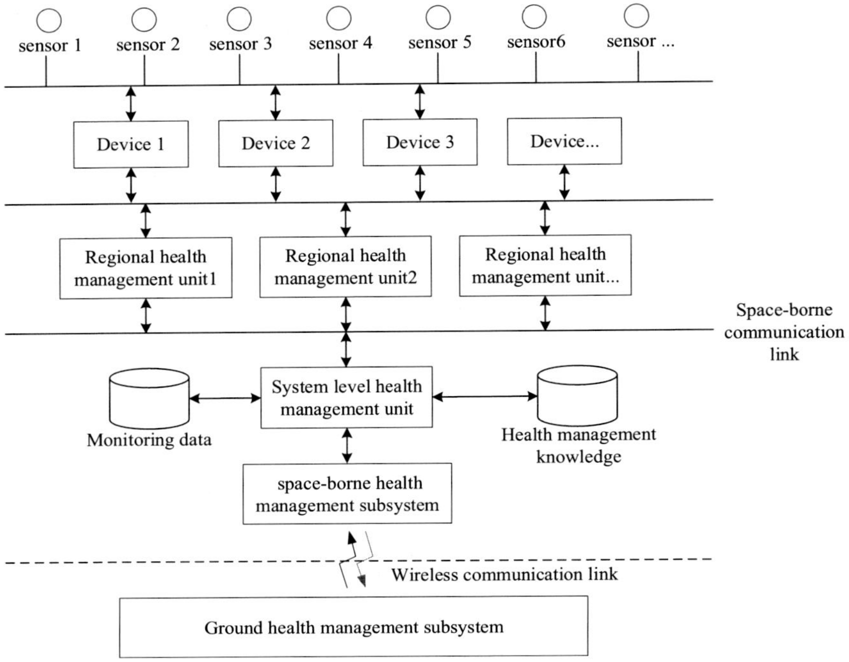

Figure 10.3 The space-borne health management system.

It is mainly composed of several relatively independent subsystems such as the structural system, propulsion system, and electronic system. Each system has a relatively independent health management system according to its own characteristics. The integration of these health management independent subsystem together constitutes the integrated health management system of the spacecraft.

1. *Structural health management.* By laying or embedding sensors such as strain, temperature, acceleration, vibration, ultrasound, chemistry, etc. on the structure of the spacecraft, the parameter changes of the structure under changing environment and working conditions are obtained. On this basis, active or passive methods are used to detect and discover dangers to the structure (stress concentration, fatigue, loosening of fasteners, crack propagation, oil and gas leakage, etc.); and evaluate the impact of the damage, take measures to mitigate the damage of the structure, or modify the task envelope to ensure the completion of the task.
2. *Mechanical devices* generally have obvious wear-out cycles, and the degradation of their performance is gradually changing, so the most

important content in the health management of mechanical components is generally fault prediction. Fault prediction is using reasonable model algorithms to make judgments about possible future failures, and predict the nature, category, degree and cause of the failures according to the current use status of the target device, combined with its operation environmental and historical data. Among them, the prediction of the remaining service life is the top priority, that is, the estimated continuous normal working time of the mechanical component from the current moment to the potential failure, which is of great significance for the adoption of the correct maintenance strategy.

3. *Electronic device health management.* The integrated electronic system has the characteristics of high complexity, multiattributes, nonlinearity, and the coupling relationship between the modules, functions, or subsystems of the system is tight. The failure transmission and the change of the symptom parameters affecting the failure of complex electronic systems are not obvious or can have sudden signals. Therefore electronic systems, unlike structures or mechanical equipment, can construct accurate physical degradation models, and it is very difficult to predict their remaining service life. The health management of electronic systems generally uses built-in test (BIT) means to obtain certain key electrical signals, and evaluate the health status of electronic equipment based on the deviation of these signals, or to detect early warning failures occurring.

Here are a few typical regional health management systems:

1. *Attitude control system.* The spacecraft attitude control system is a subsystem with a high incidence of faults, and its health management is very important for the normal operation of the spacecraft. By arranging sensor groups in the design of the attitude control system and optimizing the configuration, the changes of the state parameters of the attitude control system are obtained when the operating environment and working conditions change. New sensors, such as optical fibers, nano, and microelectronics, have many advantages, such as being easy to combine with the structure, lightweight, having small volume, small power consumption, conveniently distributed networking, and can work in harsh environments. They are widely used in the health management of the attitude control system. A state observation model of the attitude controller is constructed to evaluate the health status of the spacecraft attitude control system according to the state parameters, and active or passive methods are used to detect and diagnose potential

faults that may arise, and the harmful consequences of the faults are evaluated, then the fault maintenance plans are designed in time to minimize the safety risk of the spacecraft.

2. *Propulsion system*. Through the analysis of various working parameters and sensor information about the engine and its accessories, the diagnosis and prediction information of the engine's health status can be obtained. The general sources of health information of the engine are Engine gas path sensor, fuel system sensor, vibration sensor, structure evaluation sensor, full authority digital control (FADEC) code, engine model, logistics maintenance history, etc. Through data fusion of the various parameters and sensor information, fault location can be detected and diagnosed, fault development trend predicted, and corresponding maintenance decisions generated.
3. *Power supply system*. The normal operation of the power supply system is the cornerstone of the normal operation of the entire spacecraft. Historically, accidents damaging the entire spacecraft due to power failures have occurred. The power system of the spacecraft, especially the performance of the battery, has an obvious slow degradation process over time. By monitoring the main electrical signal parameters of the power system and the environmental changes such as temperature, the health status of the power system can be evaluated. For the most important component in the power system-the battery, the remaining life can be predicted by building an appropriate model (Fig. 10.4).

10.4.2 The ground health management system

Due to the limited storage space and computing power of the onboard computer, it is impossible to store and process a large amount of historical data, and it may be ineffective and inefficient for more complicated fault diagnosis or long-term fault prediction. The ground health management system receives the health status information and other data integrated by the onboard health management system through the telemetry channel and uses its powerful storage and computing capabilities to mine any abnormal state or development trend of the spacecraft from the massive data. Secondly, the ground health management system can make full use of historical fault information and expert experience and knowledge and can use manual assistance to complete tasks such as health assessment, fault diagnosis, fault prediction, and maintenance decision-making.

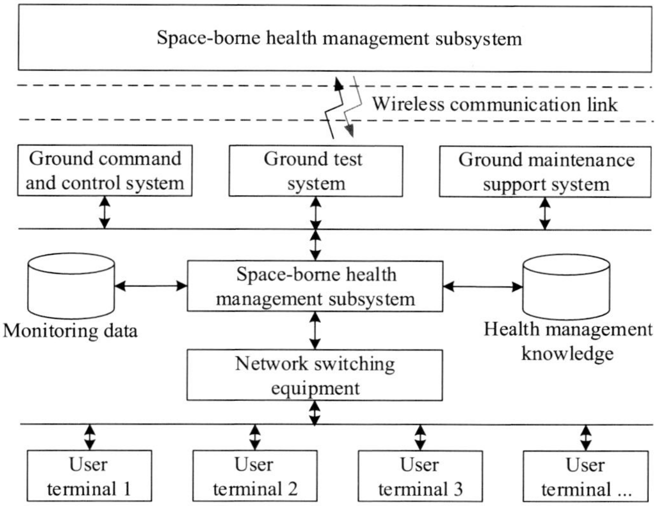

Figure 10.4 The ground health management system.

10.5 Benefits of Internet of Things to health management

As discussed before, data acquisition can be regarded as the cornerstone of the whole health management system, and sensor networks provide a technical basis for data acquisition. So what is the difference between the IoT and the sensor network? What are the benefits of applying the IoT to satellite health management? This is discussed in the following section.

10.5.1 The difference between Internet of Things and sensor network

The sensor network is a multihop self-organizing network system formed by a large number of inexpensive mini-sensor nodes deployed in the monitoring area and formed by wireless communication. Its purpose is to collaboratively perceive, collect and process the information of the sensing objects in the network coverage area and send it to the observer.

IoT is based on the computer Internet, using RFID, wireless data communication, and other technologies to construct an "IoT" covering everything in the world. In this network, objects can "communicate" with each other without human intervention. Its essence is to use radio frequency automatic identification (RFID) technology to realize the automatic identification of items and the interconnection and sharing of information through the computer Internet. RFID is a technology that allows items to "speak." In the concept of "IoT," RFID tags store standardized and interoperable information, and automatically collect them into a central information system through a wireless data communication network to realize the identification of items (commodities). Furthermore, through the open computer network, information exchange and sharing can be realized for the "transparent" management of items. The advent of the concept of "IoT" has broken previous traditional thinking. In the past, physical infrastructure and IT infrastructure have been separated: on the one hand, sensors and networks, on the other hand are data centers and computing centers; in the era of "IoT," sensors, networks and computers are integrated into a unified infrastructure. It can be said that the IoT is a fusion application of intelligent perception, intelligent recognition, ubiquitous network and pervasive computing.

Features of the IoT include:

1. comprehensive perception, that is, using sensors to obtain information about objects anytime, anywhere;
2. reliable transmission, through the integration of various telecommunications networks and the Internet, information on the object is accurately transmitted in real-time;
3. intelligent processing, using cloud computing, fuzzy recognition, and other intelligent computing technologies to analyze and process massive amounts of data and information, and implement intelligent control of objects.

To sum up, the components of the IoT and sensor networks are basically the same, including sensors, networks, data centers, etc. However, compared with the sensor network, the IoT strengthens the information exchange between the physical objects and the digital world, and also emphasizes the information processing ability of the object itself, which can provide higher-level application services for the "thing," rather than just information acquisition.

10.5.2 Internet of Things brings more benefit for health management of spacecraft

1. By using IoT technology, regional health management can be better realized. As mentioned above, the IoT emphasizes the information processing capability of the object itself, which can provide high-level application services for the "thing." It can be seen that the sensor network is used for the health management of the spacecraft, and its task is limited to the data acquisition layer of the OSA-CBM and does not have higher-level processing capabilities. By introducing the IoT into

spacecraft health management, high-level health management applications can be realized in low-level areas. In other words, each small sensing area has computing power, which can process data, and complete high-level tasks such as status evaluation, fault detection, prognostics of device in this area, rather than just collecting data to a higher-level area.

2. By using the IoT technology, we can build the information exchange mechanism between regional health management. As can be seen from the typical space-borne health management system in Fig. 10.3, the low-level regional health management system will only pass the health status information to the high-level regional health management systems, and there is no information exchange between health management systems at the same level. However, devices at the same level often have a strong coupling relationship, such as electrical connection and physical contact. An abnormal state of a certain device may cause abnormalities in nearby parts. For example, if the temperature of a certain device is too high, it may cause abnormal operation of the device in contact with it. Therefore it is necessary to conduct a "dialog" between health management systems at the same level to exchange their health status information. Compared with the sensor network, the IoT strengthens the information exchange between physical objects and the digital world by using RFID technology. Therefore applying IoT technology to health management systems can well establish the information exchange mechanism between regional health management.

10.6 Prognostics technique

Prognostics is an engineering technique that tracks the degradation path and predicts the future degradation evolution and RUL. It is a central activity of CBM, see prognostic assessment layer in OSA-CBM. There are a great variety of available techniques for prognostics evaluation, which can be divided into data-driven methods and model-based methods. Data-driven methods use information from observed data to identify the health signal and predict future health status, including neural network, support vector machine, and also some big data mining method. Since the data-driven methods depend on the trend of the data, they are powerful in predicting short-term RUL. However, they are not good enough at dealing with noise and predicting long-term RUL.

Model-based methods combine a degradation model with observed data to predict the RUL of devices. The effective of the model-based methods is not entirely determined by the observed data, but also related to the degradation model. Therefore the model-based methods are more suitable for long-term prediction, and it is more sensible to choose model-based approaches for devices with high reliability. This chapter mainly introduces the prognostics method based on the empirical model by taking the solenoid valve as an example.

10.6.1 Problems for model-based prognostics

To achieve the goal of model-based prognostics, several problems need to be considered:

1. *Defining the degradation signal.* As the indicator of the health status of components, the change of degradation signal must be able to reflect the degradation status of different stages of components truly and accurately. In addition, the degradation signal must be easily monitored by sensors in engineering practice, so that the prognostics method can be applied to practical engineering. Generally, the construction of health factors is based on the physical understanding of components, that is, what is the degradation mechanism of components, what kind of faults may occur in the future, which parameters will change in the degradation process, etc. Based on the above considerations, the signals that can reflect the degradation process of components and are easy to monitor are selected as the basis of constructing health indicator, and then these signals are mapped into a comprehensive index by certain methods.
2. *Setting up degradation model.* The degradation model is a parameterized model that reflects the degradation process. When the parameters are determined or known, the model can calculate the change of the degradation state value with time. However, in most cases, it is almost impossible to establish a physical model that can accurately describe the potential authenticity of the degradation process. Therefore empirical degradation models based on observed degradation signals are often developed. The degradation test can provide the measurement of the degradation signal, including the measurement under the original condition and the failure condition.
3. *Dealing with the uncertainties.* Since the core content of prediction is to predict the future degradation evolution and RUL, there are many

sources of uncertainty that affect the prediction, such as changes between units, random operating loads or environmental stresses, measurement errors, and model errors. Therefore when making predictions, it is necessary to deal with related uncertainties. In fact, it is impossible to completely eliminate these uncertainties, but it will be of great benefit to express, quantify and propagate these uncertainties in the forecasting process.

10.6.2 Methodology of model-based prognostics

As mentioned earlier, one of the main problems of model-based prognostics is to establish a degradation model and propose an empirical degradation model. In many engineering applications, the degradation process can be represented by a linear process with linear drift. In other words, the mathematical model of degradation can be expressed by assuming that degradation accumulates at a certain rate. Therefore a simple linear model is used to explain the basic method of fault prediction.

$$d_k = d_{k-1} + b \tag{10.1}$$

where d is the degradation state, and b is the degradation rate.

To calculate RUL, the fault threshold needs to be defined with the maximum acceptable degradation state. When the degradation exceeds the fault threshold, the component is regarded as a fault. RUL can be defined as

$$R\left(k_p\right) = k_{EOL} - k_p \tag{10.2}$$

where R represents the RUL, and k_p is the operation cycle when the measurements of degradation signals terminate and the prediction process begins. k_{EOL} is the operation cycle at the end of life (EOL), when the prediction degradation curve meets the single threshold level line. Note that the value of RUL is nondeterminate, and its value evolves when k_p changes. To deal with uncertainties, $R(k_p)$ is assumed as a random value instead of a fixed value. The objective of prognostics is to calculate the probability distribution function (PDF) of RUL, simplified as $f_R(r)$.

To achieve the objective, the methodology of prognostics (Tang, Xiao, & Hu, 2020) is presented in Fig. 10.5.

1. d_k, b_k Assuming that k is a random value, it is represented by a probability distribution. As mentioned earlier, the degradation process can be assumed to be an incremental process with a certain rate. Since

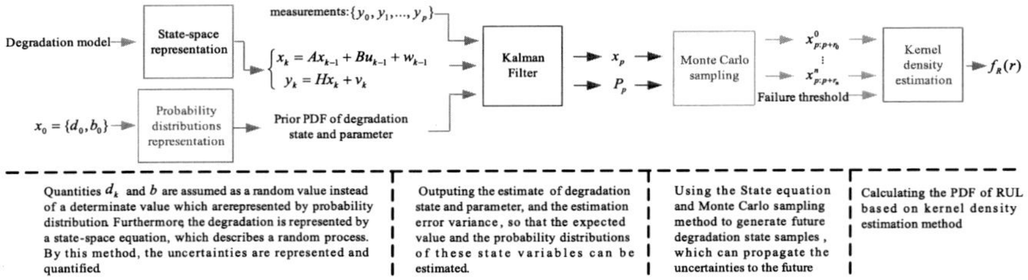

Figure 10.5 The methodology of model-based prognostics.

each degradation increment is affected by various factors such as material, assembly, and degradation stress, according to the central limit theory in statistics, it can be reasonably assumed that the change of each individual degradation increment is approximately a normal distribution. From this point of view, the degradation process is a random-effect Wiener process, which has been reported in many relevant studies. Then the model represented by Fig. 10.1 can be modified to,

$$\begin{cases} d_k = d_{k-1} + b_{k-1} + w^d_{k-1} \\ b_k = b_{k-1} + w^b_{k-1} \end{cases} \tag{10.3}$$

where d_k is the degradation value. b_k is the degradation rate, a non-determinate parameter, which allows for variation from unit to unit due to the diverse material properties of the different units and their production processes or handing condition. w^d_k represents the uncertainty of corresponding degradation process for each SV, and w^b_k is the process uncertainty of degradation rate. w^d_k and w^b_k are assumed to be invariable with the state of the system, and follow the Gaussian distribution. Here, the specification of x_k and w_k are introduced as follow,

$$x_k = \begin{bmatrix} d_k \\ b_k \end{bmatrix}, w_k = \begin{bmatrix} w^d_k \\ w^b_k \end{bmatrix} \tag{10.4}$$

where, $w_k \sim N(0, \ W)$, and W is the covariance matrix of process uncertainties.

Filtering technology under the Bayesian framework is an effective uncertainty management tool, and has a wide range of applications in the field of prediction. The most typical methods include Kalman filtering or its extended form, and particle filtering. Kalman filtering is the optimal estimation method for linear models and Gaussian distributions, and particle filtering is the optimal estimation method for nonlinear models and nonGaussian distributions. Therefore the Kalman filter is chosen to explain the method. The purpose of Bayesian filtering is to construct a state probability density function based on measurement knowledge. The Bayesian filtering problem includes two basic elements: state vector and measurement vector. The knowledge of these two vectors is transformed into a state model and a measurement model. For Kalman filtering, these two models can be expressed as follows.

$$\begin{cases} y_k = Hx_k + v_k \\ x_k = Ax_{k-1} + Bu_{k-1} + w_{k-1} \end{cases} \tag{10.5}$$

where y_k denotes the observation of degradation at time, x_k is the state vector, represented as Eq. (10.4). The quantity v_k is supposed to represent the measurement uncertainties and is assumed to be invariable and follow the Gaussian distribution, formulated as $v_k \sim N(0,\ V)$, where V is the covariance matrix. The measurement model not only captures the measurement uncertainty but also captures the additional uncertainties such as model error. w_k is the process uncertainties, represented as Eq. (10.4). u_k denotes the input of a linear system, but there is no input in this case, so B is set as 0. Accordingly, the state equation and measurement equation for KF can be expressed as

$$\begin{cases} y_k = \begin{bmatrix} 1 & 0 \end{bmatrix} x_k + v_k \\ x_k = \begin{bmatrix} 1 & 1 \\ 0 & 1 \end{bmatrix} x_{k-1} + w_{k-1} \end{cases} \tag{10.6}$$

Therefore

$$\begin{aligned} A &= \begin{bmatrix} 1 & 1 \\ 0 & 1 \end{bmatrix}, B = 0, H = \begin{bmatrix} 1 & 0 \end{bmatrix} \\ W &= \begin{bmatrix} w_k^d & 0 \\ 0 & w_k^b \end{bmatrix}, V = v_k \end{aligned} \tag{10.7}$$

2. Due to the uncertainty, the degradation signal has strong noise. Therefore the true state cannot be accurately known, but the expected value and probability distribution of these state variables can be estimated by KF. Based on the optimal criterion of minimum mean square error, the KF algorithm uses state equations and state previous

estimates and state observations to update state variable estimates. The procedure is as follows:

Algorithm: Kalman filtering for estimating degradation state and parameter.

1. Initialization of variables. The initial value of degradation state and parameter is obtained by linear fit.
2. Project state at the next moment using state equation, $\hat{x}_k^- = A\hat{x}_{k-1} + Bu_{k-1}$.
3. Project the error covariance at the next moment, $P_k^- = AP_{k-1}A^T + W$.
4. Calculate Kalman gain, $K_k = P_k^- H^T \left(HP_k^- H^T + V\right)^{-1}$.
5. Update estimate of state with measurement, $\hat{x}_k = \hat{x}_k^- + K_k\left(y_k - H\hat{x}_k^-\right)$.
6. Update estimate of error covariance, $P_k = (I - K_k H)P_k^-$.

The output of the Kalman Filter is the estimation of the expected state $\hat{x}_k$, including the degradation state and parameters, and the estimation error covariance P_k. Thus the random value of state x_k is normally distributed, and its mean value is $\hat{x}_k$, and its variance is P_k. When the operation cycle comes to the prediction beginning time k_p, the random state x_p follows a normal distribution, represented as $x_p \sim N\left(\hat{x}_p, P_p\right)$, where $\hat{x}_p$ and P_p refer to the estimated mean value and variance of state at the prediction beginning time k_p, respectively. The PDF of degradation state and parameter at time k_p estimated by KF are considered as the initialization of the prediction process.

3. The degradation state and parameters follow normal distributions, and the state equation is linear, so the future degradation state also follows a normal distribution. However, the ultimate objective is not to estimate the future degradation state and its distribution, but to estimate distribution of RUL. The equation of calculating the RUL is a nonlinear function [see Eq. (10.2)], as a consequence, the $R(k_p)$ may not follow a normal distribution, and the PDF of RUL cannot be calculated by analytical method directly. In such a case, Monte Carlo sampling is adopted to propagate the uncertainties. To obtain initial samples for prediction, the Monte Carlo method is used to draw samples of the degradation state and parameter from the Normal distribution $x_p \sim N\left(\hat{x}_p, P_p\right)$ generated by KF at time k_p. Then the future degradation evolution paths $x_{p:p+r_n}^n$ are calculated based on the state space equation in Eq. (10.5), where n denotes the nth sample. If the number of samples is N, there are N degradation evolution paths in total.

4. The RUL for each degradation evolution path can be calculated by Eq. (10.3) after all future degradation evolution paths are obtained.

Therefore there are N number of calculated RULs in total. By using kernel density estimation, the PDF of RUL $f_R(r)$ can be estimated with the N calculated RUL.

10.6.3 An example of prognostics for solenoid valve

Solenoid valves (SVs) are electromechanical devices used to control the flow of gas or liquid. SVs are widely used in transportation, industry, agriculture, aerospace, tourism, and living facilities. Because the solenoid valve has the advantages of a simple structure, easy to use, low price, fast action, small power, and light appearance, they are widely used in the field of automation as actuators and are key components in engineering systems. If they crash, it may cause the system to fail, or even cause a catastrophic event. In view of the wide application and important role of SVs, research on their reliability is of great significance. Therefore it is necessary to estimate the RUL of SVs online.

To obtain the degradation data of SV, a degradation experiment was designed. In the experiment, parameters such as drive current and drive temperature were monitored. These parameters cannot be directly obtained by the computer, and need to be detected and converted into voltage by a special sensor, obtained by the AD converter.

1. *Current sensing*. A Hall current sensor (WHB-LSP5S2) is designed to obtain current. The Hall detection technology overcomes the shortcomings of transformers and can measure AC signals of various frequencies up to 100 kHz to obtain the original current curve without distortion. Hall current sensor is a noncontact sensor, easy to use. WHB-LSP5S2 can measure a maximum current of 2 A, the output voltage (2.5−2 V) is linear with the current, is equal to or less than 0.1%, and the response time is less than 1us.
2. *Temperature sensing*. The E-type PT100 thermocouple was used to measure the working temperature of SVs during the cycle. Within a given temperature range, the resistance of the thermocouple increases linearly and can be converted into a voltage increase by the thermocouple converter, so that the voltage can be used to give accurate temperature readings.
3. *Test design*. Testing means generating stimulation and capturing responses from SVs. To generate stimulation and obtain responses, we used the PXI test equipment by National Instruments, which is rich in its functions and power and can provide fast test solutions.

SVs only need DC power, but the connection and disconnection of DC power need to be performed automatically. Therefore we used a delay switchboard and an IO card in the PXI device and used the signals generated by the IO card to control the on and off of the delay switch. The IO card was PXI6541, which is a 50 MHz digital waveform generator/analyzer. Through the NI-HSDIO driver, it can be interfaced with logic to generate a digital signal with a certain frequency and duty cycle to control the relay switch. The duty cycle is defined as the ratio of the "on" time to the total start-up period.

In addition to stimulation, the acquisition is also necessary. After the sensor converts the response parameters to voltage, it needs AD to collect data. The PXI 6224 card is selected as the PXI analog input signal acquisition DAQ module, with 32 single-ended analog input channels (16 bits, 250 kS/s), and NI-MCal calibration technology was used to improve the measurement accuracy. After the SVs ran for a certain period, the coil resistance was measured to see if the coil was short-circuited. For this, PXI 4065 was used as the DMM card. Fig. 10.6 shows the block diagram and hardware used in the experiment.

For each solenoid valve, the experiment was performed according to the process shown in Fig. 10.7, with the detailed procedure as follows:

1. *Configuration*. Before starting the test, several operating conditions need to be configured. For AD, the sampling rate, number of samples, and physical channel list were set. For IO, its configuration is a bit complicated because it controls the operating frequency and duty cycle of the solenoid valve. First, a digital waveform, including its digital logic and time interval needs to be defined, to determine the operating frequency and duty cycle. For example, if we define a digital waveform, the digital logic is "0,0,0,0,0,1,1,1,1," where the "0" logic will be connected to a DC power supply with a time interval of 0.01 S, with the operating frequency 10 Hz and the duty cycle 0.75. Then, a repeat count of the dynamic IO should be defined to determine the number of cycles the solenoid valve runs in one measurement. For example, if the repeat count is 100, the solenoid valve performs 100 switching cycles for one measurement.
2. *Measure coil resistance*. Before turning on the DC power supply, first, measure the coil resistance.
3. When starting the dynamic IO, the solenoid valve will enter the on-off cycle.

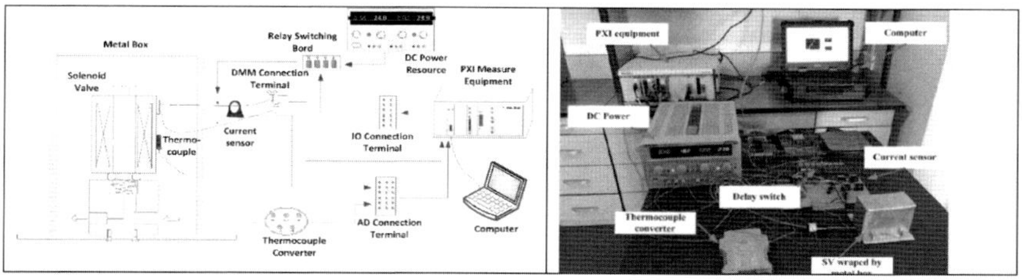

Figure 10.6 The block diagram and hardware used in our experiment.

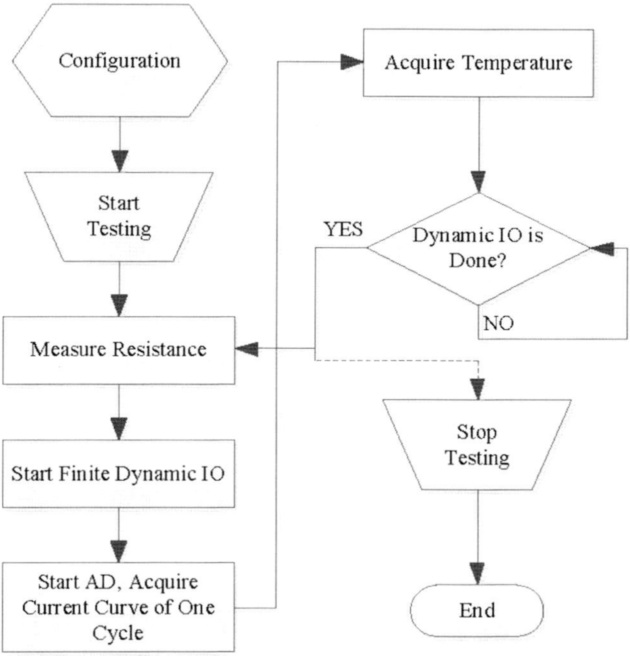

Figure 10.7 The procedure of degradation experiment.

4. *Obtain the current curve.* According to the configuration, a certain number of samples to be collected, and to be done within a complete cycle.
5. *Measure the temperature.*
6. *Wait for the completion of dynamic IO.* If not exist, return to step (2). This process uses NI LabView for design, which is a visual

programming language system design platform and development. Its biggest advantage is that it provides rapid development ability. Using graphical programming language and instrument driving, you can quickly perform the experiment and automatically save the measurement data collection in TDMS format. A graphical user interface (GUI) to configure the operating conditions and observe the slow changes in parameters were designed. A total of 8 solenoid valves were tested under a normal temperature environment but wrapped with a metal box, and the controlling cycling frequency of 5 Hz. They were tested at two different DC voltages (18 and 24 V), and two different duty cycles (50% and 80%).

When all SVs failed, the experiment stopped. After the experiment, the current curves of different duty cycles were extracted from the monitoring results as shown in Fig. 10.8. Due to limited space, the results of SV#6 and SV#8 were applied in the next analysis. Fig. 10.8 shows that the shape of the dynamic current waveform changes regularly with the increase of the duty cycle, and the contact moves slowly to the right, which means that the response time gradually becomes larger.

The dynamic current curve reflects the whole process of iron core movement. When the performance of the solenoid valve deteriorates, the movement process of the iron core changes, resulting in distortion of the current curve. Therefore it is reasonable to use the dynamic current curve as an index to evaluate the SV health status. The degree of distortion of the dynamic current curve can characterize the deterioration state of the SV. For example, in the limit state, if the moving iron

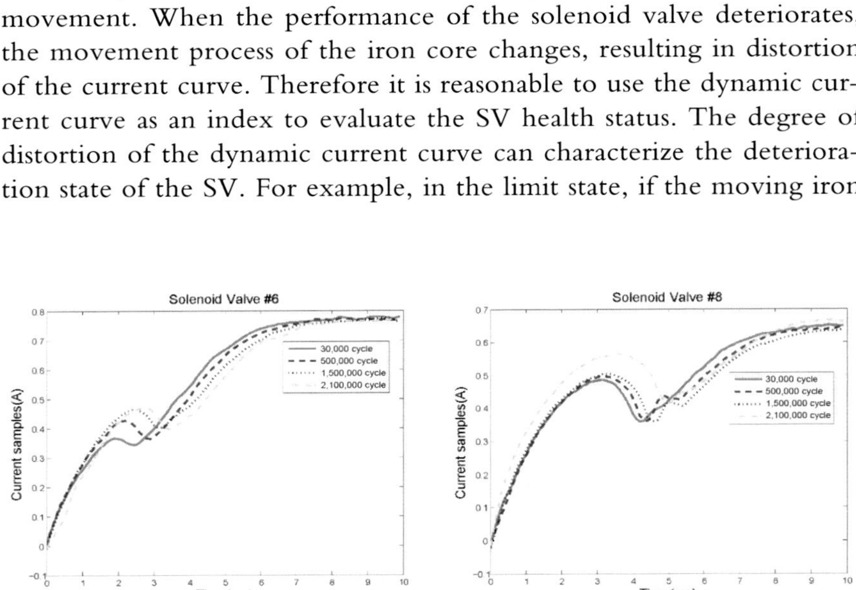

Figure 10.8 The current curves with different working cycles.

core cannot be moved for some reason, such as being stuck, the current core will increase monotonously without the current falling process. To predict the health of the solenoid valve, the degradation signal of the solenoid valve needs to be determined first. Defining a degradation signal means that a mathematical indicator that reflects the health or degradation status of SVs needs to be constructed. As mentioned earlier, in practical applications, the most important thing is that the signal is easy to monitor online. Therefore signals such as electromagnetic force, response time, and core displacement are not feasible. In this case, the influence of the drive current is considered, because the noncontact current sensor can easily monitor the drive current. In addition, the current curve contains a lot of information about the performance of the SV, as mentioned earlier.

Here, the degree of distortion of the dynamic drive current curve is defined as a degradation phenomenon. In order to measure the distortion of the dynamic drive current curve, we define a template dynamic current waveform at the beginning of the work, symbolized by I_{temp},

$$I_{\text{temp}} = [i_1, i_2, i_3, i_4, \ldots, i_n] \tag{10.8}$$

where i_n denotes the nth sample of the template dynamic current waveform. The template dynamic current waveform represents the dynamic current waveform when the SV is operating in the optimal state, that is, the SV does not deteriorate at this time. SVs can be considered to be in the best working state when shipped from the factory, so the dynamic current waveform collected at this time can be regarded as the template dynamic waveform, approximately.

The dynamic current waveform at the kth working cycle is defined as,

$$I_k = [i_{k1}, i_{k2}, i_{k3}, i_{k4}, \ldots, i_{kn}] \tag{10.9}$$

where i_{kn} denotes the nth sample of the dynamic current waveform at the kth operation cycle. When the SV works in the kth working cycle, the dynamic current waveform of the kth working cycle is obtained, and its distortion degree represents the degradation state of the SV in the kth working cycle. The greater the difference between the dynamic current waveform of the kth working cycle and the dynamic current waveform of the template, the more severe the distortion of the dynamic current waveform of the kth working cycle and the more serious the degree of deterioration of the solenoid valve. To measure the difference between the dynamic current waveform of the kth duty cycle and the template waveform, the Euclidean distance between them is defined as follows,

$$d\left(I_{\text{temp}}, I_k\right) = \sqrt{\sum_{m=1}^{n} (i_{km} - i_m)^2} \tag{10.10}$$

Therefore the Euclidean distance is defined as the degraded signal of an SV, and the Euclidean distance is used to indicate the degree of distortion of the dynamic driving current curve of the *k*th duty cycle. The trend in Fig. 10.9 shows that the distance gradually becomes larger, which means that the degradation state gradually deteriorates as the operating cycle increases. Finally, due to the failure of the SV, the current changes drastically, and there is a sudden surge. The distance value before and after the sudden increase is 0.7–0.9.

The specific process of applying the fault prediction method mentioned above to the SV is shown in Fig. 10.10.

For the same SV sample, different predicted start times and the predicted degradation evolution are very different. For example, the predicted degradation curve of SV#6 in the early stage is quite different from the actual degradation curve, but the predicted curve in the later stage is very close to the actual degradation curve. This is because the degradation rate of the solenoid valve is different at different stages, and the degradation information contained in the training data set is different at different prediction start times. This means that the prediction effect of this method has a strong relationship with the prediction start time. The closer the prediction start time is to EOL, the higher the prediction accuracy.

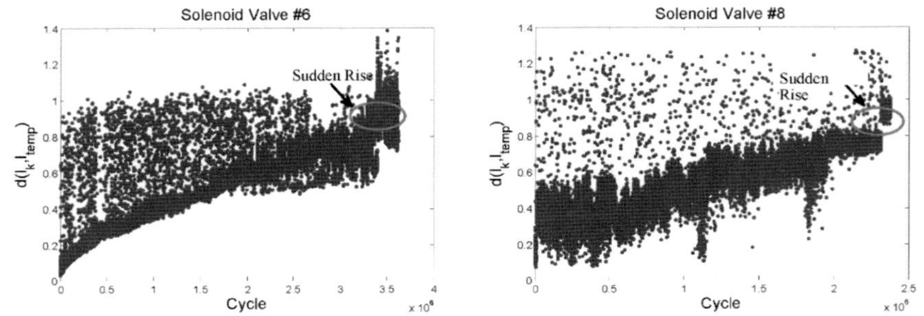

Figure 10.9 The distortion degree of dynamic driven current curve versus time.

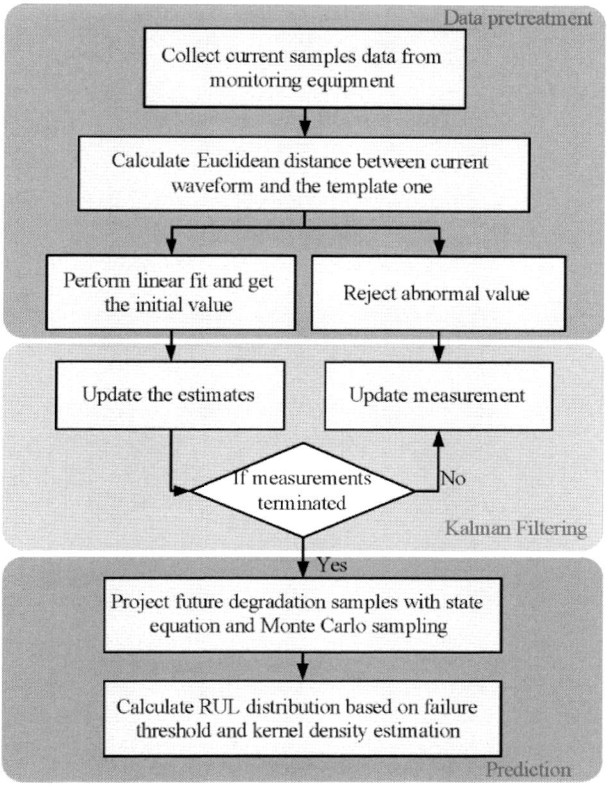

Figure 10.10 The process of prognostics for solenoid valves.

As mentioned earlier, it is very important to deal with the uncertainty of SVs predictions. KF (essentially a Bayesian estimation method) can be used to estimate the variance of the health state and the state degradation rate at the beginning of the prediction. Therefore compared with commonly used methods, the KF method can quantify the uncertainty in degradation prediction. Using the Monte Carlo method, the uncertainty captured by KF can be propagated to future degradation predictions. The gray background in Fig. 10.11 is a set of predicted degradation samples generated by the Monte Carlo method. In addition, the kernel density estimation is used to estimate the PDF of EOL, which is represented by the black curve in Fig. 10.11. As a prediction result, EOL PDF is much more reasonable than a single EOL value, because it is a more valuable reference for the maintenance of complex systems.

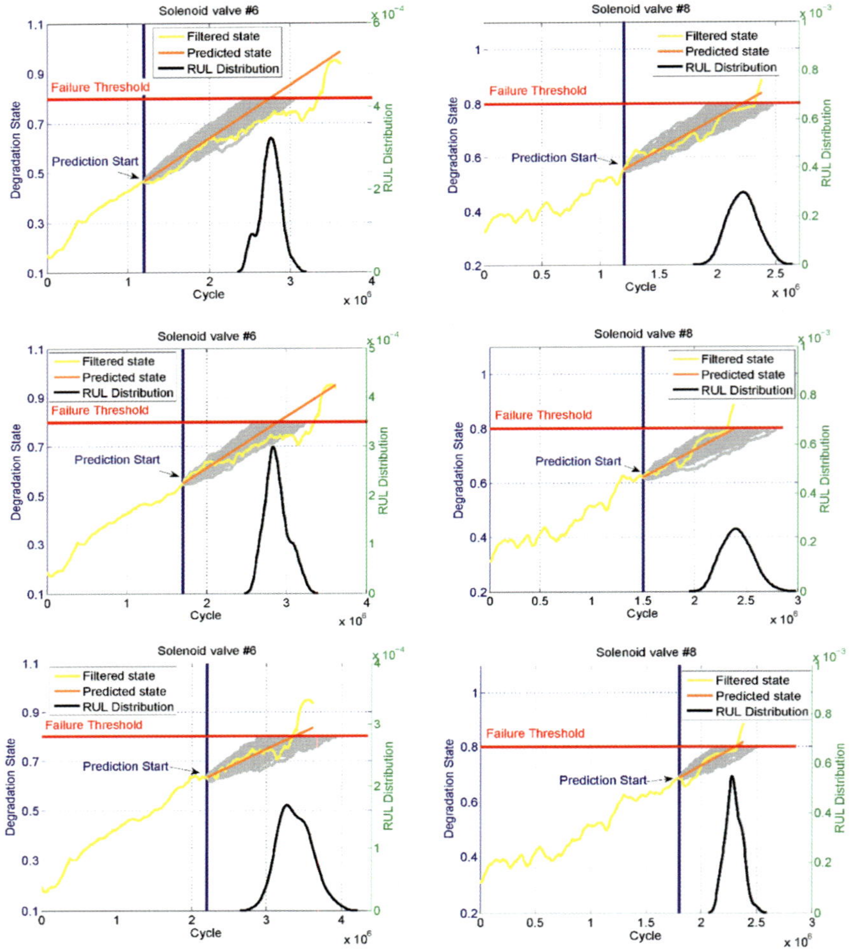

Figure 10.11 The result of prognostics for solenoid valves.

References

Akin, D. & Sullivan, B. (2001). A survey of serviceable spacecraft failures. In *Proceedings of the AIAA space conference & exposition*.

Al-Zaidy, A. M., Hussein, W. M., Sayed, M., & El-Sherif, I. (2018). Data driven models for satellite state-of-health monitoring and evaluation. *International Journal of Robotics and Mechatronics*, p.1.

Belcastro, C. M. (2006). Aviation safety program: Integrated vehicle health management technical plan summary. NASA technology report, 1−53.

Gordon, B., & Aaseng, G. (2001). Blueprint for all integrated vehicle health management system. In *Proceedings of the IEEE twentieth digital avionics systems conference* (pp. 1−11).

Hu, C., Youn, B. D., & Wang, P. (2019). *Case studies: Prognostics and health management (PHM). Engineering design under uncertainty and health prognostics. springer series in reliability engineering*. Cham: Springer.
ISO 13374-1. (2003). *Condition monitoring and diagnostics of machines—Data processing, communication and presentation—Part 1: General guidelines.*
ISO 13374-2. (2007). *Condition monitoring and diagnostics of machines—Data processing, communication and presentation—Part 2: Data processing.*
Kim, N. H., An, D., & Choi, J. H. (2017). *Introduction. Prognostics and health management of engineering systems*. Cham: Springer.
Kim, S. Y., Castet, J. F., & Saleh, J. H. (2012). Spacecraft electrical power subsystem: Failure behavior, reliability, and multistate failure analyses. *Reliability Engineering & System Safety*, 55–65.
Li, J., Shi, J. X., Li, S., & Liao, Q. Y. (2019). The method of health management on satellite constellation network. In *Proceedings of the twenty-first international conference on advanced communication technology (ICACT)*.
Lu, L. P., Liu, J., & Zou, D. P. (2012). Open system architecture for condition-based maintenance. *East China Electric Power*.
Malik, S., Rouf, R., Mazur, K., & Kontsos, A. (2019). The industry internet of things (IIoT) as a methodology for autonomous diagnostics, prognostics in aerospace structural health monitoring. *Structural health monitoring*.
Scandura, P. A., (2005). Integrated vehicle health management as a system engineering discipline. In *Proceedings of the twenty-fourth digital avionics systems conference*.
Tafazoli, M. (2009). A study of on-orbit spacecraft failures. *Acta Astronautica*, 195–205.
Tang, X., Xiao, M., & Hu, B. (2020). Application of kalman filter to model-based prognostics for solenoid valve. *Soft Computing*, 5741–5753.
Vogl, G. W., Weiss, B. A., & Helu, M. (2016). A review of diagnostic and prognostic capabilities and best practices for manufacturing. *Journal of Intelligent Manufacturing*.
Yang, L., Li, X., Feng, C., Sun, Z., & Zhang, T. (2019). *Research on autonomous health management and reconstruction technology of satellite*.
Ye, Y., Yang, Q., Yang, F., Huo, Y., & Meng, S. (2020). Digital twin for the structural health management of reusable spacecraft: A case study. *Engineering Fracture Mechanics*.

Index

Note: Page numbers followed by "*f*," "*t*," and "*b*" refer to figures, tables, and boxes, respectively.

D

E

K

L

M

N

T

U

V

W

X

Y

Printed in the United States
by Baker & Taylor Publisher Services